OPTICAL PROPERTIES OF SOLIDS

OPTICAL PROPERTIES OF SOLIDS

LECTURES AND SEMINARS PRESENTED AT
THE FIFTH CHANIA INTERNATIONAL CONFERENCE,
HELD IN CHANIA, CRETE, GREECE,
JUNE 30-JULY 10, 1969

Edited by

E. D. HAIDEMENAKIS

Executive Director
International Center for Advanced Studies and
Faculty of Sciences, University of Paris

GORDON AND BREACH, SCIENCE PUBLISHERS
NEW YORK • LONDON • PARIS

Library of Congress Catalog Card Number: 72-112780

Editorial office for Great Britain:
Gordon and Breach, Science Publishers Ltd.
12 Bloomsbury Way
London W.C. 1

Editorial office for France:
Gordon & Breach
7-9 rue Emile Dubois
Paris 14e

Distributed in Canada by:
The Ryerson Press
299 Queen Street West
Toronto 2B, Ontario

Printed in the United States of America

PREFACE

The volume in hand contains the proceedings of the international conference on "The Optical Properties of Solids" which was held in Chania from June 29 to July 10 of this year, under the auspices of the International Center for Advanced Studies. Topics which were discussed in the form of lectures, seminars and panel discussions included theoretical and experimental reviews and recent results on interband transitions, excitons, submillimeter wavelength spectroscopy, new energy band calculations, polarons, the use of sinchrotron radiation, atomic vibrations, fourier spectroscopy, many-body theory, magnetic crystals, photoemission, modulation techniques, light scattering, electron-photon-phonon interactions, and relativistic phenomena in semi-conductors.

For the last three years eleven such meetings, known as the "Chania Conferences" have taken place on the island of Crete and have attracted leading authorities from academic and governmental institutions, international organizations and industrial and business concerns, from approximately thirty-five countries. These conferences emphasize critical discussions of the state of the art and forecasts of future trends and problems in topics ranging from biology, physics and space, to computers in urban problems, technology assessment, energy resources and birth control, and investment and business strategies. They have become well known for bringing together select groups of the most distinguished representatives from the international scientific, industrial and business communities to discuss new methods and models for the improvement of man's place in society, in informal and pleasant surroundings. The hospitality of the local people, the town's magnificent old harbor, the small beach where Zorba the Greek was filmed, ancient and Minoic cities, or the vacation sites of Zeus and Europa, are being found to be ideal conditions to think, to plan and to create.

The international Center for Advanced Studies is a non-profit international organization which serves both as a meta-university and as an advisory body to industry, business and government. It is devoted to the advancement of knowledge and to the expansion of the frontiers of scientific and humanistic learning through a high quality of international exchange and cooperation. By means of select specialized, interdisciplinary and future residence research and exchange programs, the Chania Center helps bridge the existing gaps between man's knowledge of the universe and of himself, as well as those between nations or groups of nations. By promoting lasting friendships and socioeconomic and technological partnerships, it contributes to international policy in an informal way, and to the development and growth of the participating countries.

We are extemely grateful to Allis Chalmers, Gordon and Breach Publishers, Hellenic Tourism and Conferences Offices, I B M World Trade Corporation, National Bank for Industrial Development, Mr. J.G. Nicholas, Pancretan Association of America, Philips' Industries, and to Castel Wines, Efthimiadis Lines, Mamidakis Gas Company, Mr. and Mrs. M. Michaelis, National Bank of Greece, Olympic Airways, Dr. R.W. Pringle, and to the peolpe of Chania, for their support and cooperation which made the 1969 program possible. Very special thanks to Mrs. Haidemenakis for her heroic patience and kindness.

E.D. Haidemenakis

CONTRIBUTORS

F. Bassani
Istituto di Fisica, Universita di Pisa, Italy

B. Bosacchi
Istituto di Fisica dell'Universita, Milano, Italy

K. J. Button
Francis Bitter National Magnet Laboratory, Massachusetts Institute of Technology, Cambridge, Massachusetts

T. C. Collins
Aerospace Research Laboratories, Wright-Patterson Air Force Base, Ohio

J. Devreese
Faculty of Science, University of Antwerp and Solid State Physics Department S. C. K. - C. E. N., Mol, Belgium

R. N. Euwema
Aerospace Research Laboratories, Wright-Patterson Air Force Base, Ohio

R. Evrard
Department of Theoretical Physics, University of Liège, Belgium

R. Haensel
Institute for Experimental Physics, University of Hamburg Hamburg, Germany

F. A. Johnson
Royal Radar Establishment, Malvern, England

R. Kaplan
Naval Research Laboratory, Washington, D. C.

E. Kartheuser
Department of Theoretical Physics, University of Liège, Belgium

A. Lucas
Department of Theoretical Physics, University of Liège, Belgium

S. Lundqvist
Chalmers University of Technology, Goteborg, Sweden

C. Mavroyannis
National Research Council, Ottawa, Canada

T. Moriya
Institute for Solid State Physics, University of Tokyo, Japan

P. O. Nilsson
Chalmers University of Technology, Göteborg, Sweden

J. Schmit
Department of Experimental Physics, University of Liege, Belgium

B. O. Seraphin
Michelson Laboratory, China Lake, California

D. J. Stukel
Aerospace Research Laboratories, Wright-Patterson Air
Force Base, Ohio

R. F. Wallis
Naval Research Laboratory, Washington, D. C.

W. Zawadzki
Institute of Electron Technology, Polish Academy of Sciences,
Warsaw, Poland

CONTENTS

OPTICAL PROPERTIES OF SOLIDS

I

INTERBAND TRANSITIONS AND OPTICAL PROPERTIES *

Franco Bassani
Istituto di Fisica - Universita di Pisa - Italy

ABSTRACT

Relevant features of the theory of interband transitions are reviewed in connection with the structure of the optical excitation spectrum.

The theory is shown to apply to some layer compounds, particularly to graphite, but is shown to be inadequate for the excitation spectrum of alkali-halides, where the electron-hole interaction is a strong term in the total Hamiltonian.

A general procedure which deals with the case of strong electron-hole interaction and complicated band structures is mentioned and briefly outlined.

I. INTRODUCTION

The purpose of the present lectures is to discuss the connection between the electronic states of a crystal and its optical properties.

The most direct connection is established by considering the transition probability between electronic occupied states and empty states, such transitions conserve the $\underline{k}$ vector and are called interband transitions. The theory of interband transitions has been very successful to interpret the optical properties of metals and of semiconductors with large dielectric constants and small energy-gaps. However this single theory neglects the interaction between the electron

* - Based on work performed under the auspices of the C. N. R. through the G.N.S.M.

in the excited states and the hole which is left in the valence band. Such interaction is known to produce exciton states in the gap, but may have important effects also in the region with frequency higher than the energy gap. If the band structure is quite complicated the simplest effective mass approximation may not be appropriate and the problem of taking properly into account the eléctron-hole interaction is still an open problem.

In section II a general review of the phenomenological optical constants and of their experimental determination is given.

In section III the general theory of interband transitions is discussed and relevant features associated to the critical points in the band structure are discussed.

In section IV we give examples where the interband transition theory applies and explains the optical properties. We concentrate mostly on graphite because there electron-hole interaction can be neglected and the strong anysotropy can be of use to compare theory and experiment.

In section V we discuss the case of the alkali-halides, where the electron-hole interaction is expected to be very important and the band structure quite complicated. It is pointed out that the effective mass theory may not be appropriate in this case and a more general procedure is outlined. It is not possible in two lectures to give a detailed account of such a fundamental subject and to do justice to the authors who have contributed to this field. I have only tried to make remarks on the present state of the art and to indicate some open problems.

II. PHENOMENOLOGICAL STUDY OF THE OPTICAL FUNCTIONS

The optical properties of a solid are best represented by the real and imaginary part of the dielectric function (1), which appears in Maxwell equations,

$$\varepsilon(\underset{\sim}{q},\omega) = \varepsilon_1(\underset{\sim}{q},\omega) - i\,\varepsilon_2(\underset{\sim}{q},\omega) \quad (1)$$

where $\underset{\sim}{q}$ is the momentum and $\hbar\omega$ the energy transferred to the medium by the radiation field. This dielectric function is related to the other optical constants (1). In the case of the refractive index $N = n - i\ k$, the relation is:

$$\varepsilon(\underset{\sim}{q},\omega) = N^2(\underset{\sim}{q},\omega) = n^2 - k^2 - i\,2nk \tag{2}$$

In the case of the absorption coefficient $\alpha\ (q,\omega)$ the relation is:

$$\alpha = \frac{2k\omega}{c}, \text{ and}$$
$$\varepsilon_2 = \frac{nc\alpha}{\omega} \tag{3}$$

in the case of the conductivity σ

$$\varepsilon_2 = \frac{\sigma}{\omega} \tag{4}$$

In general the phenomenological theory gives the connection among optical constants and reduces to two the number of independent ones. Their knowledge completely characterizes the dielectric properties of the medium when we consider substances with magnetic permeability $\mu = 1$. Furthermore, if we know only one optical constant as function of the frequency we can derive all the others because of the Kramers-Kronig relations which connect real and imaginary part of any response function which obeys the causality condition (2). In the case of the dielectric function the Kramers-Kronig relation is

$$\varepsilon_1 = (\underset{\sim}{q},\omega) = 1 + \frac{2}{\pi}\int_0^\infty \frac{\omega'\varepsilon_2(q,\omega')}{\omega'^2 - \omega^2}\,d\omega' \tag{5}$$

From the above expression we can obtain useful sum rules by considering the limits $\omega \to 0$ and $\omega \to \infty$. In the first case we have

$$\varepsilon_1(q,0) = 1 + \frac{2}{\pi}\int_0^\infty \frac{\varepsilon_2(q,\omega')}{\omega'}\,d\omega' \tag{6}$$

In the second case, since at high frequency the electrons behave like free electrons, for which

$$\varepsilon_1(\underset{\sim}{q},\omega) = 1 - \frac{\omega_0^2}{\omega^2} \tag{7}$$

with $w^2 = \frac{4\pi n e^2}{m}$, (7')

equation (5) becomes:

$$\varepsilon_1(\underset{\sim}{q}, w \to \infty) = 1 - \frac{2}{\pi}\int_0^\infty \frac{w'\varepsilon_2(w', q)\,dw'}{w^2} \quad (8)$$

By comparing (7) and (8) we obtain:

$$\frac{2}{\pi}\int_0^\infty w\,\varepsilon_2(w, \underline{q})\,dw = \frac{4\pi e^2}{m}n \quad (9)$$

We recall that in the last formula n is the total density of the electrons which make transitions induced by the electromagnetic field, and m is their free electron mass. From the above phenomenological equations it appears that the quantity of basic interest is $\varepsilon_2(\underset{\sim}{q}, w)$. This quantity, as all other optical constants, will be a tensor in the case of noncubic materials, and when this tensor is reduced to its principal axis at most three independent components will result. For a uniaxial material we have two components $\varepsilon_{2\parallel}$ and $\varepsilon_{2\perp}$, the subscript indicating that the electric field is parallel or perpendicular to the optical axis. To investigate the optical properties in terms of electronic effects we need only to consider the case of $\underset{\sim}{q}$ very small which corresponds to the dipole approximation on the radiation field.

The main experimental methods by which the optical constants can be determined are:

(a) Optical absorption measurements, by which one determines directly the absorption coefficient $\alpha(\omega)$. These have been the traditional measurements used to determine the energy gaps in semiconductors by the absorption threshold. In bulk the measurements are very difficult to perform above threshold because the absorption coefficient becomes extremely large due to the large electron concentration. Measurements are sometimes performed on thin films.

(b) Reflectivity measurements, by which one measures the ratio $|R|^2$ of reflected to the incident intensity. The reflectivity amplitude at normal incidence is related to the indices of refraction by the Fresnel formulas

$$R = \frac{n - ik - 1}{n - ik + 1} = \frac{N - 1}{N + 1} \tag{10}$$

and the use of the dispersion relations allows a determination of n and k (3). One can also use the reflectivity at an angle of incidence with light polarized in the plane of incidence. For anysotropic uniaxial materials the formula in this case is:

$$R_{\underline{p}} = \frac{N_{\perp} N_{\parallel} \cos\vartheta - (N_{\parallel}^2 - \sin^2\vartheta)^{1/2}}{N_{\perp} N_{\parallel} \cos\vartheta - (N_{\parallel}^2 - \sin^2\vartheta)^{1/2}}, \tag{11}$$

and very precise measurements as function of the angle can allow a determination of the optical constants at a given frequency (4).

(c) Photoelectron measurements, by which one measures the kinetic energy of the electrons emitted from the crystal as function of the photon energy. This technique is more indirect because it involves the electron mean free path in the crystal (5).

(d) Electron energy loss measurements by which one measures the energy lost by fast electrons at small scattering angles (6). The energy loss probability corresponding to momentum transfer $\underline{q}$ and to energy transfer $\hbar\omega$ for an anysotropic material is given by (7)

$$W(\underline{q}, \omega) = -\frac{8\pi e^2}{\hbar} \operatorname{Im}\left[(q_i \, \varepsilon_{ij} \, q_j)^{-1}\right], \tag{12}$$

which in the isotropic case reduces to the usual formula (6)

$$W(q, \omega) = -\frac{8\pi e^2}{\hbar q^2} \operatorname{Im}\frac{1}{\varepsilon} \tag{13}$$

Also to the electron energy loss (12) one can apply dispersion relations, which have the form (7)

$$\mathrm{Re}\, f(\underset{\sim}{q},\omega) = \frac{1}{\pi} P \int_{-\infty}^{\infty} \frac{\mathrm{Im}\, f(q,\omega')}{\omega'-\omega} d\omega' + 2\,\mathrm{Re} \sum_i \frac{R_i(\Omega_i)}{\omega-\Omega_i}, \tag{14}$$

where $R_i\ (\Omega_i\ )$ indicates the residue at any poles, and f $(\underset{\sim}{q},\omega)$ indicates the function $(q_i\ \varepsilon_{ij}\ q_j\ )^{-1}$, which in the case of uniaxial material is $(\ q_p^2\ \varepsilon_\perp + q_c^2\ \varepsilon_{\parallel}\ )^{-1}$. Making use of formula (12) and of the dispersion relation (14) the optical constants can be computed from energy loss measurements as function of frequency (7).

III. THEORY OF INTERBAND TRANSITIONS AND GENERAL CONCEPTS

The optical constants are related to the electronic structure when we consider frequencies higher than the phonon frequencies. If we suppose that the electronic states are stationary states (one-electron approximation) and we use Koopman's theorem (neglect electron-hole interaction), we can relate the dielectric function to the electronic states of the crystal. We obtain (8):

$$\varepsilon_2(0,\omega) = \frac{8\pi^2\hbar^2}{\omega^2}\left(\frac{e}{m}\right)^2 \sum_{cv} \int \frac{d^3k}{(2\pi)^3} \left|\underline{e}\cdot M_{vc}\right|^2 \delta\left(E_c(\underline{k}) - E_v(\underline{k}) - \hbar\omega\right),$$

$$\text{with} \quad \underline{e}\cdot \underline{M}_{vc} = \underline{e}\cdot \int_{V\infty} d^3\underline{r}\ \psi_c^*(\underline{k},\underline{r})\ \underline{\nabla}\ \psi_v(\underline{k},r) \quad , \tag{15'}$$

where the versor $\underline{e}$ indicates the polarization direction of the electric field and E_v $(\underline{k})$ and E_c $(\underline{k})$ indicate the occupied valence bands and the empty conduction bands respectively. The study of the dielectric function with this formula may establish the connection between phenomenological constants and the microscopic electronic structure of the crystals given by the electronic energies E $(\underline{k})$.

It has been noticed long ago that the matrix element M_{vc} for allowed transition is a slowly varying function of $\underline{k}$, while the term

$$J_{vc} = \int \frac{d^3k}{(2\pi)^3}\, \delta\left(E_c - E_v - \hbar\omega\right), \qquad (16)$$

which originates from the energy conservation, may be strongly dependent of $\underline{k}$ in the vicinity of the critical points where (8)(9)

$$\nabla_k \left(E_c - E_v\right) = 0 \qquad (17)$$

Consequently a structure in the optical constant $\mathcal{E}_2(\omega)$ may arise from the joint density of states J_{vc} at critical points in $\underline{k}$ space. The critical points are those values of $\underline{k}$ where condition (17) is satisfied, and in whose proximity an expansion of $E_c - E_v$ in quadratic powers of $\underline{k}' = (\underline{k} - \underline{k}_0)$ is possible. The structure in joint density of states, and consequently in $\mathcal{E}_2(\omega)$, depends strongly on the dimensions of $\underline{k}$ space and can be classified as follows:

a) In one dimension we have singularities with a δ-like asymmetric shape. In fact

$$\int \delta\left(E_0 \mp \delta k'^2 - \hbar\omega\right) dk' = \frac{1}{\sqrt{\alpha}\sqrt{\pm(E_0 - \hbar\omega)}} \qquad (18)$$

b) In two dimensions we have two cases. A critical point M_0 with

$$\int \delta\left(E_0 + \alpha_x k_x'^2 + \alpha_y k_y'^2 - \hbar\omega\right) d\underline{k}' \simeq \frac{C}{\sqrt{\alpha_x \alpha_y}}\, \vartheta\left(\hbar\omega - E_0\right) \qquad (19)$$

and a step function-like discontinuity at $\hbar\omega = E_0$.

A point M_1, with

$$\int \delta\left(E_0 + \alpha_x k_x'^2 - \alpha_y k_y'^2 - \hbar\omega\right) \delta\underline{k}' \simeq \frac{C}{\sqrt{\alpha_x \alpha_y}}\, \lg \frac{R}{|\hbar\omega - E_0|}, \qquad (20)$$

and a logarithmic singularity at $\hbar\omega = E_0$.

c) In three dimensions we obtain more gentle discontinuities. For a point M_0 with positive coefficients of the quadratic expansion of $E_c - E_v$ the behaviour is of the type

$$\sqrt{\hbar\omega - E_0}\ . \qquad (21)$$

For a point M_1 with a negative coefficient in the quadratic expansion, the behaviour is like a constant R on one side of E_0 and like $(R - \sqrt{E_0 - \hbar\omega})$ on the other side. Graphic illustrations of all cases described above are given in references (3) and (8).

The existence of the discontinuities described in case c) has been shown to produce a structure in the optical constants of a number of semiconductors starting with Ge, and Si (9). In particular the existence of logarithmic singularities described above in case b) for saddle points has given an explanation of the sharp peaks observed in graphite and other strongly anysotropic materials (10), for which it is a good approximation to assume that the planar layers do not interact with one another and the energies do not depend on k_z. Furthermore, it is the δ -like singularity in the one dimensional density of states described above in case a) which explain the existence of sharp peaks in correspondence to Landau levels in a magnetic field (11). In fact in a magnetic field the $\underline{k}$ vector remains a good quantum number only in the direction of the magnetic field and the energy levels near a general critical point become (12):

$$E = E_0 + \left(n + \frac{1}{2}\right)\hbar\omega_c + \alpha_H k_H^2 \pm \mu_0 H g, \qquad (22)$$

with a one dimensional dependence on $\underline{k}$ and a cyclotron frequency ω_c associated to the effective masses in the direction perpendicular to H.

Another important effect in the optical constants can be obtained by considering the value of the matrix elements $|\underline{e} \cdot \underline{M}_{vc}|$ at a critical point. While if the transition is allowed this matrix element is practically a constant in the vicinity of the critical point, for some transitions $|\underline{e} \cdot \underline{M}_{vc}|$ may be zero at the critical point, in which case the expected structure does not appear. Furthermore, if the crystal is not isotropic we may have the case where $\underline{e} \cdot \underline{M}_{vc}$ is zero for a direction of polarization of the incident height and is different from zero for a for a different polarization. This produces a strong anysotropy in the optical constants which can be experimentally observed (4).

In the presence of an electric field the expression for $E_2(\omega)$ is modified because the field gives energy to the electrons, and the density of states as function of frequency is modulated (13). A basic point is that the modulation introduced by the electric field occurs only for transitions at the critical points while at a general point in the same approximation there is no effect (14).

Of course the above picture cannot explain all the excitation spectrum of solids because we have neglected correlation effects and we have used Koopman's theorem. We think it is appropriate to use the above simplified picture only for small gap materials, or when the dielectric function is very large and the electron-hole interaction can be considered a small perturbation and may be neglected in studying the basic excitation spectrum. In general this will not be the case, and we will have to consider the effect of the electron-hole interaction on the band to band transitions. Part of the correlation can be included in the explicit expression for the electron-hole interaction.

Cases for which the simple interband transition model is sufficient to explain the basic excitation spectrum are those of most semiconductors like Ge, Si etc., and of a large number of layer compounds. We will not describe the cubic semiconductors because they have been already extensively discussed (3), but we will rather concentrate on the layer compounds, and will study the simplest of such materials as an example.

Cases for which the simple interband transition model is not sufficient to explain the basic excitation spectrum are those of solid rare gases and alkali-halides crystals, and we will discuss some features of these materials and present some basic concepts to be used in any attempt to interpret their optical spectra.

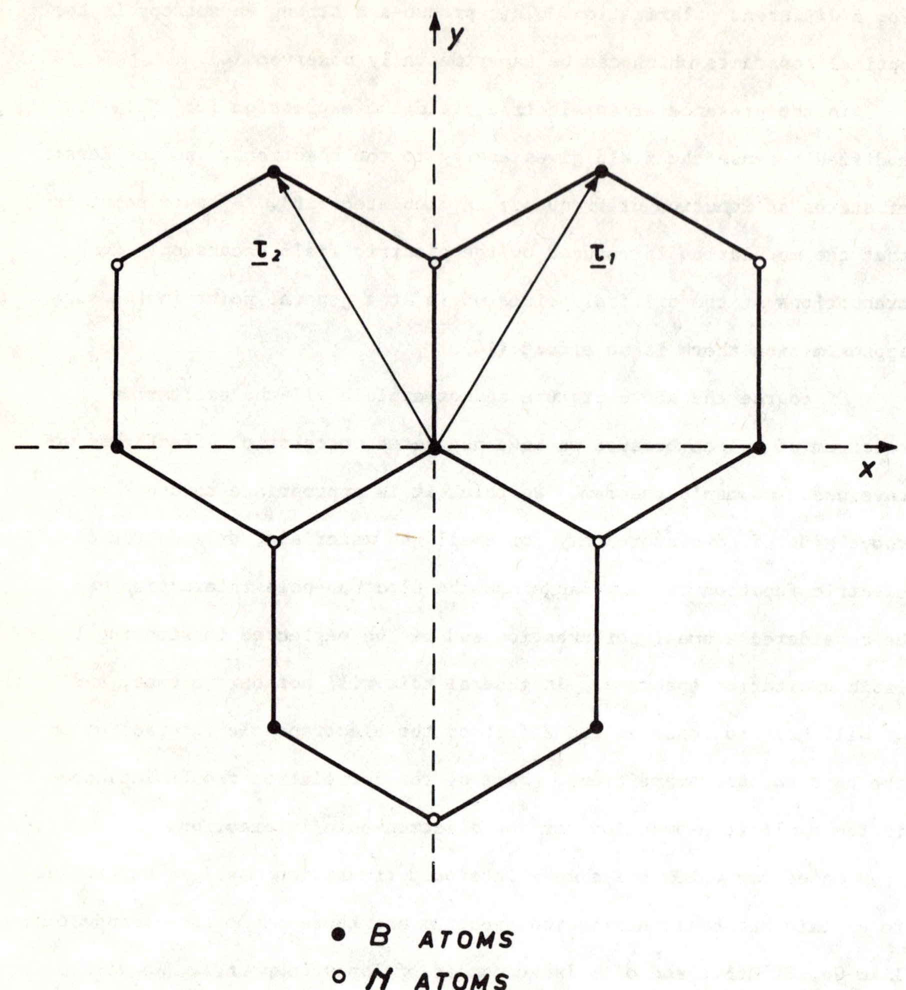

Fig. 1. Positions of the atoms in the layers of BN and graphite. The translational vectors are indicated. In graphite the two atoms in the unit cell are both C atoms.

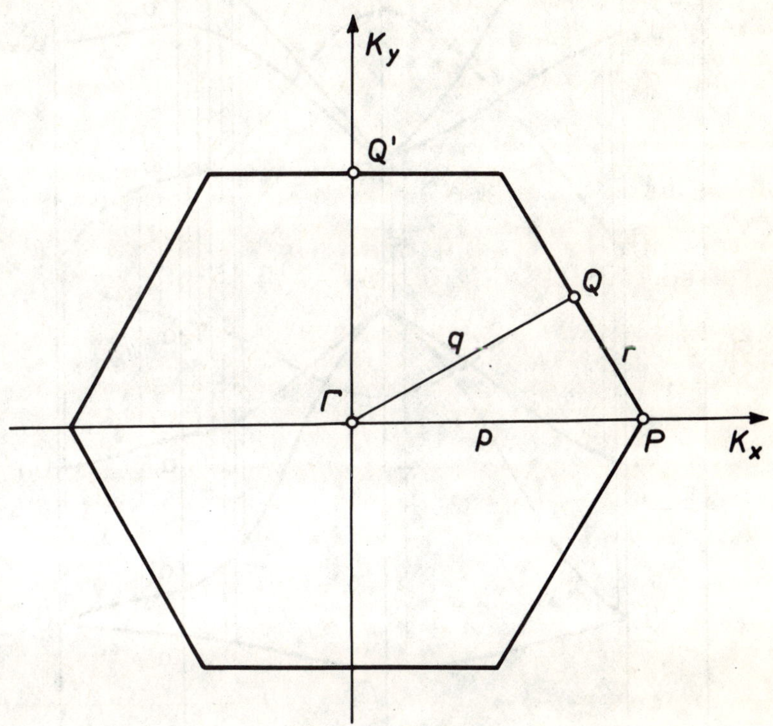

Fig. 2. Brillouin Zone for two-dimensional BN and graphite.

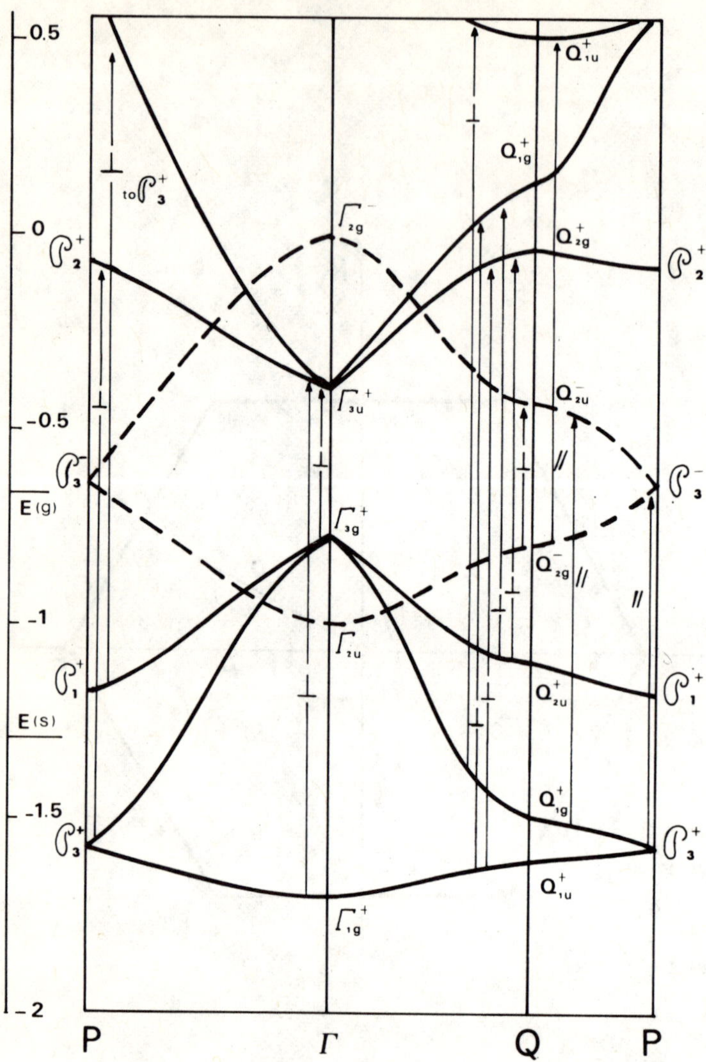

Fig. 3. Sketch of the band structure of graphite (from ref. 10) in the two-dimensional approximation. The broken lines indicate π-bands and the continuous lines indicate σ-bands. The allowed transitions at the critical points are indicated for both polarization of light. The positions of the atomic states are also reported.

IV. OPTICAL PROPERTIES AND BAND STRUCTURE OF BN AND GRAPHITE

To give an example which illustrates the above concepts we will consider the case of the simplest and very similar lamellar compounds graphite and BN, which display very nicely the critical point singularities and the anysotropy in the optical constants. They are composed of planar layers of atoms disposed as indicated in fig. 1 with basic translational vectors

$$\underline{\tau}_1 = a\left(\frac{1}{2}, \frac{\sqrt{3}}{2}, 0\right)$$

$$\tau_2 = a\left(-\frac{1}{2}, \frac{\sqrt{3}}{2}, 0\right).$$

There are two atoms per unit cell, which are different for BN, and become equal in C. The separation between the layer is so large that one may neglect the interaction among them and use the model of a two-dimensional lattice. The Brillouin Zone is presented in fig. 2, where the locations in $\underline{k}$ space of the critical points are indicated as P, Γ, and Q.

The electronic states of the valence bands and of the lowest conduction bands have been recently computed by making use of the tight-binding approximation in a semiempirical way (10) (15). The two-center overlap and potential integrals have been reduced with respect to the values would result from atomic orbitals, to take into account the distortion produced by placing the atoms into the lattice. Though the reduction factor is somewhat arbitrary, the general scheme is still meaningful and a band structure emerges which displays interesting features associated with saddle-point singularities at the point Q of the Brillouin Zone and with selection rules for transitions with light polarized along the c-axis and perpendicular to it. For the case of graphite the band structure and the associated selection rules are given in fig. 3. For the case of BN the corresponding band structure

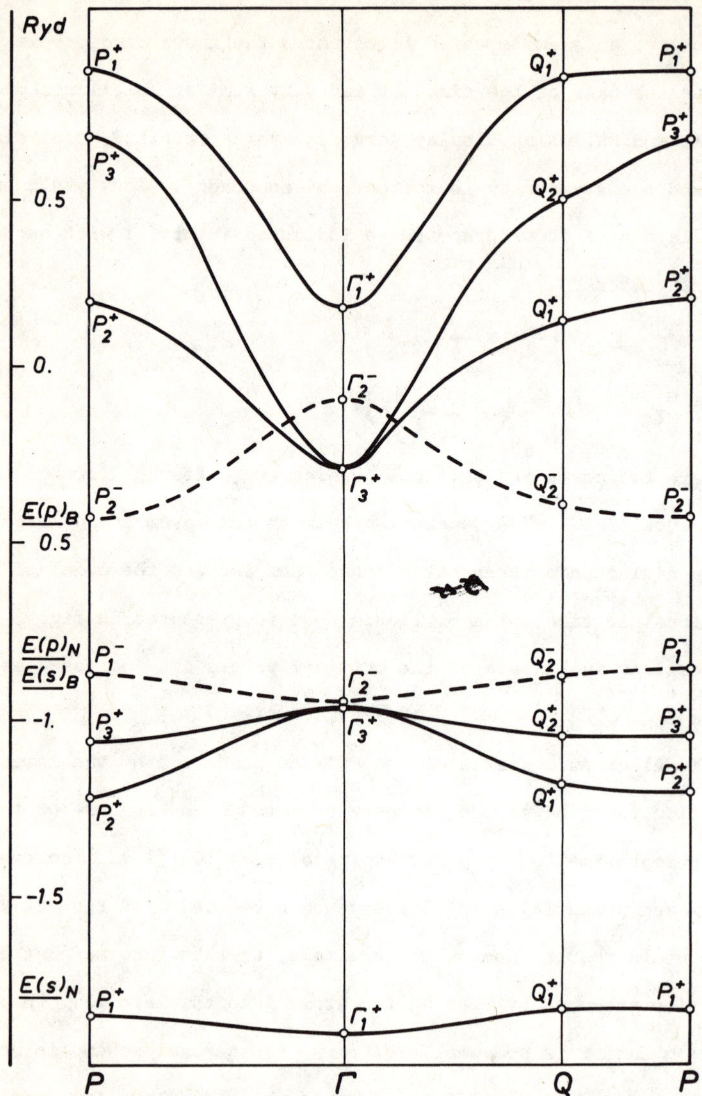

Fig. 4. Sketch of the band structure of BN in the two-dimensional approximation (from ref. 15). The broken lines indicate π-bands and the continuous lines indicate σ -bands. The positions of the corresponding atomic states are also given.

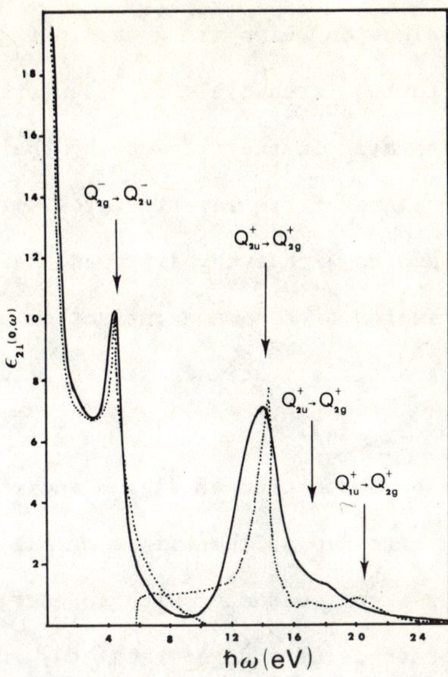

Fig. 5. Values of $\varepsilon_{2\perp}(\omega)$ in graphite. The dotted line indicates the values computed from the band structure of Fig. 3. The continuous line gives the experimental values of $\varepsilon_{2\perp}(\omega)$ (from ref. 17.)

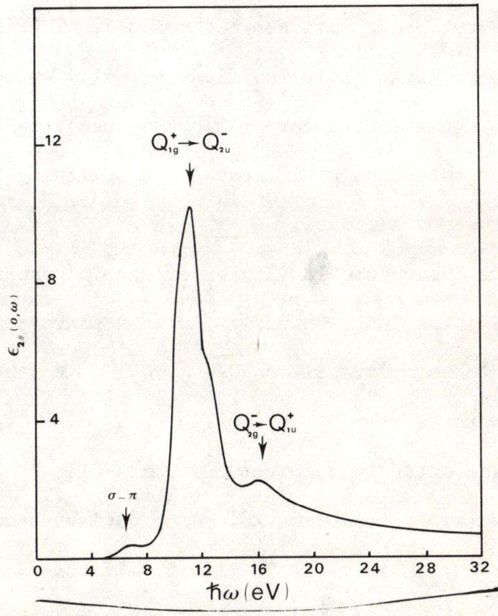

Fig. 6. Sketch of $\varepsilon_{2\parallel}(\omega)$ for graphite with indication of the relevant saddle-point transitions. The results between 2 eV and 5 eV are taken from reference (4) and those between 6 eV and 30 eV from reference (17).

and the associated selection rules are given in fig. 4. The states are indicated according to the irreducible representations of the space group, and can be separated in the all zone by the parity with respect to reflection on the plane of the layer; states with odd parity are called π states and states with even parity are called σ states. Only transitions between states of the same reflection parity contribute to $\varepsilon_{2\perp}(\omega)$ and only transitions between states of opposite reflection parity contribute to $\varepsilon_{2\parallel}(\omega)$.

One can observe by inspection of fig. 3 and fig. 4 that graphite is a semiconductor with zero gap so that light of all frequencies is absorbed, while BN is a semiconductor with an energy gap at the point P. This is due to the presence of a degeneracy at P_3^- in graphite, which is removed in BN because of the absence of inversion symmetry.

From fig. 4 we can notice that the presence of the saddle point transition $Q_2^- \rightarrow Q_2^-$ just above the gap at $P_1^- \rightarrow P_2^-$ will produce a sharp peak at about 6 eV, just above threshold. This has been observed experimentally by Choike (16), who finds actually two peaks, separated by about 1 eV. Doni and Pastori (15) have been able to interpret the existence of the two-peaks by introducing a splitting of Q_2^- due to the interaction between the planes.

From figure 3 and formula (15) we expect in graphite a smooth decrease of $\varepsilon_{2\perp}$ as function of ω, with a sharp peak at about 5 eV due to the saddle point transition $Q_{2g}^- \rightarrow Q_{2u}^-$. Another peak will appear in $\varepsilon_{2\perp}$ due to the σ - σ transitions $Q_{2u}^+ \rightarrow Q_{2g}^+$ at about 14 eV. This was actually observed experimentally by Taft and Phillipp (17) with reflectivity measurements. A comparison between the computed and the measured $\varepsilon_{2\perp}$ is given in fig. 5. If the electric field is in the direction of the c-axis, fig. 4 indicates that the resulting dielectric function $\varepsilon_{2\parallel}$ will be zero up to the weak transition

$B_1^+ \rightarrow B_3^-$ at about 6 eV, and will have a sharp peak in correspondence to the allowed saddle-point transition $Q_{1g}^+ \rightarrow Q_{2u}^-$ at about 11 eV, and a second one in correspondence to the allowed transition $Q_{2g}^- \rightarrow Q_{1u}^+$ at about 16 eV. This situation is actually found experimentally as indicated in fig. 6 where the imaginary part of the dielectric function in the direction of the c-axis is sketched as function of ω. The values of $\varepsilon_{2\parallel}$ have been obtained (7) (18) by applying formula (12) and dispersion relation (14) to experimental electron energy loss data of Zeppenfeld (19).

V. REMARKS ON THE BAND STRUCTURE AND OPTICAL PROPERTIES OF ALKALI-HALIDE CRYSTALS

The pioneering work on the band structure of alkali-halide crystals is due to Ewing and Seitz (20) and Howland (21). Their results indicate that the valence electrons are tightly bound and can be represented as electrons of the ionic shells, slightly perturbed by the other ions of the lattice. The main effect of the other ions of the lattice is to produce a shift of all the free-ion eigenvalues by the quantity $\pm\alpha_M e^2/r_0$, where α_M is the Modelung constant r_0 the lattice parameter, and the sign $\pm$ refers to positive and negative ion eigenvalues respectively. A secondary effect is to give a $\underline{k}$- dependence to the states of the valence bands, and consequently a band width, which however turns out to be rather small (of the order of 1 eV and smaller for all alkali halides). The highest valence bands corresponds to the highest p state of the halogen ion. It has symmetry Γ_{15} at $\underline{k} = 0$, it splits into two states of symmetry, X_4 and X_5 at $\underline{k} = \frac{2\pi}{a}(1,0,0)$ and into two states of symmetry L_3 and L_2 at $\underline{k} = \frac{2\pi}{a}(\frac{1}{2},\frac{1}{2},\frac{1}{2})$. The top of the valence band turns out to be Γ_{15} in most calculations, but other calculations give a slight maximum in the Σ direction (110) or in the Δ direction (100). The states $^3\Gamma_{15}$, $^2X_5$ and $^2L_3$ are split by spin orbit interaction into the states $^4\Gamma_8^- + {}^2\Gamma_6^-$, $^2X_7^- + {}^2X_6^-$ and $^2L_5^+ + {}^2L_6^-$ respectively.

The magnitude of the spin-orbit splitting is very nearly the same as that of the corresponding atomic state at Γ (slightly larger because of the normalization correction on the wave-function). At X and L it is about 2/3 of the value at Γ

In order to have a guideline for interpreting the excitation spectra of these materials a knowledge of the electronic structure of the conduction band is required. A considerable number of calculations have been recently performed on the conduction band structure by the Orthogonalized Plane Wave Method (22), by the Kohn Rostoker variational method (22), and by the Argumented Plane Wave Method (24). All of the above procedures make use of a model potential which includes an average exchange contribution estimated by the Slater approximation; they solve the Schrödinger equation for an electron in this potential both for the valence and the conduction states. Though some of the results obtained may be of physical significance for the optical spectrum, nevertheless it appears that in this type of substance the approximation of using the the same model potential for valence and conduction states is not justified. The conduction electron is an extra electron added to the lattice and sees a lattice potential of closed shell ions, so that it tends to be spread over a large distance. If the interaionic separation becomes large the conduction electron goes into the states of the alkali atom, while the valence electron goes into the states of the halogen ion. The difference between the potential of the conduction electron and that of the valence electron is apparent in the exact Hartree-Fock formulation of the problem but is not included when one considers the average exchange potential. While in metals and in small gap materials this difference may not be large, we think that in the case of the alkali halides it may be a very important element in establishing the conduction band structure. For the above reasons we have proposed a scheme which is of a semiempirical nature, but can be used easily to obtain results for the band structure of a large number of alkali-halides and to compare significant trends of the band structure between different compounds (25). The scheme consists in constructing a semiempirical pseudopotential appropriate to the conduction states as the lattice as sums of pseudo-potentials of the free ions. It is along the line suggested by Heine and Habarenkov for the case of metals and semiconductors. One assumes:

$$V_p(\underline{r}) = \sum_v \left[V_p(\underline{r} - \underline{R}_v) + V_p^+\left(\underline{r} - \underline{R}_v + \frac{a}{2}\right)\right] , \qquad (23)$$

where V_p and V_p^+ are the pseudopotentials of the appropriate negative and positive ions, a is the lattice parameter, and R_v the translations of the f.c.c. lattice. While the potentials along would be strongly attractive, the pseudopotentials are generally weak because the repulsion from core and valence electrons is included by the orthogonality condition. The pseudopotential V^- for instance does not give any bound states to an extra electron and the potential V^+ gives as bound states the spectro-scopy terms of the optical electron in the alkali atom. A suitable form for the ionic pseudopotentials has been given by Giuliano and Ruggeri (26) for the alkali ions and can be extended to the halogen ions, it consists in writing

$$V_p(\underline{r}) = \sum_\ell V_\ell(|\underline{r}|) P_\ell , \qquad (24)$$

$\underline{P}_\ell$ being the angular momentum projection operator, and

$$V_\ell(\underline{r}) = \mp \frac{e^2}{r + \frac{A_\ell}{r^2}} - \frac{N_\ell^2 e^{-\alpha_\ell r}}{r} \qquad (24')$$

with N atomic number, and A_ℓ and α_ℓ appropriate parameters. This pseudopotential has the correct behaviour at the origin and at infinity and contains a repulsive effect of the occupied states at a distance within the ionic radius. For d-states and higher which do not have inner states of comparable symmetry α_ℓ is such smaller than for s and p

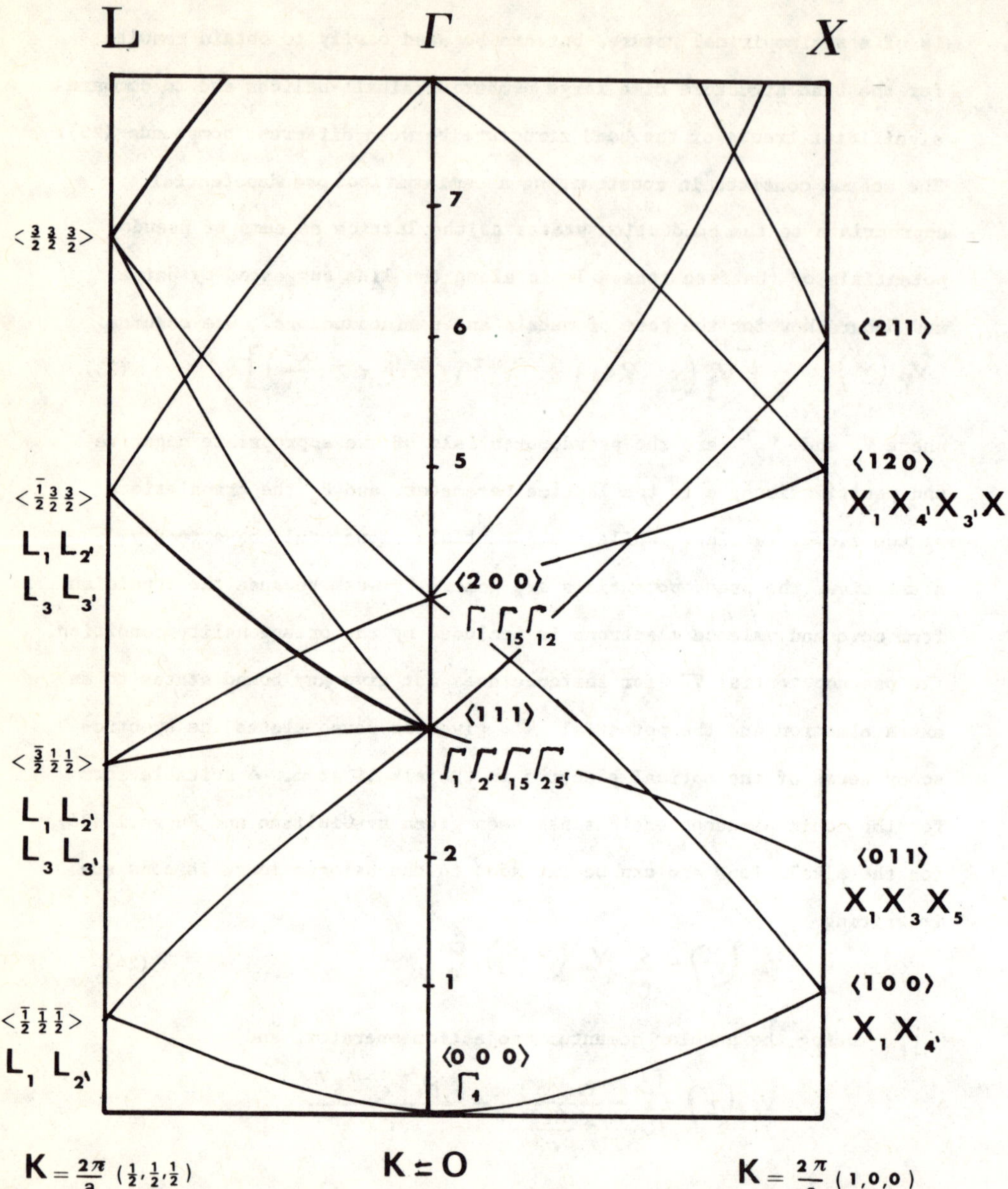

Fig. 7. States of the "empty lattice" of an alkali-halide crystal. The crystal symmetries of the states are indicated for the lowest eigenvalues. The accidental degeneracies will be split by the crystal potential.

states α_ℓ is so large that it has no effect on the band structure and the A_ℓ are chosen to give the spectroscopic terms in the case of the alkali ions and are taken as a fixed parameter in the halogen ion case.

The above described crystal pseudopotential has small matrix elements between plane waves and consequently it can be treated as a perturbation on the "empty lattice" formed with a constant potential, where the conduction electrons are represented by plane waves. By argument of perturbation theory we can see what to expect as the sequence of the energy levels of the conduction band. The lowest states of the empty lattice are given in fig. 7 and their crystal symmetries are indicated. We expect at the center of the Brillouin Zone the sequence $\Gamma_1 \ll \Gamma_{25}' < \Gamma_2', \Gamma_1, \Gamma_{15}$. The reason why Γ_{25}' is the lowest among the states which follow Γ_1 is due to the fact that Γ_{25}' is a d-like state and it feels a more attractive potential. For the same reason the sequence at X is $X_1 < X_3 <$ X_4', X_1, X_5, X_3 being the d-like state. At the point L we have the sequence $L_1 < L_2' < L_3$, L_1, L_3'. The difference between the results we obtain with this procedure and the results of other more involved calculations is that our states are more free-electron like and the states X_3 and Γ_{25}' are not lowered as much with respect to the other states. The trend from one substance to the other can also be anticipated on physical grounds. As we go from LiCl to KCl the state X_3 (d-like symmetry) becomes lower relative to Γ_1 (s-like), while the states L_1 and L_2' increase. In NaCl and KCl the absolute minimum of the conduction band is Γ_1 while in LiCl it is probably L_1. In all alkali-halides X_3 is a secondary minimum with large effective mass, well pronounced in KCl and somewhat uncertain in NaCl.

Various attempts have been made to interpret the excitation spectrum of alkali-halides on the basis of band structures of this type and this analysis is still under way. The experimental evidence is that one cannot interpret the existing structure in terms of interband transitions, because the peaks are too sharp and one has to take into account the electron-hole interaction. Such interaction produces exciton states in the gap and may produce exciton states originating from higher minima in the conduction band. Phillips (9) has first suggested that saddle-points in the conduction band may also give exciton states in resonance with the continuum. Baldini and Bosacchi (27) have attributed the strong exciton peak above ionization to the minimum at X_3, which has d-like symmetry. Another minimum exists at L_2', but transitions to it from the valence state would be forbidden in first order. Transitions from the core states, as observed by Haensel (28) with sincrotron radiation, should result in similar excitons, but the selection rules are different depending on the specific core state.

The interpretation of the structure which is observed experimentally is complicated by the fact that in this type of crystals the electron-hole interaction is a strong one because of the small dielectric constant and consequently bound and resonant states may be associated to the overall band structure rather than to a single critical point. In fact, the equation one must solve to determine the excitation spectrum, in the hypothesis that the valence band is completely flat, is

$$\left(E(\underline{k})-E\right)\varphi(\underline{k})+\int d\underline{k}'\,\mathcal{U}(\underline{k},\underline{k}')\,\varphi(\underline{k}')=0 \tag{25}$$

where $E(\underset{\sim}{k})$ stands for $E_c(k) - E_v$,

$$\mathcal{U}(\underline{k},\underline{k}')=\frac{1}{(2\pi)^3}\int \psi^*(\underline{k},\underline{r})\,U(\underline{r})\,\psi(\underline{k},\underline{r})\,d\underline{r} \tag{25'}$$

is the Bloch transform of the electron-hole interaction $\mathcal{U}$ (r), and $\varphi(\underset{\sim}{k})$ is the envelop function obtained by expanding the wave function in the complete set of Bloch states

$$\Phi_E(\underline{r})=\int d\underline{k}\,\varphi(\underline{k})\,\psi(\underline{k},\underline{r}). \tag{25''}$$

If we take plane waves instead of Bloch functions and use a quadratic expression for E(k) we obtain by a Fourier transform the effective mass approximation of Kohn and Luttinger and the Wannier excitons if the potential is Coulomb like. This may be a poor approximation however if $\mathcal{V}(r)$ is not a slowly varying potential and if the critical points in the band structure are not well separated. The general solution of the equation (25) is of the form:

$$\varphi(\underline{k}) = \delta(\underline{k} - \underline{k}_E) - \int \frac{d\underline{k}'\, \mathcal{V}(\underline{k}, \underline{k}')\, \varphi(\underline{k}')}{E(\underline{k}) - E - i\epsilon} \tag{26}$$

where ϵ is an infinitesimal which insures that the boundary conditions are automatically satisfied for scattering states. In a recent paper (29) we have shown that it is still possible to demonstrate the existence of bound states in the gap and resonant states in the continuum in correspondence to minima of the reduced zone, provided one can satisfy the condition that $\mathcal{V}(\underline{k}, \underline{k}')$ when $\underline{k}$ and $\underline{k}'$ are in the region of $\underline{k}$-space belonging to different minima is much smaller than $\mathcal{V}(\underline{k}, \underline{k}')$ when $\underline{k}$ and $\underline{k}'$ are in the region of $\underline{k}$ space relative to the same minimum. In this case the absolute minimum gives the usual excitons and the higher minima may give the resonant excitons. The resonant excitons are characterized by a finite lifetime, which has to be large compared to the transition time. If the above condition is not satisfied, as may well be the case in alkali-halides, then to determine the eigenvalues E one should solve a Fredholm type secular equation:

$$\mathrm{Det} \left| \delta_{nn'} - \int \frac{d\underline{k}'\, f_n(\underline{k}')\, f^*_{n'}(\underline{k}')}{E(\underline{k}') - E} \right| = 0 \tag{27}$$

where

$$f_n(\underline{k}) = \frac{1}{(2\pi)^{3/2}} \int d\underline{r}\, \psi(\underline{k}, r) \sqrt{-\mathcal{V}(\underline{r})}\, g_n(\underline{r}), \tag{27'}$$

$g_n(\underline{r})$ being the function of an arbitrary complete set. The real values of E give the bound states and the complex values of E, when (27) is properly extended into the complex E plane, may give resonances if the imaginary part is negative and small compared to the real part. The all band structure contributes to the integral and consequently may influence the results.

In some cases it may not be surprising to find that the overall effect of the band structure in the integral appearing in equation (27) is only slightly different from what would result if the lattice potential were not present at all; in particular this will tend to be true as E becomes large. In case there are stationary and resonant excitons it is likely that they absorb most of the oscillator strength of the transition probability and one may be unable to observe transitions to the scattering states which would correspond to the interband transitions.

REFERENCES

1. See for instance:
J.A. Stratton; Electromagnetic Theory, McGraw-Hill, 1941
F. Seitz; The Modern Theory of Solids, Chapter XVII, McGraw-Hill, 1940
F. Stern; Elementary Theory of the Optical Properties of Solids - Solid State Physics - Edited by F. Seitz and D. Turnbull vo. 15, 299, (1963)

2. L.D. Landau and E.M. Lifshitz; Electrodynamics of Continuous Media, Pergamon Press. (1960)

3. D.L. Greenaway and G. Harbecke; Optical Properties and Band Structure of Semiconductors, Pergamon Press, 1968

4. D.L. Greenaway, G. Harbecke, E. Tosatti and F. Bassani; Phys. Rev. 178, 1340 (1969)

5. P.O. Nilsson; Photoemission and Optical Studies of Metals and Alloys, this same volume

6. H. Reather; Solid State Excitations by Electrons in Springer Tracts in Mod. Physics 38, 84 (1965)

7. F. Bassani and E. Tosatti; Physics Letters 27A, 446 (1968)
E. Tosatti; Il Nuovo Cimento B, to be published

8. See for instance:
F. Bassani; Band Structure and Interband Transitions, in: Proceedings of the International School "E. Fermi" Course XXXIV, page 33 (1966)

9. J.C. Phillips; The Fundamental Optical Spectra of Solids, in Solid State Physics, Edited by F. Seitz and D. Turnbull vol. 18, 55 (1966)

10. F. Bassani and G. Pastori-Parravicini; Il Nuovo Cimento B, 50 pag. 95 (1967)

11. L.M. Roth and P.N. Argyres; Magnetic Quantum Effects, Chapter 6 of Semiconductors and Semimetals vol. I, Edited by R.K. Willardson and A.C. Beer, Academic Press 1966

12. A. Baldereschi and F. Bassani; Proceedings of the IX International Conference on the Physics of Semiconductors, pag. 280 Moscos (1968)

13. See for instance:
B.O. Seraphin; Modulated Reflectance in the Study of Band Structure, this same volume
D.E. Aspnes and N. Bottka; Electric-Field Effects on the Dielectric Function of Semiconductors and Insulators, from Semiconductors and Semimental, Edited by R.K. Willardson and A. Beer vol VI, Academic Press, New York (1969)

14. F. Aymerich and F. Bassani; Il Nuovo Cimento 48B, 358 (1967)

15. E. Doni and G. Pastori-Parravicini; Il Nuovo Cimento B, to be published

16. W.J. Choyke; private communication, to be published.
The author is indebted to Dr. Choyke for sending his data prior to publication

17. E.A. Talf and H.R. Philipp; Phys. Rev. 138A, 197 (1965)

18. E. Tosatti and F. Bassani; Il Nuovo Cimento, to be published

19. K. Zeppenfeld; Zeits. Phys. 211, 391, (1968)

20. D.H. Ewing and F. Seitz; Phys. Rev. 50, 760 (1936)

21. L.P. Howland; Phys. Rev. 109, 1927 (1958)

22. A.B. Kunz; Phys. Rev. 175, 1147 (1968)

23. Y. Onodera and Y. Toyozawa; J. Phys. Soc. Japan 22, 833 (1967)

24. P.D. De Cicco; Phys. Rev. 153, 931 (1967)

25. F. Bassani and S. Giuliano; Il Nuovo Cimento B, to be published

26. S. Guiliano and R. Ruggeri; Il Nuovo Cimento B, 61 53 (1969)

27. G. Baldini and B. Bosacchi, Phys. Rev. 166, 863 (1968) also B. Bosacchi, Excitons in Alkali-Halides, this same volume

28. R. Haensel, Optical Properties of Alkali-Halides and Solid Rare Gases by Syncrotron Radiation, this same volume

29. F. Bassani, G. Iadonisi and B. Preziosi, Phys. Rev., to be published

II

INVESTIGATION OF OPTICAL PROPERTIES OF ALKALI HALIDES AND SOLID RARE GASES BY SYNCHROTRON RADIATION

R. Haensel

Physikalisches Staatsinstitut, II. Institut

für Experimentalphysik der Universität Hamburg

ABSTRACT

Optical absorption spectra of inner shell transitions in alkali halides and solid rare gases have been obtained by the use of synchrotron radiation in the photon energy range 10 to 260 eV. For the alkali halides the results are compared with those in the fundamental absorption region, for the solid rare gases with the corresponding atomic transitions. The discussion of the results includes band structure calculations as well as exciton theory. In the experimental part some remarks are made on the characteristics of synchrotron radiation.

I. INTRODUCTION

Alkali halides and rare gas solids are formed from ions or atoms respectively which have completely filled atomic shells, and whose electrons, therefore, are closely bound to the nucleus. As a result there is a big energy gap between the occupied valence band states and the unoccupied conduction band states. Because of the big band gap these solids are transparent in the visible and have the onset of optical absorption in the vacuum ultraviolet. There are two reasons for the rapid growth of interest in the optical behaviour of these wide band gap materials during the last few years. One reason is the development of experimental techniques which has allowed more and more refined experiments. The second reason is the development of band structure calculations, which are now available for nearly all alkali halides and solid rare gases. Both developments are closely connected, for on the one hand the band calculations stimulated the experimentalists to compare experimental results with theoretical predictions. On the other hand the theoreticians used experimental values (e.g. of band gap energy) to adjust certain parameters in their calculations. Because of the big binding energy of the valence electrons the wavefunctions are highly localized (in the alkali halides at the halogen ion) and can be theoretically treated like the core states by a tight binding approximation. The valence bands are relatively flat and their width is much smaller than the band gap energy. The conduction band states are composed of the unoccupied atomic states; because of the overlapping of the wavefunctions of neighbouring atoms they are spread over both the negative and positive ions. As the band calculations give results on the energy eigenstates of an electron in the conduction band with a filled valence band, they describe only interband transitions correctly. In large band gap materials with a low dielectric constant however, transitions play a big role where the hole, left by the excited electron, is not completely screened by the surrounding valence electrons but forms bound states together with the excited electron. These states are called excitons.

Excitons may be described with the help of two different models[1]: The Frenkel picture describes the excitons as being built up of atomic wavefunctions. The Wannier-Mott picture describes it as a hole surrounded by the electron at a distance of the order of the lattice constant. The Wannier-Mott excitons from hydrogen-like eigenstate series, which converge towards certain singularities in the continuum conduction band. In our case an intermediate picture between these two boundary models has to be applied: Baldini[2] found that in valence band transitions of the solid rare gases, the first exciton is very close to the first atomic transition in the gas, but he also saw exciton series which are especially clear in solid xenon. In alkali halides exciton series have also been seen e.g. by Fischer and Hilsch.[3] Band calculations covering the conduction band states are available for LiF[4,5], LiCl[6], LiBr[7], LiJ[8], NaCl[6,9,10], NaBr[7], NaJ[8], KCl[6,11-13], KBr[7], KJ[14], RbCl[7], RbBr[7], CsJ[15,16], and for Ar[17-19], Kr[20], and Xe[21].

A review of the theoretical and experimental problems involved in the investigation of optical properties of alkali halides and solid rare gases may be found in papers by Knox and Teegarden[22] and Baldini.[2]

II. VALENCE BAND AND CORE SHELL TRANSITIONS

The bulk of experimental material of optical properties of alkali halides and solid rare gases is available on transitions from the valence band. The valence band is built up of p-symmetric states. As the lowest part of the conduction band is mainly s-symmetric, the onset of optical absorption mainly shows structure due to transitions between these states including the excitons. As different transitions may have initial states at different points of the valence band, the optical absorption structure in the fundamental absorption region does not directly give a picture of the conduction band structure but of the combined valence- and conduction-band structure. Since the valence states have p-like symmetry wavefunctions concentrated at the halogen ion, the result is that only transitions into s- and d-symmetric final states, whose wavefunctions have a considerable weight at the halogen ion, can be observed.
To get information on the other states in the conduction band, transitions from the core levels on both ions with different symmetry have to be studied additionally. Transitions from p-symmetric core states are also interesting. As the core bands

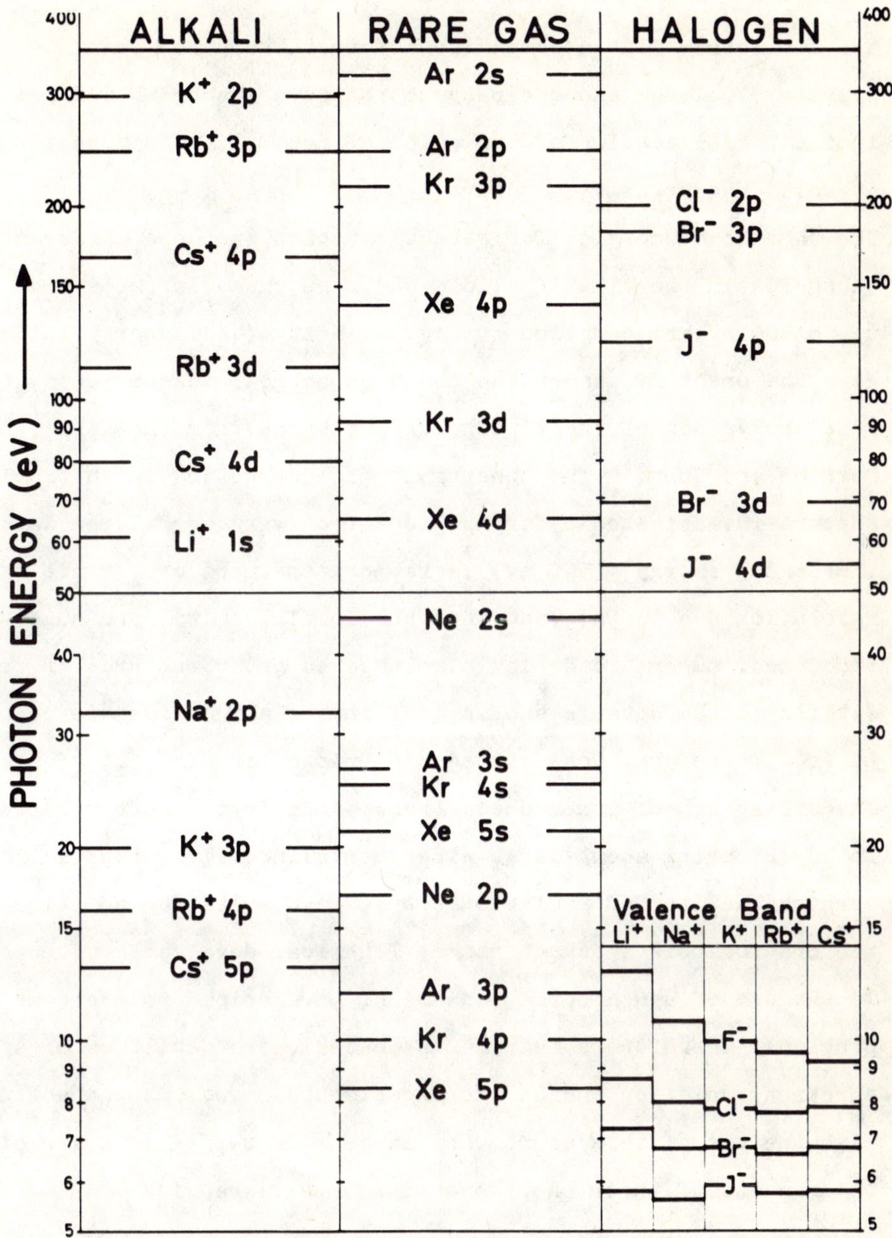

Fig. 1 Threshold energies for transitions from the different occupied shells in the alkali halides and rare gases. Each compound is indicated for the valence band transitions. For core levels the dependence of the chemical composition is small and therefore has been neglected.

are completely flat a comparison of the absorption from the p-symmetric valence and the p-symmetric core states should help in the identification of the point of the Brillouin zone scheme at which the transition occurs.

The threshold energies for core excitation from the different subshells in the alkali halides and rare gases in the energy range up to 500 eV are compiled in Fig. 1. Most of the energy values give the onset of absorption based on optical measurements, the rest are values of the binding energy taken from the tables of Bearden and Burr.[23] The observable fine structure in absorption spectra investigated by transitions from very deep lying levels (threshold energy > 500 eV) lacks more and more obtainable energy resolution due to the fact that the initial states are Auger-broadened. Auger broadening also smeares out structures in transitions in the outer s-shells[24] of the alkali halides (except Li^{+}1s). Therefore these states are omitted in Fig. 1.

Investigations of inner shell transitions in the energy range 10 to 500 eV bring about large experimental problems, especially concerning the lack of radiation sources and suitable radiation detectors for this spectral range. Technical developments, especially the use of synchrotron radiation, have helped to overcome these problems. Therefore some remarks on the characteristics of synchrotron radiation and on the experimental problems connected with the use of this source should be made before a review of the results of inner shell transitions will be given.

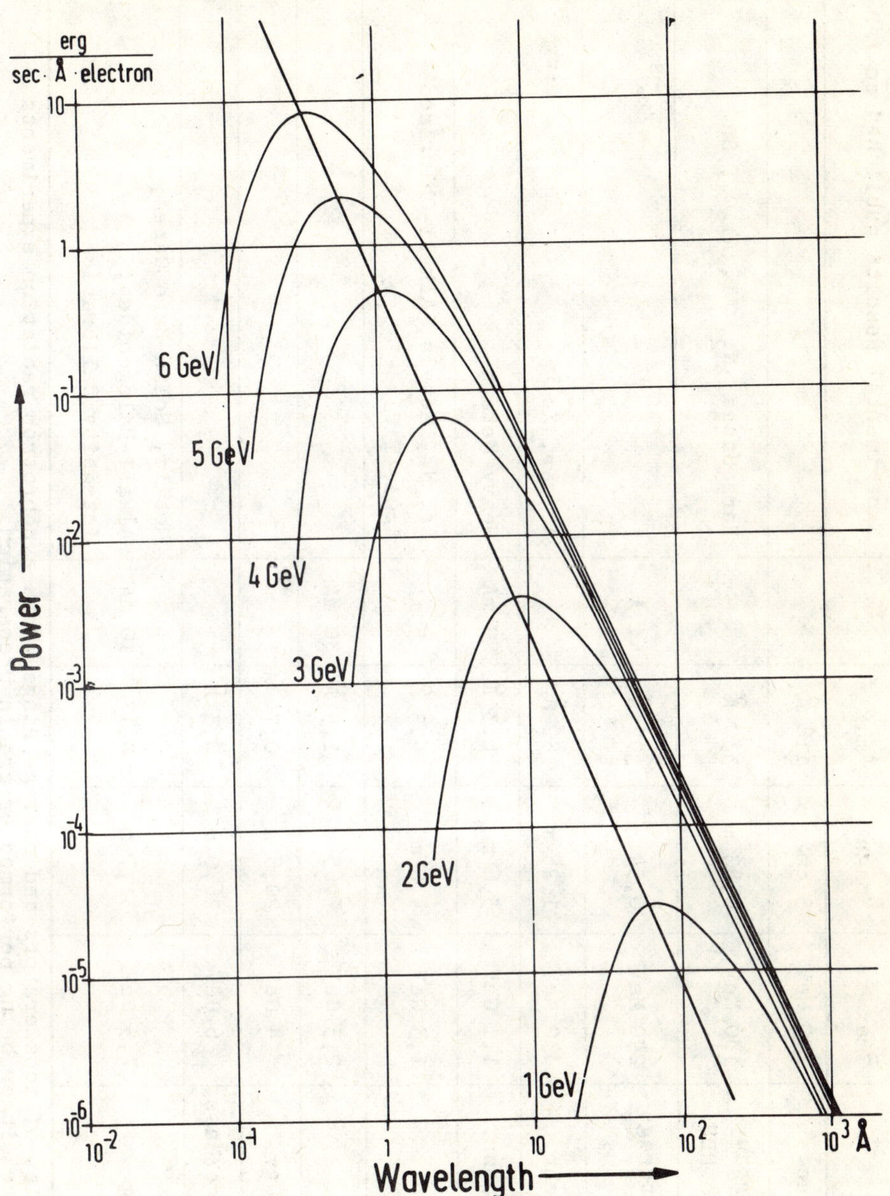

Fig. 2 The spectral distribution of the synchrotron radiation emitted by electrons of different energies at DESY.

Machine	E_{max}	R_{magn}	λ_{max}	I	Experimental Results published so far
Glasgow Linac	100 MeV	5 cm[≠]	118 Å		
National Bureau of Standards NBS	180 MeV	0,83 m	330 Å		Rare Gases, N_2, Al, Al_2O_3, Sn
Wisconsin Storage Ring	240 MeV	0,54	92 Å	~1 A	
Glasgow Synchr.	330 MeV	1,25 m	82 Å		Ar, Kr, Xe
Frascati Synchr.	1,1 GeV	3,6 m	6,3 Å	9 mA	Heavy Metals
Tokyo Synchr.	1,3 GeV	4,0 m	4,3 Å	5 mA	Ar, N_2, Metals, Alkali-Chlorides, AgCl, TlCl
Bonn Synchr.	2,3 GeV	7,65 m	1,5 Å	30 mA	
NINA Daresbury	4 GeV	20,8 m	0,76 Å		
CEA Cambridge/Mass.	6 GeV	26,0 m	0,3 Å		
DESY Hamburg	7,5 GeV	31,7 m	0,17 Å	10 mA	Metals, Solid Rare Gases, Alkali-Halogenides, Intensity Calibration

Table I: List of accelerators and storage rings where synchrotron radiation experiments are either being performed or are in preparation.

[≠] in a 70 kΓ supercondicting magnet

III. SYNCHROTRON RADIATION

Synchrotron radiation is emitted by electrons centripetally deflected in bending magnets of high energy cyclic accelerators and storage rings. The radiation has a continuous spectral distribution extending from the radio wave region up to the ultraviolet or x-ray region dependent on the electron energy. Fig. 2 shows, as an example of spectral distribution the characteristics of the synchrotron radiation at the 7,5 GeV electron synchrotron DESY in Hamburg. The intensity is much higher than that of other sources especially of gas discharge continua and thus makes synchrotron radiation most suitable for absorption, reflection- and photoyield measurements in gases and solids. The high degree of polarization of the radiation is also very useful for solid state investigations. Table I summarizes the different accelerators where synchrotron radiation experiments have been performed or are in preparation. The basic parameters of the different accelerators (maximum energy E_{max}, the magnetic radius R_m, λ_{max} giving the spectral maximum, and the circulating current I) are included together with remarks on the experimental results published so far. The theoretical basis for synchrotron radiation may be found in textbooks.[25,26] An extended development of the theory as based on the classical electrodynamics by Schwinger[27] and on the quantum mechanics by Sokolov et.al.[28] has been given. General descriptions of the characteristics of synchrotron radiation at different accelerators can be found in references[29-31].

As the synchrotron radiation has a very small angular dispersion around the instantaneous electron flight direction it is usually taken from a beam pipe mounted tangentially on the electron orbit in a bending magnet and led to the spectrometers several meters apart from the tangential point. In Hamburg and Tokyo, where the data, to be reviewed in the next chapter, have been taken, different spectrometers in normal and in grazing incidence have been used. At DESY photomultipliers have been used for radiation detection, whereas in Tokyo photographic films have been used. The latter does not allow exact intensity measurements, so that the evaluation of quantitative absorption coefficients is difficult. On the other hand they register the whole spectrum simultaneously and therefore are insensitive to current fluctuations in the accelerators which need intensity normalization procedures during scanning of the absorption spectrum with photomultipliers. These current fluctuations are a special problem in synchrotrons but will not play an important role in storage rings, such as the one in Stoughton, where alkali halide experiments are being performed by a group from the University of Illinois.[32] At accelerators in the GeV (10^9 eV) range the spectrum of synchrotron extends to the x-rays (in the 100 keV-region), thus causing background radiation. This background radiation can be suppressed by inserting a mirror between the source and the spectrometers, which only reflects wavelengths longer than a limiting wavelength which depends on the entrance angle. Beside this straylight, caused by the high energy background radiation, the continuous spectral distribution gives rise to background caused by light which is reflected from the grating in higher orders. This higher order light can be suppressed by the choice of gratings having an appropriate reflection characteristic (entrance angle, blaze angle) and by the use of different filters.

IV. EXPERIMENTAL RESULTS FROM INNER SHELL TRANSITION MEASUREMENTS

IV.1 Initial state with p-symmetry.

As can be seen from Fig. 1 the core shells with the lowest threshold energy are the outer shells of the alkali ions. Cs^+5p has its threshold at 13 eV, Rb^+4p at 16 eV, K^+3p at 20 eV, and Na^+2p at 32 eV.

Transitions from the K^+3p-shell. Reflection measurements for investigation of transitions from the K^+3p-level have been performed by synchrotron radiation on KCl, KBr and KJ.[33] Blechschmidt et.al.[33] used the multi-angle reflection method to evaluate n and k, as well as ε_1, ε_2 and $|Im\frac{1}{\varepsilon}|$ in the energy range 12 to 30 eV. At this range synchrotrons are not the only radiation sources available, but with synchrotron radiation it is easy to scan the photon energy spectrum continuously with high resolution over a wide range. In addition measurements in parallel- and perpendicular-polarized light allow for checking consistency and accuracy of the data obtained. Stephan et.al.[35-36] using a multiline capillary discharge radiation source have made reflection measurements for 15° angle of incidence and a subsequent Kramers-Kronig analysis for an evaluation of KF, KCl and KBr in the energy range 5 to 45 eV. The results of both groups for KCl and KBr differ slightly in quantitative values but agree completely in the shape of the spectra as compared one to another and with results obtained from electron energy loss experiments made by Creuzburg[37] and Keil.[38] In the KF-results[34] a repetition of the valence band structures can clearly be seen in the $K^{+}3p$-transitions. In KCl and KBr the Γ_1- and X_3-excitons can clearly be seen at the beginning of each transition. The similarity of the rest of the absorption structure is not as great as in KF. There could be reasons for these differences: a) The initial states of valence band- and $K^{+}3p$-transitions are on different ions and the final state wavefunctions in the conduction band of KCl and KBr might be more inequally distributed on both ion types as in KF. b) The spin-orbit splitting of $K^{+}3p$ is small ($\sim$0.27 eV) and the double peaks in absorption have not been resolved by Stephan et.al.[34-36], but Keil in KBr[38] and Blechschmidt et.al. in KJ[33] see some indication of it in the Γ_1-exciton. The valence band on the other side of KCl is split by 0.1 eV and of KBr by 0.5 eV, thus giving, in the case of KBr, double peaks in the valence band absorption, which would have to be unfolded before a comparison with the $K^{+}3p$-transition could be made, a problem which is difficult because of interaction between the transitions from the two valence band parts.

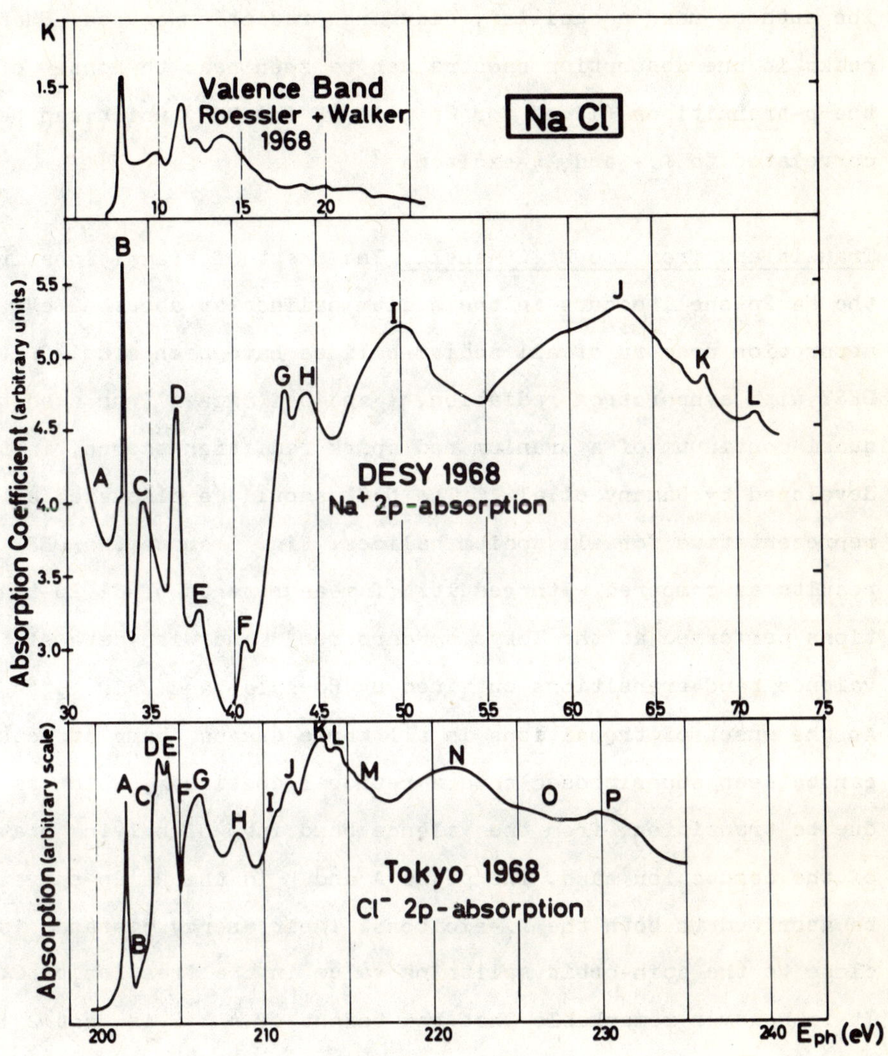

Fig. 3 Comparison of the absorption structure of NaCl for transitions from the valence band, Na^+ 2p-level and Cl^- 2p-level.

<u>Transitions from the $Rb^{+}4p$ and $Cs^{+}5p$-shell.</u> The $Rb^{+}4p$ and $Cs^{+}5p$-transitions have mainly been studied by Saito et.al.[39] The authors used a capillary discharge radiation source. Sharp peaks in the absorption spectra can be seen near the onset of the p-transitions (13 eV for Cs^{+}, 16 eV for Rb^{+}) which can be correlated to Γ_1- and X_3-excitons.[7]

<u>Transitions from the $Na^{+}2p$-shell.</u> The onset of transitions from the $Na^{+}2p$-shell occurs in the sodium halides at about 32 eV. The absorption spectra of all sodium-halides have been studied at DESY with synchrotron radiation,[40] and bei Sagawa[41] who used the quasi-continuum of a uranium rod spark radiation source, initially developed by Damany et.al.[42] The NaCl should be discussed as representative for all sodium halides. Fig. 3 shows the DESY results as compared with results of measurements of $Cl^{-}2p$-transitions performed at the Tokyo synchrotron[43] and with data of valence band transitions obtained by Roessler and Walker.[44] At the onset of transitions in all three curves sharp structures can be seen superimposed onto a residual continuum, which is due to transitions from the valence band into high lying states of the conduction band. The peaks A and B in the $Na^{+}2p$-curve can be ascribed to both the Γ_1-excitons. Their energy distance is close to the spin-orbit splitting value in the free ion of 0.2 eV. It is however remarkable that the height of A : B is not 2 : 1 as would be expected from the statistic contribution of the $2p_{3/2}$ and $2p_{1/2}$ shell, but 0,6 : 1 in NaF, 0,3 :1 in NaCl, 0,47 : 1 in NaBr and 0,19 : 1 in NaJ. This inversion of the

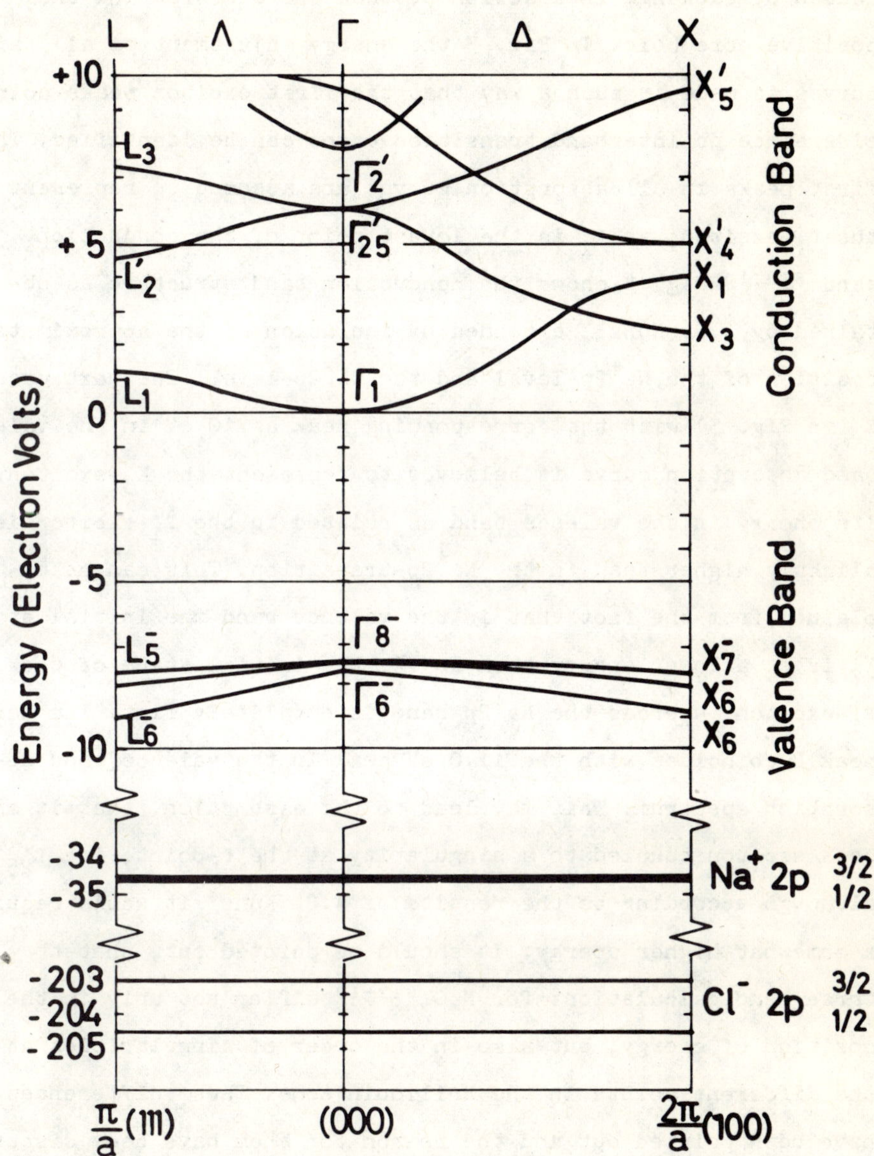

Fig. 4 Bandstructure of NaCl after A.B. Kunz[6] with inclusion of the position of Na^+ 2p and Cl^- 2p. The zero-point of the energy scale has arbitrarily been positioned at the bottom of the conduction band.

oscillator strengths is explained by Onodera and Toyozawa[45] as caused by exchange interaction between the electron and the positive core hole. In Fig. 3 the energy adjustment of all three curves is made in such a way that the first exciton peaks coincide since no interband transition edges can be identified. The first peaks in all absorption curves are assumed to represent the Γ_1-exciton, as Γ_1 is the lowest point of the conduction band.[6,9,10] Fig. 4 shows the conduction band structure as obtained by A.B. Kunz[6], extended by inclusion of the approximate position of the Na^+2p-level and the Cl^-2p-level. The next peak C (in Fig. 3) with the corresponding peak at 10 eV in the valence band absorption curve is believed to represent the X_3-exciton. Its energy in the valence band as related to the Γ_1-exciton is slightly higher than in the Na^+2p-transition. This can be explained from the fact that in the valence band the initial state X_7' is at a lower energy than at Γ_8', the initial state of the Γ_1-exciton, whereas the Na^+2p band is completely flat. The next peak D coincides with the 11.0 eV peak in the valence band absorption spectrum. This may lead to the assumption, that it may be a exciton coupled to a singularity at the Γ-point, i.e.Γ_{25}' although according to the results of A.B. Kunz[6] it would require a somewhat higher energy. It should be pointed out, that the three band calculations for NaCl[6,9-10] differ not only in the position of energy, but also in the order of singularities at the different points in the Brillouin zone. These differences have been pointed out and the reason for them have been discussed by A.B. Kunz et.al.[46]

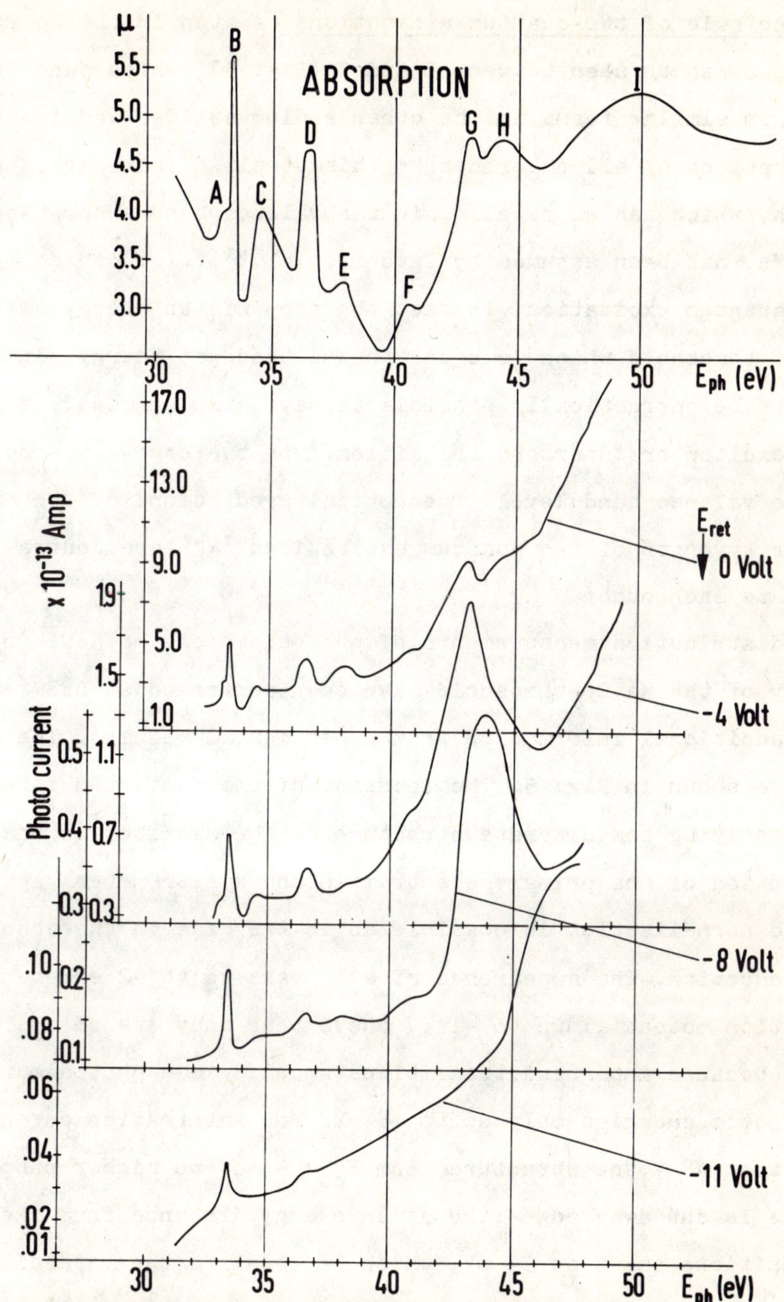

Fig. 5 Photoemission of NaCl in the vicinity of Na^+ 2p-transitions for different retarding potentials.

<u>About the role of two-quantum excitations.</u> A step in the Na^+ 2p-absorption can be seen between F and G (Fig. 3), which can also be seen in similar forms in the other sodium halides and in the 2p-absorptions of all chlorides.[43] This steplike increase of absorption, which has no parallel in the valence band absorption structure, has been assumed by Iguchi et.al.[43] to be caused by double quantum excitation. In fact the step has an energy distance from the threshold which is equal to the band gap energy. It would therefore be energetically possible to have simultaneously excited one exciton or interband transition from the core- and one from the valence band level. Theoretical predictions of the oscillator strength of two quantum excitations[47,48] are contradictory to each other.

Energy distribution measurements of photoelectrons of NaCl in the vicinity of the Na^+ 2p-threshold have been performed at DESY[49] to get additional information from experimental results. The results are shown in Fig. 5. The increase of the continuum intensity underlying the discrete structure can be ascribed to the distribution of the primary spectrum in the spectrometer, against which no normalization is possible as is the case in the absorption curve reduction. The appearence of all peaks up to 42 eV even at retardation potential up to -11 V shows that they are of excitonic nature, because interband transitions should yield photoelectrons with kinetic energies only up to $\sim$8 eV. For retardation potentials higher than -8 V the structure from E_{ph}= 43 eV to higher photon energies is cut away edge-like at an energy distance from the 2p-transitions onset at 34 eV, which is equal to the retarding potential. This indicates, that here one-electron transitions play the main role since in the case of two-electron transitions both electrons would have low E_{kin} and therefore would already be suppressed by lower retarding potentials. Peak G at 43.0 eV is suppressed not so much by small retarding potentials as the peaks at higher photon energies. This shows that it is lying at an energy just below the onset of inelastic sctattering. Thus in our opinion, the double-exciton- or double-electron-excitation can be excluded as an explanation for the absorption step near 43 eV.

IV.2 Initial state with s-symmetry

Transitions from the Li^+ 1s-shell. The investigation of initial states which have no p-symmetry gives us the possibility to study complementary states in the conduction band. The onset of transitions from the Li^+ K-shell is at about 60 eV. The absorption behaviour of these transitions has been studied at DESY for all lithium halides.[50] The theoretical band calculations of the lithium halides[5-8] indicate that the L_1-point is very close to the Γ_1-point. In LiCl and LiBr it is slightly above, in LiF and LiJ it is even below Γ_1, thus making L_1 the lowest point in the conduction band. This has an important result because only where L_1 is the lowest point is it an M_o-point,[51] to which normal excitons can be coupled. The even parity of L_1 is only valid if the origin of the Brillouin zone is in the halogen ion. For the origin in the alkali ion the parity is changed, a fact which is valid for all points at L.[4,52]

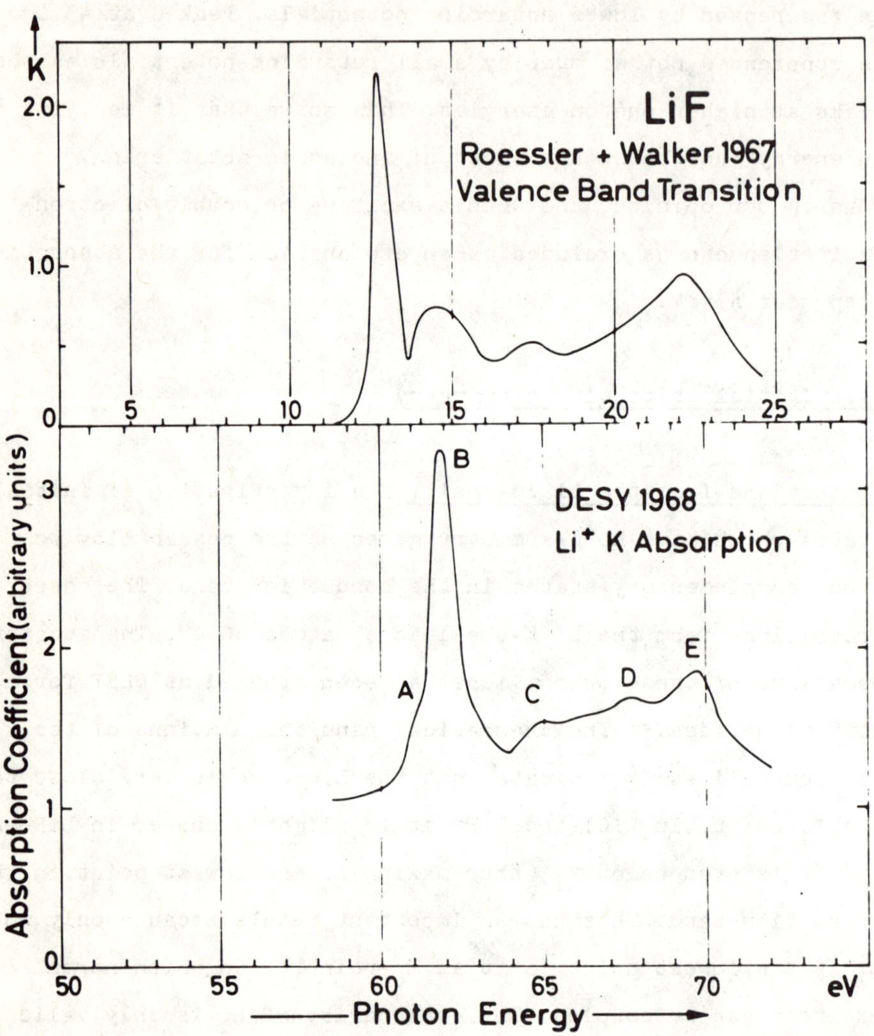

Fig. 6 Comparison of LiF valence band and Li^+ K-absorption after Kunz, Miyakawa and Oyama.[5]

Comparison of the experimental data on Li^+1s-transition with valence band transition is mainly possible for LiF, where, in the fundamental absorption region, data are available up to 25 eV from Roessler and Walker,[53] whereas for the other Li-halides data are only available near the threshold.[54] The band structure of LiF has been calculated by Kunz, Miyakawa and Oyama,[5] who also from the theoretical point of view, made a comparison of the experimental valence-band and Li^+1s-transition data (see Fig. 6) also including electron loss data.[37] The first peak in the valence band data is ascribed to the L_1-exciton, L_1 being the lowest point of the conduction band. As the wavefunctions of L_1 are highly concentrated around the halogen ion, transitions from the 1s-level of Li^+, although allowed by the optical selection rules, are very weak and show up in the 1s-transitions only as a small shoulder A. It should be noted that LiJ[8], also having the L_1 lower than Γ_1, shows a similar shoulder at the onset of a higher absorption peak.[50]
Then the 1s-transitions show a prominent absorption peak B, which is ascribed[5] to an exciton at the p-symmetric X_4'-point but it should be pointed out, that the energy difference A - B is much smaller in the experiment, than should be expected from theory. It should be noted, that, according to Moore[55], the excitation energy of the free Li^+ ($1s^2 \rightarrow 1s2p$) is around 61,3 eV and therefore very close to A and B. The other structures may be either excitons or interband transitions, but their assignement in view of the band structure is difficult. However, it is remarkable that, except for peak D the rest of the structure has a similar

shape, despite the fact that both transitions come from states of different symmetry, localized at different ion types. An important proof of the energy adjustment of the two curves in Fig. 6 would be a measurement of the LiF-emission, which would directly show the energy difference between the valence band and the Li^+ K-level. The transition should occur around 50 eV, but no results of emission measurements on LiF came to the authors knowledge.

Another check could be in the form of an independent measurement of the binding energy of the electrons in the different shells by the ESCA-method.[55a] Although some examples are given in reference[55a] of alkali halides electron binding energies, no values are found for the lithium halides.

Transitions from the Ne 2s- and Ar 3s-shell. Up to now we were only concerned with alkali halides. In solid rare gases measurements of inner shell transitions give the opportunity to directly compare the same transitions in the gaseous and solid state in contrast to the alkali halides, where no transitions from core states have been studied on vapours. In solid rare gases the weak van-der-Waals binding forces make one expect that there are major similarities in the absorption spectra, not only as far the position of absorption structures is concerned, but also with regard to the shape of the lines. For many years the synchrotron of the National Bureau of Standards has been used for extensive measurements of the autoionization levels of solid rare gases.[56] Different types of absorption profiles have been identified and in collaboration with Fano and Cooper[57] ascribed to the interaction of the discrete states with the underlying continuum. In Ne 2s-transitions[58] assymetric and in Ar 3s-transitions[59] "window" type absorption profiles have been found. Although some evidence of assymetric line shape has also been seen in different solids[51], it seemed especially interesting to see if and to what extend characteristic line shapes in gases could also be seen in the solidified rare gases. First results obtained at DESY[60] indicate, that the assymetric absorption lines in Ne 2s can also clearly be seen in the solid, somewhat broadened and shifted by 3 eV to higher energies. In Ar 3s one "window" type line has also been seen.

IV.3 Initial state with d-symmetry

Transitions from the Kr 3d-shell. The band calculations for solid Kr[20] indicate that the lowest parts of the conduction band mainly consist of s- and d-symmetric states. The inversion of parity at L does not occur in the monoatomic solid. It is therefore interesting to investigate, to which points of the Brillouin zone excitons are coupled which may be found in the absorption spectrum. Elliot[61] has shown that two types of exciton transitions exist. The first are the "allowed" exciton series, formed between initial and final states, where direct optical transitions are allowed by the selection rules (the normal case for the p-symmetric valence band and the lowest s- and d-symmetric parts of the conduction band). The second are the "forbidden" excitons, formed between states, where direct transitions are not allowed. Here all Rydberg series

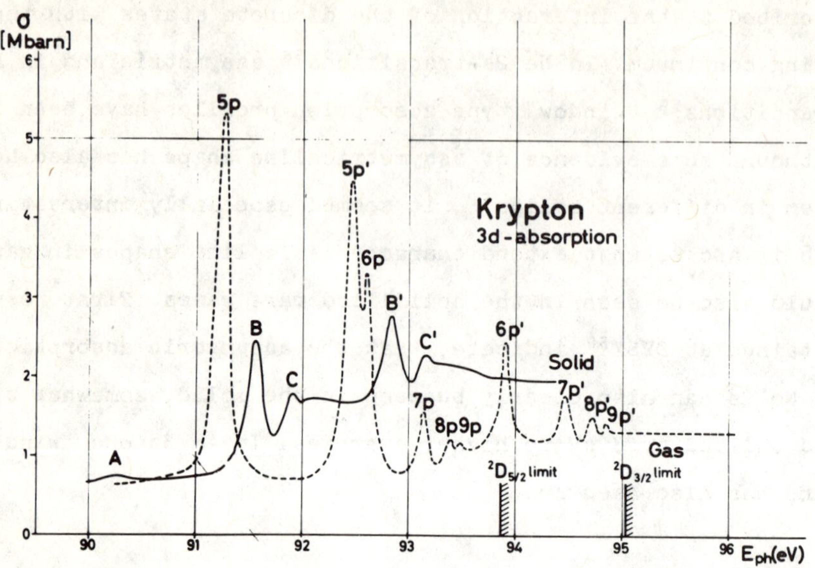

Fig. 7 Absorption of Kr in the vicinity of the 3d-threshold in solid and atomic state.

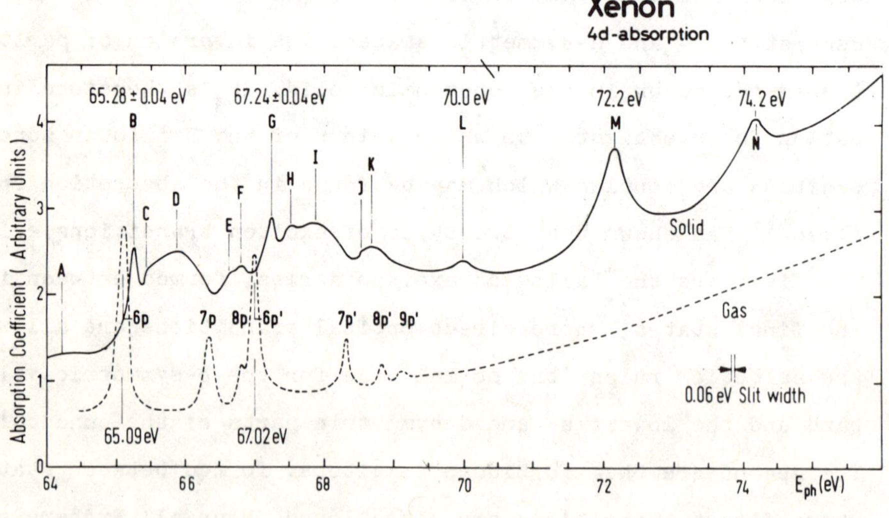

Fig. 8 Absorption of Xe in the vicinity of the 4d-threshold in solid and atomic state.

members, except n=1 can be formed. The classical experimental example is the exciton series in Cu_2O.[62]
Fig. 7 shows the absorption structure of solid and gaseous Kr in the region of the gas absorption lines.[63] The energy of the Kr gas lines has already been evaluated by Codling and Madden[64] and served as energy calibration for the DESY measurements. In solid Kr many peaks can be seen (also above the ionization limit in the gas) which can be correlated in pairs, the corresponding partners always having an energy difference of about 1.22 eV, the spin orbit splitting energy of the 3d-shell.[64] At the onset a weak peak A can be seen, superimposed onto a continuum absorption of valence electron transitions into high lying states of the conduction band. Then a very high peak B at 91.6 eV with its partner B' at 92.8 eV follows, which are very close to the first members of the gas transitions. Because of their shape and their energy position lines B, C and B', C' are supposed to be members of exciton series. As no transitions into the n=1 exciton of a exciton series coupled to the s-symmetric Γ_1-point are allowed from the d-symmetric initial state it is assumed that B and B' are the n=2 members, C and C' the n=3 members of exciton families converging to the Γ_1 point at 92,2 eV and 93,4 eV.
In this way A could be the forbidden n=1 exciton. The binding energy B in the exciton series

$$E = E_o - \frac{B}{n^2} \qquad \text{with} \qquad B = \frac{\mu}{\varepsilon^2}$$

would be 2.5 eV and 2.3 eV which is somewhat higher than 1.73 eV and 1.52 eV for the corresponding valence band excitons.[65] Assuming B and C to be the n=1 and n=2 members then B would have the small value of 0.5 eV. Another argument for the existence of a "forbidden" exciton series can be seen by comparing the oscillator strengths of B, B' and C, C' which are about 3:1. This is much closer to 2.6:1 theoretically predicted for $\frac{n=2}{n=3}$ of a "forbidden" series[61], than 8:1 theoretically predicted for $\frac{n=1}{n=2}$ of an "allowed" series.

Transition from the Xe 4d-shell. A similar but admittedly less unambiguous identification of a forbidden exciton series can be made for Xe 4d[66] (Fig. 8), where A is also the forbidden n=1 exciton, B and G are the n=2 excitons and C and H the n=2 excitons. The latter are not so well separated as in Kr, since more structures are in the immediate neighbourhood. Thus we would be made to assume that in Xe higher conduction bands are closer to the bottom of the conduction band (the s-band with Γ_1 as the lowest point) than in Kr, which would be similar to the alkali halides, where, with increasing Z the higher conduction band are attracted by the s-band. The band calculations of Kr[20] and Xe[21], however, show very similar shapes as far as the distance of the s- and d-bands are concerned.

Beyond the first exciton series Xe shows a rich structure, where two peaks in the spin-orbit-splitting distance can always be correlated. These structures may be explained as excitons or (especially at higher energies) as interband transitions, but an unambiguous correlation is difficult. For further investigations of the solid rare gases an extension of Baldini's valence band transition measurements[65] beyond 14 eV is highly desirable.

ACKNOWLEDGMENTS

The DESY experiments on which this paper is based have been performed together with Dr. C. Kunz of DESY and Mr. G. Keitel, Mr. P. Schreiber and Dr. B. Sonntag of the University of Hamburg. I would like to thank them for the stimulating and very pleasant collaboration, which made these experiments possible. For the communication of results of the potassium halide measurements prior to publication I would like to thank Mr. D. Blechschmidt, Mr. R. Klucker and Dr. M. Skibowski of the University of Munich. I am also grateful to the directors of the II. Institut für Experimentalphysik der Universität Hamburg and the Deutsches Elektronen-Synchrotron for their continuous interest and support of the synchrotron radiation group.
Thanks is also due to the Deutsche Forschungsgemeinschaft for its financial support.

Literature:

1. R.S. Knox, Theory of Excitons (Academic Press, New York and London, 1963)
2. G. Baldini, Phys.Rev. 128, 1562 (1962)
3. F. Fischer and R. Hilsch, Nachr. Akad. Wiss. Göttingen, II. Math.-Physik Kl. 8, 241 (1959)
4. D.E. Ewing and F. Seitz, Phys.Rev. 50, 760 (1936)
5. A.B. Kunz, T. Miyakawa and S. Oyama (to be published)
6. A.B. Kunz, Phys.Rev. 175, 1147 (1968)
7. A.B. Kunz, Phys.Stat.Sol. 29, 115 (1968)
8. A.B. Kunz, (to be published)
9. T.D. Clark and K.L. Kliewer, Phys.Letters 27A, 167 (1968)
10. C.Y. Fong and M.L. Cohen, Phys.Rev.Letters 21, 22 (1968)
11. L.P. Howland, Phys.Rev. 109, 1927 (1958)
12. S. Oyama and T. Miyakawa, J.Phys.Soc.Japan 21, 868 (1966)
13. P.D. De Cicco, Phys.Rev. 153, 931 (1967)
14. Y. Onodera, M. Okazaki and T. Inui, J.Phys.Soc.Japan 21, 2229 (1966)
15. Y. Onodera, J.Phys.Soc. Japan 25, 469 (1968)
16. U. Rössler (to be published)
17. R.S. Knox, J.Phys.Chem.Sol. 9, 265 (1959)
18. R.S. Knox and F. Bassani, Phys.Rev. 124, 652 (1961)
19. L.F. Mattheis, Phys.Rev. 133, A 1399 (1964)
20. W.B. Fowler, Phys.Rev. 132, 1591 (1963)
21. M.H. Reilly, J.Phys.Chem.Sol. 28, 2067 (1967)
22. R.S. Knox and K. Teegarden in Physics of Color Centers, ed. W.B. Fowler (Academic Press New York and London, 1968)
23. J.A. Bearden and A.F. Burr, Rev.Mod.Phys. 39, 125 (1967)
24. D.H. Tomboulian in Handbuch der Physik XXX, ed. S. Flügge (Springer-Verlag, Berlin, Göttingen, Heidelberg, 1957) p. 294
25. J.D. Jackson, Classical Electrodynamics (John Wiley & Sons, Inc. New York and London, 1962)
26. R.P. Feynman, R.B. Leighton and M. Sands, The Feynman Lectures on Physics, Vol. I. chap. 34-3 (Addison-Wesley Publishing Comp., 1965)

27. J. Schwinger, Phys.Rev. 75, 1912 (1949)
28. A.A. Sokolov and J.M. Ternov, Synchrotron Radiation (Akademie-Verlag, Berlin, 1968)
29. K. Codling and R.P. Madden, J.Appl.Phys. 36, 380 (1965)
30. R. Haensel and C. Kunz, Z.Angew.Phys. 37, 3449 (1966)
31. R.P. Godwin in Springers Tracts in Modern Physics, Vol. 51; ed. G. Höhler (Springer-Verlag 1969) (to be published)
32. F.C. Brown, A. Fujita and Ch. Gähwiller (private communication)
33. D. Blechschmidt, R. Klucker and M. Skibowski (to be published)
34. G. Stephan and S. Robin, C.R. Acad.Sci. 267, 1286 (1968)
35. G. Stephan and S. Robin, Opt.Comm. 1, 40 (1969)
36. G. Stephan, E. Garignon and S. Robin, C.R.Acad.Sci. 268, 408 (1969)
37. M. Creuzburg, Z.Phys. 196, 433 (1966)
38. P. Keil, Z.Phys. 214, 266 (1968)
39. H. Saito, S. Saito, R. Onaka and B. Ikeo, J.Phys.Soc. Japan 24, 1095 (1968)
40. R. Haensel, C. Kunz, T. Sasaki and B. Sonntag, Phys.Rev. Letters 20, 1436 (1968)
41. T. Sagawa and S. Nakai, J.Phys.Soc. Japan 26, 1427 (1969)
42. H. Damany, J.-Y. Roncin and N. Damany - Astoin, Appl.Opt. 5, 297 (1966)
43. Y. Iguchi, et.al., Sol.State Comm. 6, 575 (1968)
44. D.M. Roessler and W.C. Walker, J.Opt.Soc.Am. 58, 279 (1968)
45. Y. Onodera and Y. Toyozawa, J.Phys.Soc. Japan 22, 833 (1967)
46. A.B. Kunz, W.B. Fowler and P.M. Schneider (to be published)
47. T. Miyakawa, J.Phys.Soc. Japan 17, 1898 (1962)
48. J.C. Hermanson, Phys.Rev. 177, 1234 (1969)
49. R. Haensel, G. Keitel, C. Kunz, G. Peters, P. Schreiber, and B. Sonntag (to be published)
50. R. Haensel, C. Kunz and B. Sonntag, Phys.Rev. Letters 20, 262 (1968)

51. J.C. Phillips, in Solid State Physics Vol. XVIII, ed. F. Seitz and D. Turnbull (Academic Press, New York and London, 1966) p. 56
52. P.M. Scop, Phys.Rev. 139, A 934 (1965)
53. D.M. Roessler and W.C. Walker, J.Phys.Chem.Sol. 28, 1507 (1967)
54. K. Teegarden and G. Baldini, Phys.Rev. 155, 896 (1967)
55. C. Moore, NBS Circular 467, I (1948)
55a K. Siegbahn et.al. ESCA Atomic, Molecular and Solid States Structure Studied by Means of Electron Spectroscopy (Almqvist & Wiksells Boktryckeri AB, Uppsala, 1967)
56. R.P. Madden and K. Codling, in Autoionization; Astrophysical, Theoretical, and Laboratory Experimental Aspects, edited by A. Temkin (Mono Book Corporation, Baltimore, Md., 1966) p. 129
57. U. Fano and J.W. Cooper, Rev.Mod.Phys. 40, 441 (1968)
58. K. Codling, R.P. Madden and D.L. Ederer, Phys.Rev. 155, 26 (1967)
59. R.P. Madden, D.L. Ederer and K. Codling, Phys.Rev. 177, 136 (1969)
60. R. Haensel, G. Keitel, C. Kunz and P. Schreiber (to be published)
61. R.J. Elliot, Phys.Rev. 108, 1384 (1957)
62. P.W. Baumeister, Phys.Rev. 121, 359 (1961)
63. R. Haensel, G. Keitel, C. Kunz and P. Schreiber (to be published)
64. K. Codling and R.P. Madden, Phys.Rev. Letters 12, 106 (1964)
65. G. Baldini, Phys.Rev. 128, 1562 (1962)
66. R. Haensel, G. Keitel, C. Kunz and P. Schreiber, Phys.Rev. Letters 22, 398 (1969)

III

EXCITONS IN ALKALI HALIDES

Bruno Bosacchi

Istituto di Fisica dell'Universita, Milano, Italy
and

Gruppo Nazionale di Struttura della Materia del C.N.R.

I. INTRODUCTION

This seminar is mainly based on the lectures by Prof. Bassani (1), who has shown the power of the band scheme in the discussion of the optical properties of semiconductors. The idea of Phillips to extend the same approach also to the alkali halides (2) has proved very stimulating. The optical spectra of these crystals have been thoroughly investigated recently, and a wealth of experimental data, in excellent agreement among them, are now available (3-7). On the theoretical side, several calculations on their band structure have recently been performed (8-17). One could have therefore the feeling that a good basis exists for the interpretation of the sharp structures of the optical spectra of the alkali halides in the region below 10 eV. However, this is probably an over-optimistic view of the problem, which is still quite open, in spite of the fact that its history goes back to the beginning of the optical spectroscopy of solids.

After a short discussion of the experimental techniques in Sect. II, and a glance to some ideas of the exciton theory in Sect. III, we shall discuss the reflectivity spectra in Sect. IV. Finally, Sect. V

Address from September 1st, 1969: Argonne National Laboratory, Argonne, Illinois, USA

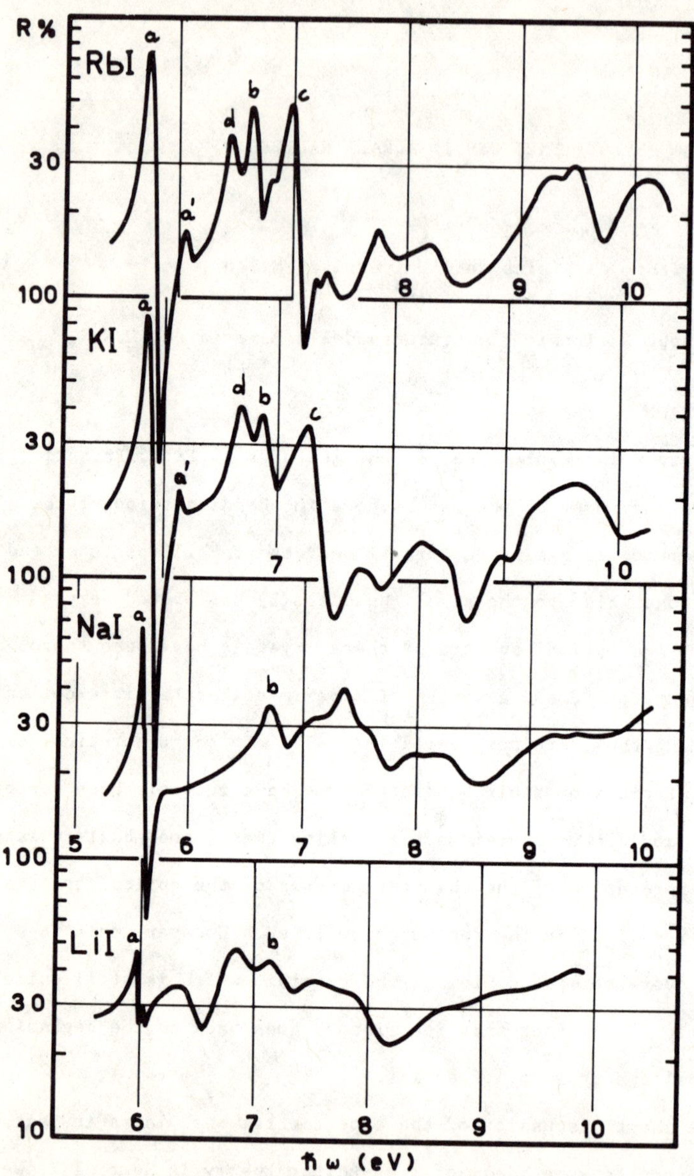

Fig. 1 - Reflectivity spectra of the iodides at 55°K

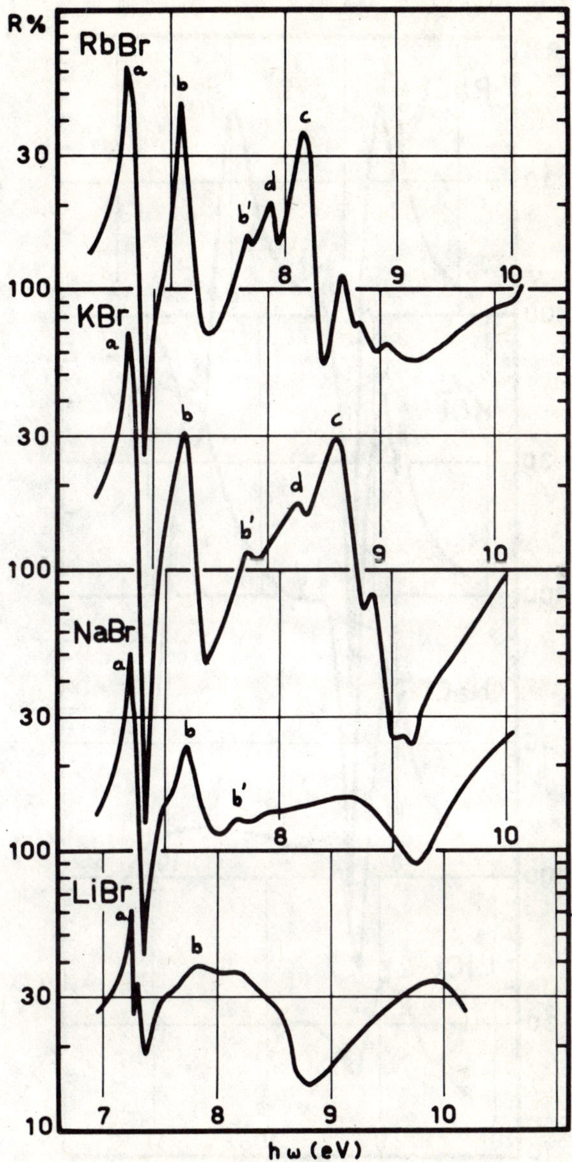

Fig.2 - Reflectivity spectra of the bromides at 55°K

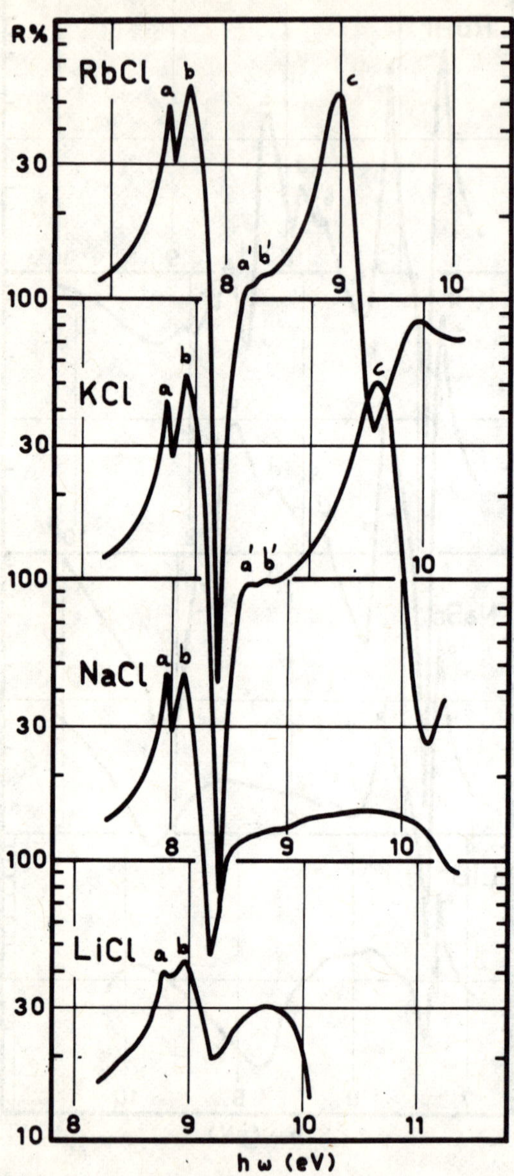

Fig.3 - Reflectivity spectra of the chlorides at 55°K

deals with some effects of the exciton-phonon interaction on the exciton lines.

II. EXPERIMENTAL

The fundamental optical properties of the alkali halides can be investigated either through the absorption in thin films, or through the reflectivity from cleaved single crystals. In the latter case one has the advantage of a better crystal structure and some details, which are characteristic of the long range order, can be detected more easily; there are, however, more experimental difficulties, since, in the ultraviolet, reflectivity is very sensitive to the unavoidable contamination of the surface, which occurs at low temperatures, even in a good vacuum, after some minutes from the cleavage.

The quantity of most interest is actually ε_2, the imaginary part of the complex dielectric constant $\varepsilon = \varepsilon_1 + i\varepsilon_2$. To obtain ε_2 two independent measurements have to be performed, for example, two high precision reflectivity measurements at different angles of incidence, as used also recently for graphite (18); however, by exploiting the connection between the real and imaginary parts of an analytic function, one can get ε_2 from a single reflectivity measurement, through a Kramers-Kronig analysis of the data (19). Often reflectivity spectra can be used directly for a qualitative discussion, and this will be done here.

The apparatus we have used has already been described (6); a double beam technique allows a rapid and direct recording of the spectra, with only minor corrections to be considered afterwards. Some modifications have been recently made in order to do the cleavage under vacuum at low temperature (7); this procedure guarantees the best conditions for the measurement and can be used also for the hygroscopic crystals.

The near normal incidence reflectivity spectra at 55°K of all the fcc iodides, bromides and chlorides are shown in Figs. 1, 2 and 3 re-

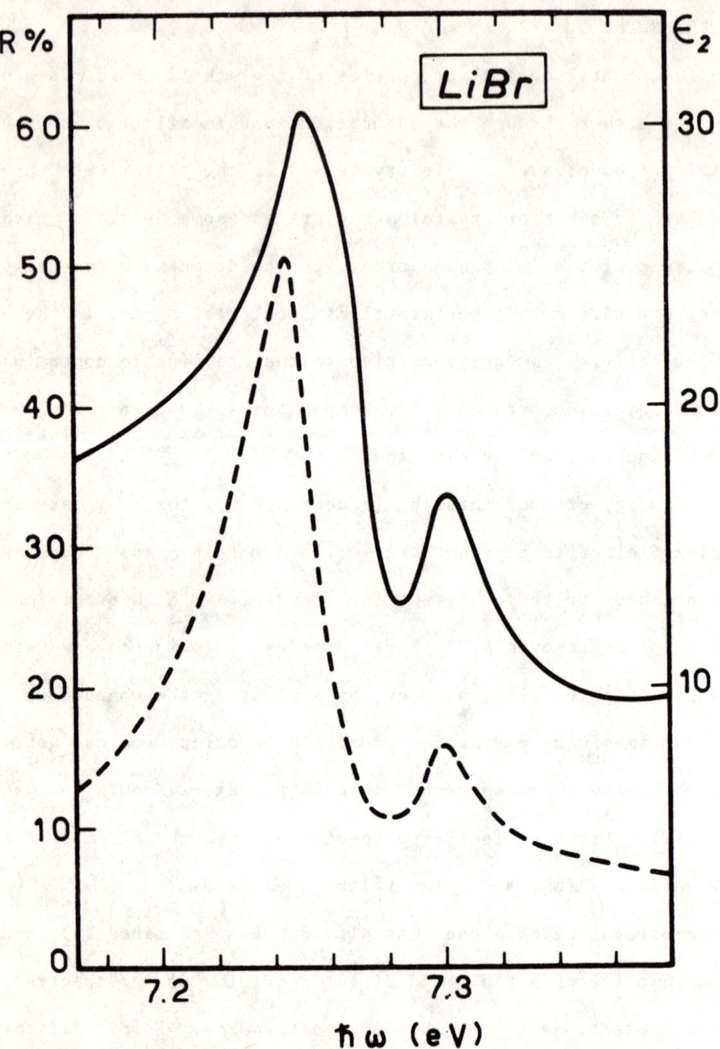

Fig.4 - Reflectivity and ε_2 spectrum of LiBr at 55°K

spectively. We have found it useful to draw them in such a way that the lowest peaks are in the same position for any sequence with the same halogen ion. Logarithmic scales have been used for the reflectivity values; some of the peaks have been labeled with the letters a, a', b, b', c and d, to simplify the discussion of Sect. IV. At our knowledge, some of these spectra (LiI, NaBr, LiBr and LiCl) have never been reported till now; the others agree very well with the previous results (4, 5, 6). According to the observation at the beginning of this section, one sees that reflectivity spectra show more resolved structures than the absorption spectra at the same temperature; they compare very well even with the absorption results at much lower temperature: for example, the structures indicated with a' and b' in RbCl and KCl (Fig. 3), are absent in ref. 3. But the most striking difference is the appearance of a fine structure in the lowest exciton lines of some iodides and bromides, when very high resolution is employed. Since these details are scarcely visible in Figs. 1 and 2, we have reported in Fig. 4, as an example, the reflectivity and ε_2 spectrum of LiBr in the region of interest (20). At 55°K we have found these structures only in LiI, LiBr and NaI, where the lowest exciton is split, and in NaBr, where it has a shoulder on the high energy side; no structure has been resolved for the remaining crystals (7). However, lowering the temperature, new structures appear; preliminary results would show that in LiI, at liquid helium temperature, a third peak is resolved on the high energy side of the exciton a (21); in the case of KI, at the same temperature, the peak a' shows a structure, on the high energy side, composed of at least three peaks, each other separated by ~ 16meV (21). These structures in the exciton lines are a completely new and unexpected result for the alkali halides: they will be discussed in Sect. V.

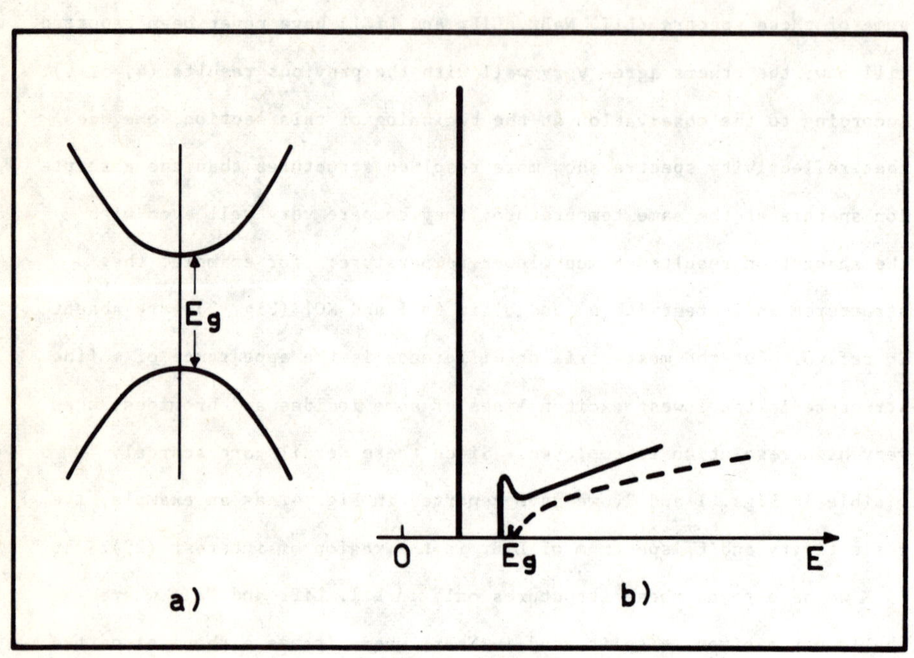

Fig.5 - b) Absorption spectrum for the band model shown in a)

III. EXCITONS IN THE BAND SCHEME

It has been shown (22) that the shape of the optical spectrum in the band scheme is essentially determined by the joint density of states. This function has always a number of sharp singularities (minimum M_0, saddle points M_1 and M_2, maximum M_3), which can be seen in the spectrum and localized in the Brillouin Zone, thus correlating the band structure to the optical excitation spectrum. The above picture, however, is valid only in the one electron approximation, and can be drastically changed by the electron-hole interaction. This is just what happens in the alkali halides, where excitonic effects play a dominant role. The interpretation is therefore not so direct as in many semiconductors. Exciton theory has been extensively treated (23,24); we shall only collect those ideas which allow one to introduce the excitons in the band scheme.

Excitons are the excited electronic states of semi-conductors and insulators. They can be visualized from two different approximations. In the Frenkel approximation, the excitation is an atomic one, slightly perturbed by the weak interaction among the crystal units. In the Wannier approximation, the excitation is an electron-hole pair, resulting from the transfer of an electron from the full valence band, to the empty conduction band, leaving a hole behind. The Coulomb attraction between the two particles, gives a quasi-particle, the exciton, with bound states below its ionization limit. We will adopt this second point of view. In its frame, an effective mass theory has been developed, in close analogy to that for the impurity states in semiconductors. The probability of transition from the ground state of the crystal to the exciton states, has been evaluated by Elliott (24). In the very simple case illustrated in Fig. 5a, spherical bands with the extreme at the same K, the direct allowed absorption near the edge is qualitatively

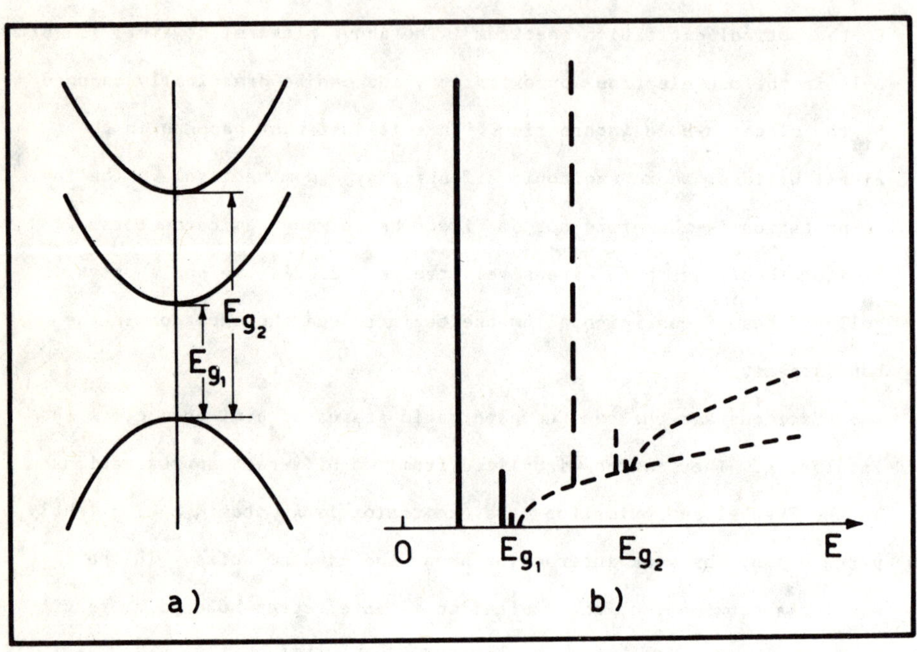

Fig.6 - b) absorption spectrum for the band model shown in a)

depicted in Fig. 5b: the spectrum below the edge is a series of hydrogenic lines, at energies $h = E_g - R/n^2$, where $R = \mu e^4 / 2h^2 \varepsilon^2$ is the binding energy, and $n=1,2,...$ Above the edge the absorption is that of an M_0 singularity, times on enhancement factor which is particularly effective close to the edge; the spectrum above the edge joins continuously to the envelope of the lines at large n below the edge. The model can be generalized as in Fig. 6a, with the results of Fig. 6b. This has been done by Phillips (25). There are now two transitions, and the exciton bound to the upper minimum falls in the continuum of the first one; it is usually referred to as a resonant state in the continuum, or excitonic resonance. Actually, such a state would be broadened by scattering processes, but, if its lifetime is sufficiently long, it can emerge from the continuum. The importance of subsidiary minima in the conduction band to give resonant states has recently attracted large attention (26) and has also received a very general formulation (27). Phillips has hypothetized also the existence of resonant structures related to saddle points, but this question is still rather controversial (25,28). We shall try to avoid it in our discussion.

IV. DISCUSSION OF THE SPECTRA

In this section we shall consider the general features of the spectra, which can be framed reasonably in the existing band schemes, but shall avoid any detailed attribution of the peaks, which would be too tentative at present. Though the energy band structure of the alkali halides has already been discussed by Professor Bassani, it is useful to recall shortly some results.

Since we are concerned with the energy region below 10 eV, the inner core states cannot be excited, and only the highest full valence band, and the lower conduction bands, are of interest. The highest valence band arises from the p-states of the halogen ions, and its

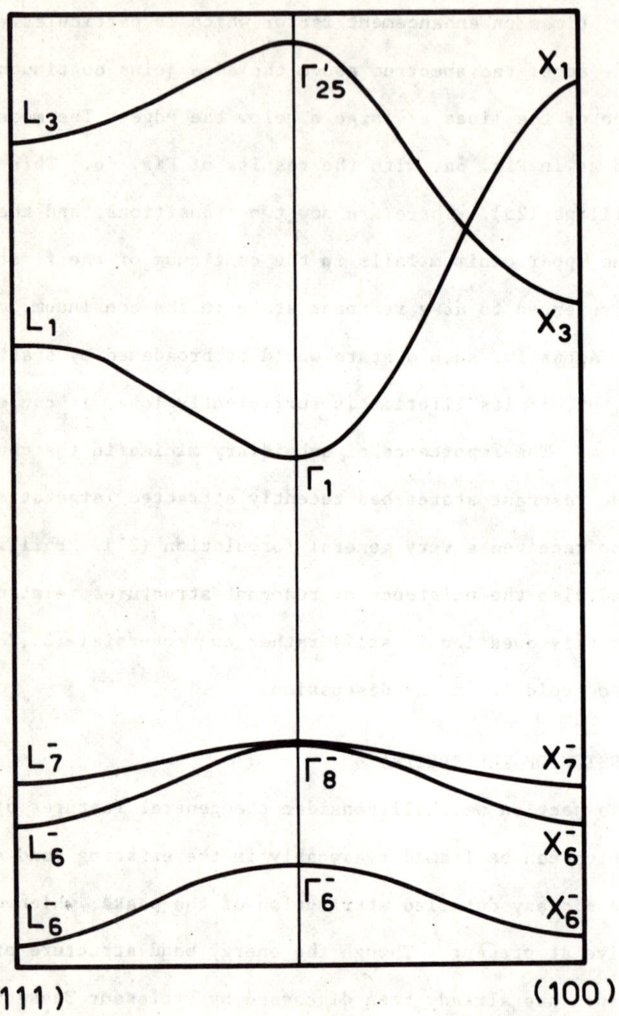

Fig.7 - Schematic band structure of an alkali halide

maximum is generally located at Γ, even if in some calculations it appears slightly displaced. As shown in Fig 7, the valence band is split at Γ by the spin orbit interaction; the magnitude of this splitting is somewhat slightly larger than that of the free halogen atom: it decreases from the iodides to the chlorides. When going to low symmetry points, the upper branch is further split by the crystal potential. Though the valence band is mainly related to the halogen ions, its width appears to increase when passing from the Rb to the Li halides (13,14, 29). The situation is much more complicated for the conduction bands, which have been extensively studied in last years (8-17). The common point of all the calculations is that the absolute minimum of the conduction band is at Γ, with the only exception of the Li halides, where it could be at L_1 (15,17). Another common point is the minimum at X_3, related to the d-states of the alkali ions, which shifts down strongly in energy, compared to the free electron like band scheme. In most calculations X_3 is even lower than X_1, causing therefore overlapping of the s and d bands; but, on this point, there is some disagreement (11, 17). Disagreement exists also about the position of the L_1 point, which has probably an energy much lower than that quoted in ref. 9 and is found in some case to be a minimum (16). We believe that at present there are too many discrepancies among the various results; this makes it meaningless to rely upon the absolute values and to compare directly a calculation and an optical spectrum: there are so many critical points in a band scheme that an attribution can always be found for every peak! What one can do is rather to exploit the systematic behaviours of the bands from crystal to crystal and to compare it to that of the experimental peaks; one can consider, for example, the trend of some peaks going from the Rb to the Li chlorides, or from the iodides to the chlorides. It is evident that, to this aim, single calculations are of little use: one needs systematic works over a whole series of alkali halides, like those of Kunz, or Bassani and Giuliano. A possible way to collect the information from such calculations is shown in Fig. 7, which shows the trend of the X_3 and L_1 points, according to ref. 13, when going from Rb to Li. More rigorously, we have reported the energy differences $(X_3-X_7^-)-(\Gamma_1-\Gamma_8^-)$ and $(L_1-L_7^-)-(\Gamma_1-\Gamma_8^-)$. One has to look for similar qualitative behaviour in the trends of the experimental structures, before a consistent assignment can be done.

In view of the general agreement in locating at Γ_1 the absolute minimum, the lowest energy interband transition is $\Gamma_8^-\rightarrow\Gamma_1$, at least in the Rb, K and Na halides. There is therefore little doubt in attributing to the exciton associated with this transition the lowest exciton peak a; a different viewpoint, however, has been expressed by Pekar (30). The situation can be different for the Li halides, where the lowest peak is perhaps associated with the $L_7^-\rightarrow L_1$ transition; indirect transition would also be possible in these cases, and this could explain the high values of the reflectivity in the tail at low energies in LiI and LiBr and perhaps also some weak structure which is visible in the same region in LiCl. The onset of the $\Gamma_8^-\rightarrow\Gamma_1$ transition is known approximately from photoconductivity and two-photon absorption measurements (31,32); one can put, immediately below this threshold, the n=2 member of the Wannier series a'; this is well visible in RbI and KI and a little less in RbCl and KCl, whereas it is hidden in the bromides and not resolved in the remaining crystals. The exciton b, related to the transition $\Gamma_6^-\rightarrow\Gamma_1$ from the spin orbit split lower branch of the valence band, is also easily identificable in chlorides and bromides. The attribution is more difficult for the iodides, where several peaks could be good candidates. In any case, since the spin-orbit mate can be only one, there are other two peaks which must be explained. An important step

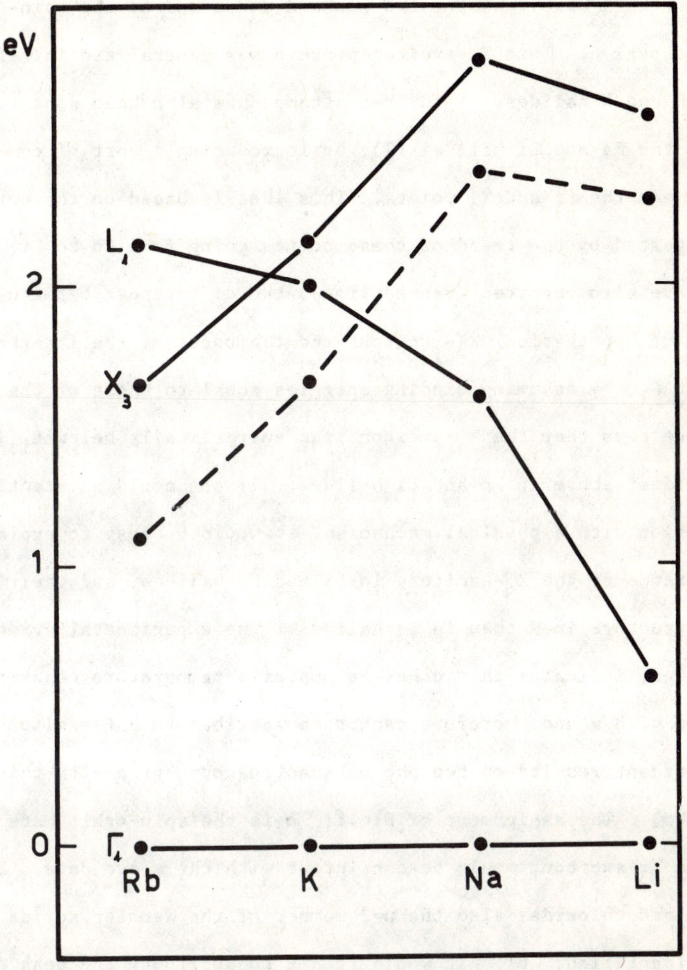

Fig. 8 - Energetic trend of the points L_1 and X_3 going from Rb to Li halides. The dashed line shows the trend of an X_3 -exciton (see text)

in this direction is the suggestion by Onodera and Toyozawa (33) that excitons are produced below the X_3 minimum; on the basis of ref. 9 they attributed to the X_3-excitons the two peaks d and b and to the spin-orbit mate of a the peak c. This X_3-exciton picture was generalized in ref.6 to all the Rb and K halides. A further attempt has also been done to extend it to the Na and Li halides (7), by introducing a sort of competition between the L_1 and X_3 points. This idea is based on the correlation suggested by the trend of these points going from Rb to Li; in Fig. 8 we have also reported (dashed line), the differences between the energies of hypothetical X_3-excitons and the onset of the interband transition at Γ_1, by assuming binding energies equal to those of the Γ-excitons. One sees that the X_3-exciton lies energetically below L_1 in Rb and K halides, above in Na and Li halides. If one could substantiate this correlation with a physical mechanism, it would be easy to explain the disappearance of the X_3-excitons in Li and Na halides, and their less sharp structure in K than in Rb halides. The experimental evidence for KI, however, indicates that c has an opposite temperature behaviour in comparison with a and therefore cannot be ascribed to a Γ-exciton (34). More recent results on two photon spectroscopy agree with this conclusion (35). The assignment of Fig.1: b is the spin-orbit mate of a and c is an X_3 exciton, would be consistent with the above data. In the bromides and chlorides also the n=2 member of the Wannier series of b is easily identified: b'. It would remain to attribute the peak d, which Kunz has recently related to the point L_1 which he finds to be a minimum in the case of KI and RbI (16), and the higher energy structures in the iodides, for which transitions from lower points of the valence band, or to higher points of the conduction bands, could be invoked. However, we prefer to avoid the introduction of further incertitudes in our discussion and to conclude with some general comments.

The band scheme does not appear so successful as in semiconductors to explain the optical properties of alkali halides. One can be confident that systematic and very detailed band calculations, possibly with well chosen potentials, will resolve soon the problem. It could also be, however, that new ideas are necessary ; for example, the simple picture sketched in Sect.III, to take electron-hole interaction into account a posteriori, can be inadequate. Also the ideas of ref. 27, discussed by Bassani in his last lecture, could throw much light on the problem. Something more can be done also from the experimental side, like measurements with different selection rules (two-photon spectroscopy or excitation of inner cores) or measurements in external fields, using the modulation techniques described by Dr. Seraphin in his lectures (36). We end reporting some preliminary results just on this point. Piezo-reflectance measurements on KI, with pressure in the direction 100, seem to cause a broadening of the exciton, possibly related to the removal of the Γ_8^- degeneracy in the valence band (37). Going up with the energies, these measurements could give useful information on the deformation potentials. Electrooptic measurements on alkali halides are not expected to be so effective in studying band structures, as they are for the semiconductors (36); however, very important information can equally be drawn from them. Preliminary measurements of transverse electro-absorption on RbI thin films, in agreement with those of Stevens and Menes (38), seem to show a shift of the Γ n=1 exciton towards the lower energies and perhaps a splitting of the n=2 exciton (39). These results appear to be in qualitative agreement with calculations by Ralph (40).

V. EXCITON-PHONON INTERACTION

Exciton-phonon interaction is of course not included in the band scheme. Usually, however, it is assumed that in the alkali halides it simply broadens the exciton lines (23). This viewpoint was based on the fact that any fine structure of the exciton lines had never been observed; moreover, the commonly quoted linewidths, desumed from thin films absorption data, appeared too large, compared to the phonon energies, to suggest any possible observation of the effect. We have found, however, that the linewidths, obtained from a Kramers-Kronig analysis of the reflectivity data on single crystals, are much narrower than anticipated (41); moreover we have also found a fine structure on some exciton lines (7,20,21). It appears therefore that the above viewpoint has to be changed.

Fig.4 shows the reflectivity and ε_2 spectra of LiBr at 55°K in the region of the lowest exciton peak. We have already discussed and ruled out in ref. 20 several possible causes of the splitting; in particular we do not believe that the observed structures are due to transitions at different points in the Brillouin Zone: they are present in several cases (7) and the order of magnitude of the splitting itself discourages an interpretation in terms of band scheme, but rather suggests that phonons play a role.

Exciton-phonon interaction has been recently the subject of several papers (42); there are two main mechanisms which can be considered. The first is a sort of dynamic Jahn-Teller instability, as discussed by Moran for the F center (43) and by Toyozawa and Inoue for the heavy impurities in alkali halides (44). In the case of an exciton, which also can be seen as a localized defect, such an interpretation relies upon the degeneracy of the valence band at Γ_8^-. This degeneracy is not present in the ground state of the crystal, since the valence band is completely full; however, it reflects itself on the exciton state. It is just this degeneracy which, being removed by the non cubic vibrational modes, could cause the appearance of structures in the absorption lines, as shown in ref. 44. This viewpoint, however, has recently been questioned by Perlin (45). The above mechanism would have to be present in all the alkali halide crystals, but only in those with a strong difference between the masses of the ions, like those where the splitting is observed, the lattice dynamics would be really effective. This Jahn-Teller instability has been invoked in ref. 20.

The second mechanism is based on the well known phonon sidebands picture (46). This viewpoint has been generally adopted to explain similar structures recently observed in other crystals (47,48). For example, Bachrach and Brown introduce zero-phonon and one-phonon lines, related to the LO mode, whereas the remaining phonons simply broaden these lines (48). There is in general good correspondence between the energies of the LO phonons and those of the splittings, which are equal or something lower. To explain this last point exciton-phonon bound states have been introduced (49).

Both the above schemes have some difficulties in our case, though the more recent results appear to be clearly in support of the latter. Indeed, it is difficult to use the Jahn-Teller mechanism, first because in the Li halides, where the effect is stronger, the Γ-exciton picture may be wrong, as seen in the previous section; but chiefly because, when going to liquid helium temperatures a third peak is resolved in LiI, approximately at an energy from the first peak which is twice that of the second (21). For the phonon sidebands mechanism the problem is that the observed splittings appear to be larger than the LO phonon energies, in contrast with the results of ref.47 and 48. One could perhaps invoke local phonon modes of higher energies, through the abrupt hardening of the local force constants when the transition takes place; but it seems impossible to explain in this way the case of NaI, where the splitting is nearly twice the LO phonon energy. On the other hand, the phonon sidebands mechanism is strongly supported by a further and very preliminary experimental results (21). At liquid helium temperature a', the n=2 Γ-exciton, shows, in the case of KI, other three equidistant peaks, separated by ~ 16meV, an energy close to that of the LO phonon; there is a close analogy with the results found for CuCl by the Strasbourg group (50). The following unifying picture seems to emerge: transitions involving few phonons appear when the exciton is sufficiently extended. The crystals where the effect has been well observed in the first exciton are indeed those with the higher dielectric constant, and therefore the larger extension of the exciton. In KI, where the dielectric constant is lower and the first exciton more strongly coupled, the structure is hardly visible in the first exciton peak and we have to go to the second term of the Wannier series, much more localized, in order to see the effect.

The results we have apoken about in this section are very preliminary; some of them have been obtained just few days ago and must still be confirmed and carefully analysed. A final report will be soon published.

Acknowledgements

The results on which this seminar has been based have been obtained in collaboration with Prof. G. Baldini ; it is a pleasure to acknowledge his guidance over the past few years. I am also very grateful to Prof. F. Bassani for very useful suggestions and continuous help and advice. I wish also to thank Miss E. Pozzi and Mr. A. Bosacchi for help with measurements and for discussions.

References

1. F. Bassani, this volume
2. J.C. Phillips - Phys. Rev. 136, A1705 (1966).

3. K.J. Teegarden and G. Baldini - Phys. Rev. 155, 896 (1967).

4. T. Tomiki - J. Phys. Soc. Japan 22, 209 (1967) (NaCl) and 22, 463 (1967) (KCl).

5. D.M. Roessler and W.C. Walker - J. Opt. Soc. Am. 57, 677 (1967) (RbI and KI) and Phys. Rev. 166, 599 (1968) (KCl and NaCl).

6. G. Baldini and B. Bosacchi - Phys. Rev. 166, 863 (1968) (RbI, KI, RbBr, KBr, RbCl, KCl and NaCl).

7. G. Baldini and B. Bosacchi, submitted for publication to Phys. Stat. Sol. (NaI, LiI, NaBr, LiBr and LiCl).

8. S. Oyama and T. Miyakawa - J. Phys. Soc. Japan 21, 868 (1966) (KCl).

9. Y. Onodera, M. Okazaki and T. Inui - J. Phys. Soc. Japan 21, 2289 (1966) (KI).

10. P. De Cicco - Phys. Rev. 153, 931 (1967) (KCl).

11. T.D. Clark and K.L. Kliewev - Phys. Letters 27A, 167 1968) (NaCl).

12. C.Y. Fong and M.L. Cohen - Phys. Rev. Letters 21, 22 (1968) NaCl).

13. A.B. Kunz - Phys. Rev. 175, 1147 (1968) (KCl, NaCl and LiCl).

14. A.B. Kunz - Phys. Stat. Sol. 29, 115 (1968) (RbBr, KBr, NaBr, LiBr and RbCl).

15. A.B. Kunz - Phys. Rev. 180, 934 (1969) (NaI and LiI).

16. A.B. Kunz - J. Phys. Chem. Sol. in press (RbI and KI).

17. F. Bassani and E.S. Guiliano, to be published (KCl, NaCl and LiCl).

18. D.L. Greenaway, G. Harbeke, F. Bassani and E. Tosatti - Phys. Rev. 178, 1340 (1969).

19. See for example, D.L. Greenaway and G. Harbeke - Optical Properties and Band Structure of Semiconductors, Pergamon Press, London 1968.

20. G. Baldini and B. Bosacchi - Phys. Rev. Letters 22, 190 (1969).

21. G. Baldini and A. Bosacchi and B. Bosacchi, to be published.

22. See for example, F. Bassani, this volume, p. F. Bassani, Proc. Int. School "E. Fermi" Course XXXIV, edited by J. Tauc (Academic Press, New York, 1966) p. 33 and also ref.19, Ch.4.

23. R.S. Knox in Solid State Physics, edited by F. Seitz and D. Turnbull (Academic Press, New York 1965), Suppl. 5.

24. R.J. Elliott in Polarons and Excitons, edited by C.G. Kuper and G.D. Whitfield (Oliver and Boyd, Edinburgh 1963) p. 269.

25. J.C. Phillips, Proc. Int. School "E. Fermi" Course XXXIV edited by J. Tauc (Academic Press, New York, 1966) p. 156.

26. H. Kaplan - J. Phys. Chem. Solids 24, 1593 (1963).
G.A. Peterson, Proc. Int. Conf. on the Physics of Semiconductors (Dunod, Paris 1964). p. 771.

27. F. Bassani, G. Iadonisi and B. Preziosi, to be published.

28. B. Velicky and J. Sak - Phys. Stat. Sol. 16, 147, (1966).
C.B. Duke and B. Segall - Phys. Rev. Letters 17, 19 (1966).
J. Hermanson - Phys. Rev. Letters 18, 170 (1967).

29. T.I. Kucher and K.B. Tolpygo, Fiz. Tverd. Tela 2, 2301 (1960) English Transl. Soviet Phys. Solid State 8, 2609 (1966) .

30. S.I. Pekar - Nuovo Cimento B 60, 291 (1969).

31. Y. Nakai and K.J. Teegarden - J. Phys. Chem. Solids 22, 327 (1961).

32. J. Hopfield and J.M. Worlock - Phys. Rev. 137, A1455 (1965).
D. Frohlich and B. Stagimus - Phys. Rev. Letters 19, 496 (1967).

33. Y. Onodera and T. Toyozawa - J. Phys. Soc. Japan 22, 833 (1967).

34. F. Fischer and R. Hilsch - Nader. Akad. Wiss. Gottingen IIa, 8, 241 (1959).

35. K. Park and R.G. Stafford - Phys. Rev. Letters 22, 1426 (1969).

36. B.O. Seraphin, this volume

37. A. Bosacchi, Thesis, University of Milan, Nov. 1969, unpublished.

38. M. Stevens and M. Menes - Phys. Letters 27A, 472 (1968).

39. E. Pozzi, Thesis, University of Milan, Nov. 1969, unpublished.

40. H.I. Ralph - J. Phys. C. (Proc. Phys. Soc.)1, 378 (1968).

41. G. Baldini and B. Bosacchi, to be published.

42. See for example, among the more recent papers:
A.S. Davydov and E.N. Myasnikov - Phys. Stat. Sol. 20, 153 (1967) and S.A. Moskalenko, M.I. Schniglyuk and B.I. Chinik - Fiz. Tverd. Tela 10, 351 (1968) English Transl. Soviet Phys. Sol. State 10n 279 (1968)

43. P.R. Moran - Phys. Rev. 137, A1015 (1965).

44. V. Toyozawa and M. Inoue - J. Phys. Soc. Japan 21, 1663 (1966).
K. Cho - J. Phys. Soc. Japan 25, 1372 (1968).

45. Yu. E. Perlin - Fiz. Tverd. Tela 10, 1941 (1968) English Transl. Soviet Phys. Solid State 10, 1531 (1969).

46. See for example, E. Mulazzi, G.F. Nardelli and N. Terzi - Phys. Rev. 172, 847 (1968).
D.B. Fitchen in Physics of Color Centers, edited by W.B. Fowler (Academic Press, New York 1968) p. 293.

47. W.Y. Liang and A.D. Yoffe - Phys. Rev. Letters **20**, 59 (1968).
W.C. Walker, D.M. Roessler and E. Loh - Phys. Rev. Letters **20**, 847 (1968).
J. Dillinger, C. Konak, V. Prosser, J. Sak and M. Zvara - Phys. Stat. Sol. **29**, 707 (1968).

48. R.Z. Bachrach and F.C. Brown - Phys. Rev. Letters **21**, 685 (1968).

49. Y. Toyozawa and J. Hermanson - Phys. Rev. Letters **21**, 1637 (1969).

50. J. Ringeisser, A. Coret and S. Nikitine in Localized Excitations in Solids, edited by R.F. Wallis (Plenum K.S. Song, ibid. p. 287.

IV

SELF-CONSISTENT ENERGY BAND CALCULATIONS USING THE OPW METHOD FOR VARIOUS EXCHANGE AND CORRELATION APPROXIMATIONS.

T. C. Collins, R. N. Euwema, and D. J. Stukel

Aerospace Research Laboratories, Wright-Patterson Air Force Base, Ohio

I. INTRODUCTION

The complexity of the Hartree-Fock exchange operator has so far stopped anyone from doing a truly first principles calculation in crystals. One has been forced to approximate this operator and in doing so, has built in an adjustable parameter. In order to better understand this we will present in Section IIa the exchange approximations which are the most applicable to crystalline calculations. They are the Slater[1], Kohn-Sham-Gaspar[2,3], and the Liberman[4] approximations. We will also give the relationship of the eigenvalues to the binding energies in each of the approximations. This, in the past, has lead to a great amount of confusion.

These exchange approximations will be tested on atomic systems and on several different crystal systems. In order to test these approximations on the crystal electronic system, one has to have a self-consistent model. Thus the self-consistent orthogonalized plane wave (SCOPW) method is outlined in Section IIb. One needs to know the approximations of the model, what constitutes a self-consistent energy band calculation, and estimates of errors built into the model and these will be brcught in in the outline. Further a description of the way each exchange approximation is handled in the band calculation is given with Liberman's term outlined in detail since this is the first band calculation using this approximation.

The results of the different exchange approximations using Kr as an example are given in Section IIIa. The results of all the exchange approximations in the crystalline system using ZnS as an example are given in Section IIIb. In Section IIIc the detailed comparison to experiment is presented with GaAs as an example. We examine the effects of hydrostatic pressure upon the calculated resúlts, and compare the results with photoemission and reflection data. In the last part of this section, the effects of correlation on the atom and crystal functions are included and the new system is brought to self-consistency. The correlation potential used is the local static potential of Carr and Maradudin[5] following the procedure of Lundqvist and Ufford.[6] The colclusions of this study and our estimates of where to go from here are given in Section IV.

II. Theoretical Background

a. Exchange Approximations

In the energy band theory of solids and in atomic theory, the Hamiltonian for the electronic system is (in atomic units)

$$\mathcal{H} = -\sum_{i=1}^{N} \nabla_i^2 - \sum_{\alpha, i} \frac{2}{|\vec{r}_i - \vec{R}_\alpha|} + \sum_{i<j} \frac{2}{r_{ij}} \tag{2-1}$$

The first term on the right side is the kinetic energy term, the second term represents the interaction of the nuclei (or nucleus in the atomic system) with the electrons, and the last term is the electron-electron interaction. One generally assumes that the wave function for this system can be approximated by an antisymmetric product of one-electron functions. These one-electron functions are determined by requiring that the total energy can be a minimum and the one-electron wave functions be orthogonal to each other, i.e. the HF method. The HF equations obtained for the one-electron wave function are

$$\mathcal{H}_1 \phi_i(x_1) + \left\{ \sum_{j=1}^{N} \int d\tau_2 \phi_j^*(x_2) \phi_j(x_2) \frac{2}{r_{12}} \right\} \phi_i(x_1)$$

$$- \left\{ \sum_{j=1}^{N} \int d\tau_2 \phi_j^*(x_2) \phi_i(x_2) \frac{2}{r_{12}} \right\} \phi_j(x_1) = \epsilon_{i,HF} \phi_i(x_1) \tag{2-2}$$

Here $\hat{H}_1$ is the sum of the kinetic energy of the electron and its potential in the field of all nuclei. One not only obtains the ϕ_i's from the above equation, but also the $\varepsilon_{i,HF}$'s. $\varepsilon_{i,HF}$ is identical to the negative of the energy that would be required to remove the ith electron from the system provided that the system did not relax after the electron is removed (Koopman's Theorem)[7]. In other words, the eigenvalues, $\varepsilon_{i,HF}$ of the HF equations are equivalent to the binding energies, B_{iHF}, where the binding energies are defined to be

$$B^{HF}_{i,HF} = \epsilon_{i,HF} = \langle i | \mathcal{H}_1 | i \rangle + \sum_{j=1}^{N} \left\{ \langle ij | \tfrac{2}{r_{12}} | ij \rangle - \langle ij | \tfrac{2}{r_{12}} | ji \rangle \right\}. \tag{2-3}$$

For atomic systems one can calculate binding energies using the HF method as long as Koopman's Theorem holds. Even if Koopman's Theorem fails, it is not too difficult to obtain the total energy,

$$E_{Total} = \sum_{i=1}^{N} \langle i | \mathcal{H}_1 | i \rangle + \tfrac{1}{2} \sum_{i,j} \left\{ \langle ij | \tfrac{2}{r_{12}} | ij \rangle - \langle ij | \tfrac{2}{r_{12}} | ji \rangle \right\} \tag{2-4}$$

for an atomic system of N electrons. One can recalculate the total energy with one electron removed and take the difference to obtain the binding energy. However when one is calculating energies of electrons in crystals, the fact that the exchange term in Eq. 2-2 is different for each interacting electron compels one to make approximations. Although exchange approximations are needed for crystalline systems and not for atomic systems, they should be valid in both types of systems.

A simplified one-electron operator which replaces the exchange operator in the HF equations was suggested by Slater.[1,8] If one multiplies and divides the exchange term of Eq. 2-2 by $\phi_i^*(X_1)\phi_i(X_1)$, one has

$$\left\{-\sum_{j=1}^{N} \frac{\int d\tau_2 \phi_j^*(x_2)\phi_i^*(x_1)\phi_i(x_2)\phi_j(x_1)\frac{2}{r_{12}}}{\phi_i^*(x_1)\phi_i(x_1)}\right\}\phi_i(x_1) = V_{ex}\phi_i(x_1) \tag{2-5}$$

To obtain Slater's exchange operator, one takes a weighted average of V_{ex} over occupied states and approximates the ϕ_i's by plane waves. Doing this one obtains

$$V_{ex}^{S}(\bar{r}) = -6\left\{\frac{3}{8\pi}\rho(\bar{r})\right\}^{1/3}, \tag{2-6}$$

where $\rho(\bar{r})$ is the electron density of the system,

$$\rho(\bar{r}) = \sum_{i=1}^{N}\phi_i^*(\bar{r})\phi_i(\bar{r}) \tag{2-7}$$

If one now uses the same approximation in the expression for the binding energy one has

$$B_{i,S}^{S} = \epsilon_{i,S} = \langle i|H_1|i\rangle + \sum_{j=1}^{N}\langle ij|\frac{2}{r_{12}}|ij\rangle + \langle i|V_{ex}^{S}|i\rangle \tag{2-8}$$

An interesting point was brought out by Rudge. If one multiplies

Slater's approximation for the weighted average exchange term in $\langle\psi|H\psi\rangle$ by $3\alpha/2$, one has the following:

$$\langle \Psi | \mathcal{H} | \Psi \rangle \approx \sum_{i=1}^{N} \langle i | \mathcal{H}_1 | i \rangle + \tfrac{1}{2} \sum_{i,j} \langle ij | \tfrac{2}{r_{12}} | ij \rangle + \tfrac{1}{2} \sum_{i} \langle i | \tfrac{3}{2} \alpha V_{ex}^{s} | i \rangle . \tag{2-9}$$

Varying $\langle\psi|H|\psi\rangle$ with respect to the orbitals one obtains

$$\mathcal{H}_1 \phi_i(x_1) + \left\{ \sum_{j=1}^{N} \int d\tau_2 \phi_j^*(x_2) \phi_j(x_2) \tfrac{2}{r_{12}} \right\} \phi_i(x_1) + \alpha V_{ex}^{s} \phi_i(x_1) = \epsilon_{i,\alpha} \phi_i(x_1) \tag{2-10}$$

If $\alpha = 2/3$ one has the Kohn-Sham[2] or Gaspar[3] exchange approximation in $\langle\psi|H|\psi\rangle$.

For any value of α, Eq. 2-9 and Eq. 2-10 are related by the variational method. Also as Rudge noted, the kinetic energy and the potential energy, as defined by Eq. 2-9, obey the virial theorem when the ϕ_i's are obtained from Eq. 2-10. One now has to decide whether to substitute in the total energy and then take the difference between the neutral atom and ion or to take the difference and then substitute into the expression for the binding energy. If the difference is taken before substitution of the exchange approximation one finds that the binding energy is no longer identical to the Kohn-Sham eigenvalues. However, if the substitution is made before the difference of the total energies is taken, the binding energy is in a crystal, almost equal to the eigenvalue. The difference of the two is proportional to $N^{-2/3}$ where N is the number of electrons in the system.

This lack of equivalence between binding energy and eigenvalues has been given the name of "Koopman Corrections," K. In this paper it is defined as

$$K^{C}_{i,A} \equiv B^{C}_{i,A} - \epsilon_{i,A} \tag{2-11}$$

where A represents the exchange approximation used to obtain the orbitals and C represents the exchange approximation used in the calculation of the binding energy.

Liberman[3] suggested also that plane waves be used for the ϕ_i in the Hartree-Fock exchange potential. Assuming that all states are filled for k less than the Fermi momentum, k_F, and all states are empty for k greater than k_F, the Hartree-Fock exchange potential reduces to the expression

$$V^{L}_{ex}(\bar{r}) = -8\left\{\frac{3}{8\pi}\rho(\bar{r})\right\}^{1/3} F(\eta) \tag{2-12}$$

where $F(\eta) = \frac{1}{2} + \frac{1-\eta^2}{4\eta} \ln\left|\frac{1+\eta}{1-\eta}\right| \quad , \eta = k/k_F$ (2-13)

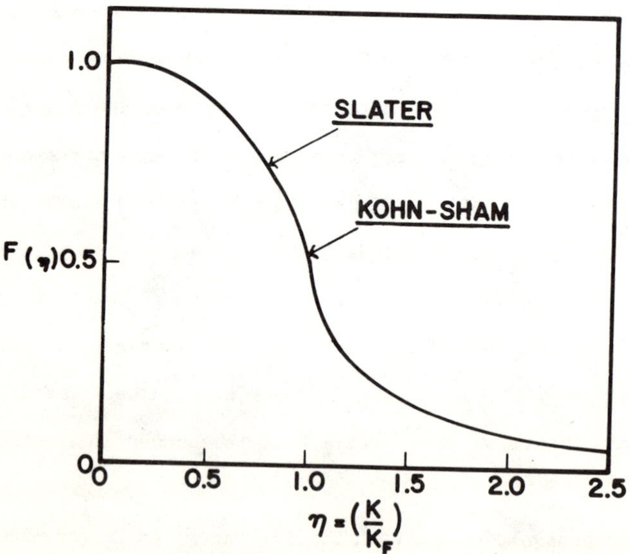

Fig 1. Function F(η) versus η.

A plot of $F(\eta)$ versus η is given in Fig. 1 along with the values corresponding to Slater's and Kohn-Sham-Gaspar's approximations. The wave vector, k, and the Fermi wave vector, k_F are taken to be

$$k_F(r) = \{3\pi^2 \rho(r)\}^{1/3} \tag{2-14}$$

and

$$k(r) = \{E - V(r)\}^{1/2} \tag{2-15}$$

Slater, Wilson and Wood[10] modified Liberman's approximation by using

$$k_F(r) = \{E_F - V_{Coulomb}(r) + 4\left[\tfrac{3}{8\pi}\rho(r)\right]^{1/3}\}^{1/2} \tag{2-16}$$

instead of Eq. 2-14. We have found it advantageous to calculate k_F both ways and always use the larger value. This approach gives results slightly closer to the HF results for Krypton.

Following Liberman's simplification when evaluating the sum over filled electron states in the Hartree-Fock binding energy expression of equation 2-8, we obtain the Liberman approximation to the binding energy:

$$B^L_{i,L} = \epsilon_{i,L} = \langle i|\mathcal{H}_1|i\rangle + \sum_{j=1}^{N} \langle ij|\tfrac{2}{r_{12}}|ij\rangle + \langle i|V^L_{ex}|i\rangle \tag{2-17}$$

b. SCOPW Calculation Model

The crystalline energy band calculations were made using modified versions of our basic SCOPW energy band programs.[11,12] Following the usual OPW procedure, the electronic states of Zn and S (for example) are divided into tightly bound core states and loosely bound valence states.

The core states are calculated using a spherically averaged crystalline potential. There must be no appreciable core overlap between nearest neighbor atoms. The valence states are expanded in a modified Fourier series,

$$\psi_{v}^{\vec{k}_0}(\vec{r}) = \frac{1}{\sqrt{\Omega_0}} \sum_{\mu} C_{\mu v} \left\{ e^{i\vec{k}_\mu \cdot \vec{r}} - \sum_{a} e^{i\vec{k}_\mu \cdot \vec{R}_a} \sum_{n,l} \Phi^{a}_{nl;\vec{k}_\mu}(\vec{r}-\vec{R}_a) A^{a}_{n,l;\vec{k}_\mu} \right\},$$

$$\vec{k}_\mu = \vec{k}_0 + \vec{K}_\mu \tag{3-1}$$

where Ω_0 is the unit cell volume, $\hat{k}_0$ is a vector in the first Brillouin zone, the $\vec{K}_\mu$ sum is over all reciprocal lattice vectors, and the sum over $\vec{R}_a$ is over all atoms in the crystal. The sum over nl is over atomic core states,

$$\Phi^{a}_{n,l,\hat{k}}(\vec{r}-\vec{R}_a) = \frac{P_{n,l}(|\vec{r}-\vec{R}_a|)}{|\vec{r}-\vec{R}_a|} Y^{l}_{0,\hat{k}}(\theta) \tag{3-2}$$

where $y^{l}_{0;\hat{k}}$ is the spherical harmonic with z-axis taken along k, and n and l label the principal and orbital angular quantum numbers of the atomic-like states. The constants $A_{nl;\hat{k}_\mu}$ are chosen so that the quantity in brackets, an OPW, is orthogonal to all core state wave functions. (See Woodruff[13] for a more complete discussion of the basic OPW formalism.)

The two requirements of no core overlap and convergence of the valence Fourier series with a reasonable number (about 250) of OPWs determines the division into core and valence states. In ZnS the Zn 4s states and the S 3s and 3p states are considered valence states. As we showed in detail in two previous papers,[11,12] core overlap is then negligible and the Fourier series has adequately converged by 229 OPWs.

The calculation is self-consistent in the sense that core and valence state energies and wave functions are calculated alternately with appropriate updating of crystalline potential and orthogonality coefficients, $A^{a}_{nl;\bar{k}_\mu}$ until the valence energies change less than 0.02 eV from iteration to iteration. In calculating the contribution to the crystalline potential made by the valence electron density, the electron density average over the Brillouin zone is estimated by a weighted average over the four high symmetry points Γ, X, L and W of the zincblende Brillouin zone. The adequacy of this four point weighting is discussed at some length in a paper comparing our results with X-ray charge densities to be published with P. M. Raccah.[14] Because the band energies are only calculated at the high symmetry points in the SCOPW method, the high symmetry point results are extended throughout the Brillouin zone using a pseudopotential interpolation procedure.[15] The pseudopotential results are used to calculate the imaginary part of the dielectric constant, ε_2. The comparison of the calculated ε_2 curves with the experimental results allows a good judgment of the accuracy of the calculated bands.

The exchange potential for the X_α case is treated in the following way. A valence wave function is written as the sum of plane wave terms, ϕ_{pw}, and the sum of terms involving the core wave functions, ϕ_{v-c},

$$\phi_v = \phi_{pw} + \phi_{v-c}.$$

The valence electron density, ρ_v, can be written

$$\rho_v = \sum |\psi_v|^2 = \sum |\phi_{pw}|^2 + \sum \{ \phi^*_{pw}\phi_{v-c} + \phi^*_{v-c}\phi_{pw} + \phi^*_{v-c}\phi_{v-c} \}$$

$$\equiv \rho_{pw} + \rho_{v-c} \qquad (3\text{-}3)$$

Throughout the self-consistent calculation, the ρ_{v-c} term is spherically symmetrized. When calculating new core states, ρ_{pw} is also spherically symmetrized about each inequivalent nucleus and $\rho^{1/3}$ is calculated. For the valence calculation, $V_{ex}(k)$ is needed. The total electron charge density is written

$$\rho^{1/3} = \{\rho^{1/3} - [\rho_{v-c} + \rho_{CORE}]^{1/3}\}_{\substack{\text{CRYSTAL} \\ \text{MESH}}} + [\rho_{v-c} + \rho_{CORE}]^{1/3}_{\substack{\text{ATOMIC} \\ \text{MESH}}} \tag{3-4}$$

The first term is calculated over an appropriate crystalline mesh of 650 points in 1/24 of the Brillouin zone and is Fourier transformed by the three dimensional generalization of the formulae

$$f_p = \sum_{j=-N}^{N} g(j) e^{2\pi i pj/(2N+1)}$$

$$g(j) = \frac{1}{2N+1} \sum_{p=-N}^{N} f_p e^{-2\pi i pj/(2N+1)} \tag{3-5}$$

The last term in Eq. 3-4, which is spherically symmetric about each atom site, is calculated in the usual way,

$$f(\bar{k}) = \sum_{a} e^{2\pi i \bar{k} \cdot \bar{R}_a} \int dr\, g(r) \frac{\sin(kr)}{kr} 4\pi r^2 \tag{3-6}$$

where the sum is over atoms in the unit cell.

The calculation involving the L exchange approximation is more involved because a separate exchange potential is needed for each state because of the occurrence of E in the relation $k^2 = (E_\mu - V(r))$. In the core calculation, ρ_{pw} is again spherically symmetrized and the calculation proceeds using the appropriate core energy in each case. For the valence state calculation, this procedure would be too cumbersone. Consequently, a weighted average of high symmetry point valence band energies is taken and this value is used for E_μ in the expression for k^2 for all valence states. A study of the effect of varying this E_v, from the bottom to the top of the valence band showed no appreciable change in the valence charge density.

It was found that the breakup of ρ in Eq. 3-4 into a smoothly varying crystal mesh part and a spherically symmetrical core part gave misleading results because of the slightly different $F(\eta)$ in the two cases (because of spherically symmetrizing ρ_{pw} in the core part). Consequently a much finer crystalline mesh of 1785 points in 1/24 of the Brillouin zone was used. The total quantity $F(\eta)\rho^{1/3}$ was evaluated at each crystalline mesh point and then Fourier transformed according to the formulae of Eq 3-5.

A complication arises because of the nonorthogonality of the L wave functions due to the different Hamiltonians used for the different states. The SCOPW method requires that the same Hamiltonian be used for core and valence calculations, or more explicitly, the variational solution of the valence problem requires that the functions $\phi^a_{nl;\bar{k}_\mu}$ in Eq. 3-1 be solutions of the valence Hamiltonian with the classic L prescription. This core density is then used to determine the updated crystalline potential. When the core states are mutually self-consistent, one final core wave function calculation is performed with the previously determined core density, but with all the E_μ taken equal to E_v. The resulting $\phi^a_{nl;\vec{k}_\mu}$ are solutions of the valence Hamiltonian and are used in Eq. 3-1 along with the correspond $A^a_{ne;\bar{k}_\mu}$.

While the preceding technique is sufficiently accurate to determine the valence electron distribution, the variation of E_μ with E_v must be taken into account when finding the final valence and conduction band eigenvalues. This final eigenvalue determination is made by freezing the self-consistent crystalline charge density, both valence and core, and then carrying the calculation from the determination of new orthogonality functions, $A^a_{nl;\bar{k}_\mu}$, through the crystal mesh exchange potential and through the valence eigenvalue calculation using each of four E_vs in $F(\eta)$. The resulting eigenvalues, when plotted as a function of E_v, are shown in Figure 2 for the case of ZnS. The intersections of the curves $E_\mu(E_v)$ with the 45 degree line $E_\mu = E_v$ then give the required L eigenvalues, i.e. the solution where $k^2 = (E_\mu - V(r))$ is used in determining $F(\eta)$ in the exchange potential.

For the crystalline binding energy calculation (in the Liberman or Slater approximation), the desired binding energies of Eq. 2-17 are related to the X_α eigenvalues by the equations

$$B^{L\,\mathrm{or}\,S}_{\mu,\alpha} = E_{\mu,\alpha} + \int \Psi^{*}_{\mu,v}(\bar{r})\left[V^{L\,\mathrm{or}\,S}_{ex} - V^{\alpha}_{ex}\right]\Psi_{\mu,v}(\bar{r})\,d\bar{r}$$

$$= E_{\mu,\alpha} + \int \left[\rho^{v}_{PW} + \rho^{v}_{v-c}\right]\left[V^{L\,\mathrm{or}\,S}_{ex} - V^{\alpha}_{ex}\right]d\bar{r}$$

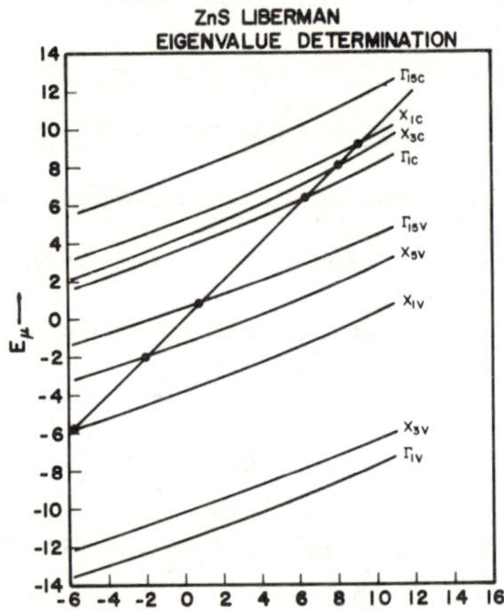

Fig 2. Eigenvalue E_μ versus value of E_v in $k = [E_v - V]^{1/2}$. The 45° line gives the correct eigenvalue.

The first integral is performed numerically over a crystalline mesh of 1485 points. The second is performed for each inequivalent atom. The appropriate E_μ is used for E_v for each valence state.

III Results

a. Atomic Systems, Krypton

Before we examine the results obtained using the approximation to the

TABLE I - Atomic Krypton. The experimental ionization energies obtained by electron emission and by optical spectra (denoted by *) are given in column 1. The relaxed RHF ionization energies are given in column 2 and the RHF eigenvalues are given in column 3. The L, α_1, α_M(.70) and $\alpha_{2/3}$ eigenvalues are given in columns 4 thru 7. The L, α_1, α_M and $\alpha_{2/3}$ binding energies calculated in the Hartree-Fock approximations are given in columns 8 thru 11. The α_1, α_M and $\alpha_{2/3}$ binding energies calculated in the Liberman approximation are given in columns 12 thru 14. The values are given in Ry.

	1	2	3	4	5	6	7	8	9	10	11	12	13	14
	EXP	I_{RHF}	E_{RHF}	E_L	E_{α_1}	E_{α_M}	$E_{\alpha_{2/3}}$	B_L^{HF}	$B_{\alpha_1}^{HF}$	$B_{\alpha_M}^{HF}$	$B_{\alpha_{2/3}}^{HF}$	$B_{\alpha_1}^{L}$	$B_{\alpha_M}^{L}$	$B_{\alpha_{2/3}}^{L}$
1s	1052.55	1055.40	1059.39	1058.59	1049.50	1039.47	1038.36	1059.85	1056.29	1059.23	1059.55	1055.10	1057.94	1058.26
2s	141.20	142.15	144.17	143.93	139.74	136.91	136.60	144.49	141.95	144.02	144.25	141.40	143.46	143.69
$2p^{1/2}$	126.94	127.50	129.76	130.12	126.94	123.87	123.54	130.15	127.42	129.63	129.87	127.36	129.57	129.82
$2p^{3/2}$	123.12	123.43	125.76	126.06	122.90	119.92	119.59	126.16	123.45	125.63	125.87	123.34	125.52	125.77
3s	21.24	21.80	22.45	22.22	20.36	19.29	19.17	22.56	21.34	22.39	22.51	20.98	22.03	22.15
$3p^{1/2}$	16.39	16.55	17.24	17.15	15.72	14.67	14.56	17.37	16.12	17.19	17.31	15.89	16.96	17.08
$3p^{3/2}$	15.73	15.94	16.63	16.53	15.12	14.09	13.98	16.75	15.52	16.57	16.69	15.29	16.34	16.46
$3d^{3/2}$	6.54	6.84	7.56	7.56	6.91	6.00	5.91	7.69	6.43	7.50	7.62	6.29	7.36	7.48
$3d^{5/2}$		6.74	7.45	7.45	6.81	5.91	5.81	7.59	6.34	7.40	7.52	6.19	7.26	7.38
4s	1.76	2.33	2.38	2.36	1.94	1.61	1.58	2.38	2.05	2.34	2.37	1.99	2.31	2.35
$4p^{1/2}$	1.078*	1.019	1.083	1.080	0.906	0.629	0.600	1.083	0.804	1.047	1.072	0.773	1.026	1.059
$4p^{3/2}$	1.029*	0.944	1.029	1.017	0.851	0.580	0.552	1.026	0.757	0.990	1.014	0.720	0.968	0.996
TE			5577.788	5577.573	5577.004	5577.561	5577.5530							

exchange term, let us compare the Hartree-Fock energies and the binding energies with experiment. The proper way, as pointed out in IIa, to calculate ionization energies for various electron levels of an atom is to subtract the total energy of the self-consistent atom from that of the self-consistent ion. This is the only correct way to handle the relaxation of the electronic wave function. Relativistic Hartree-Fock (RHF) calculations were carried to self-consistency for atomic krypton and separately for each krypton ion obtained by removing one electron from each atomic state in turn. The resulting ionization energies, obtained by subtracting total energies of atom and ion, are given in column 2 of Table I. The RHF eigenvalues for the atoms are given together with the experimental ionization energies obtained by electron emission[16] and by optical spectra[17] in Table I column 3 and 1 respectively.

The well known theorem that the relaxed ionization energy is smaller than the frozen ionization energy is illustrated by comparing the RHF eigenvalues of column 3 (which by Koopman's theorem give the ionization energies assuming frozen wave functions) with the relaxed ionization energies of column 2 obtained by subtracting total energies. Taking relaxation into account markedly improves the agreement between RHF results and experiment. The deviation is largest for the s levels. But consideration of radiative effects (the Lamb shift) would further improve the agreement. For example in Mercury the Lamb shift lowers the ionization energy of the 1s state by 38 Ry.[18] A rough estimate for krypton can be obtained from the fact that the Lamb shift goes as $N^4 q^5 mc^2$, where N is the number of electrons and q is the fine structure constant. Scaling the 38 Ry to krypton gives $\sim$ 2 Ry for the further decrease in the calculated ionization energy. This clearly is of the correct order of magnitude to explain the 3 Ry discrepancy between the relaxed RHF calculation and the experimental value. For the 4p states there is an experimental discrepancy between the electron emission results and the optical results. Since the optical results are more reliable because of the complicated electron-electron interactions that are so important at low energies for the electron emission results, Table I contains the 4p optical results. Thus one sees from the above discussion that the relaxed RHF calculation with allowance for the Lamb shift at the low end agrees exceptionally well with experiment.

The calculations using the different approximations to the exchange term are also given in Table I. The eigenvalues plus total energies are given in columns 4 through 7, and the binding energies calculated using the Hartree-Fock expression for the exchange are given in columns 8 through 11. We have also included binding energies using Liberman's approximation for the exchange term (wave functions obtained in the different approximations for this term) in columns 12 through 14. If one applied the same corrections for relaxation and Lamb shift to the α_1 (Slater) eigenvalues, it is clear that the superficial original agreement becomes much worse. While these corrections move the HF results into agreement with experiment, the same corrections move the approximation results (this includes all the approximations: Slater's α_1; the minimum total energy value, α_M; Kohn-Sham's $\alpha_{2/3}$; and Liberman's, L eigenvalues) away from experiment. It can be concluded from this that α_1 eigenvalues do not mystically account for correlation. They do accidently account for atomic core relaxation, but this is purely coincidental. Representative samples of eigenvalues and binding energies are shown in Fig. 3. The eigenvalues vary linearly with the exchange coefficient and increase from α_1 to $\alpha_{2/3}$. The $\alpha_{2/3}$ energy levels are much to compressed to match experiment. The α_1 levels are somewhat too compressed. While L and RHF energy levels are spread out the most and match experiment the closest the close agreement between the L and RHF eigenvalues is impressive.

A very sensitive test of the wave functions obtained from the various exchange approximations is to use them in the calculation of the total energy of an atom. The krypton results are given in Fig. 4 and summarized in the last row of Table I. The exchange coefficient (α_1) which gives

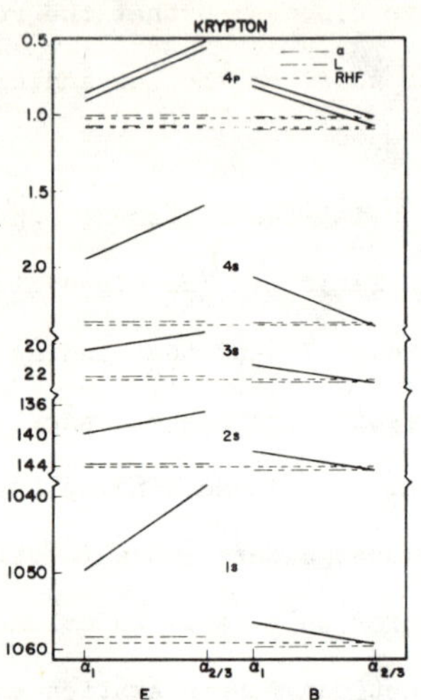

Fig 3. The variation of the eigenvalues and binding energies of Kr using the RHF, and X_α exchange term. α varies from Kohn-Sham value to Slater value. The energies are in Ry.

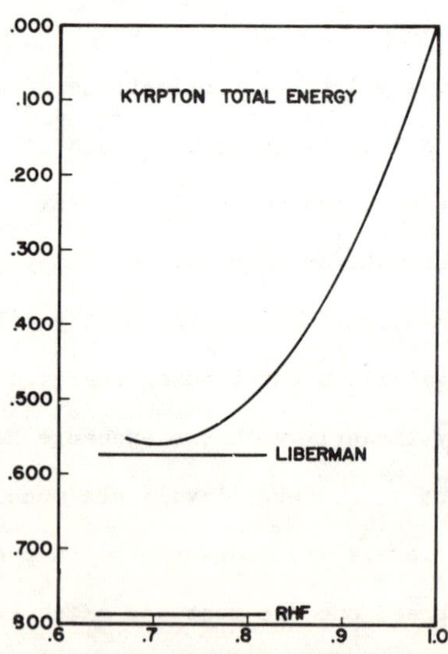

Fig 4. The total energy calculated in HF approximation using orbitals obtained from RHF solution, Liberman solutions and X_α solutions. α varies from the Kohn-Sham value to the Slater value.

the lowest total energy is 0.70. This is very close to the free electron value of 2/3. From Fig. 4 it can be seen that the L, the $\alpha_{2/3}$, and the α_M wave functions all give about the same total energy, which is still considerably above the RHF value. The difference between this energy and the RHF energy is about half that of the α_1 to $\alpha_{2/3}$ difference. So while the $\alpha_{2/3}$ value is an improvement over the α_1 value, it is an improvement by a factor of two, not be an order of magnitude.

Atomic binding energies again vary linearly with exchange coefficient. (see Table I and Fig. 3.) The variation of the upper binding energy levels is comparable to the variation of the eigenvalues, but is in the opposite direction. The $\alpha_{2/3}$ binding energies very closely match the RHF results, and hence match experiment, while the α_1 binding energies are too high. The binding energies in the L approximation agree very closely with the binding energies calculated with the proper RHF Hamiltonian, partially justifying our use in the next section of the L approximation

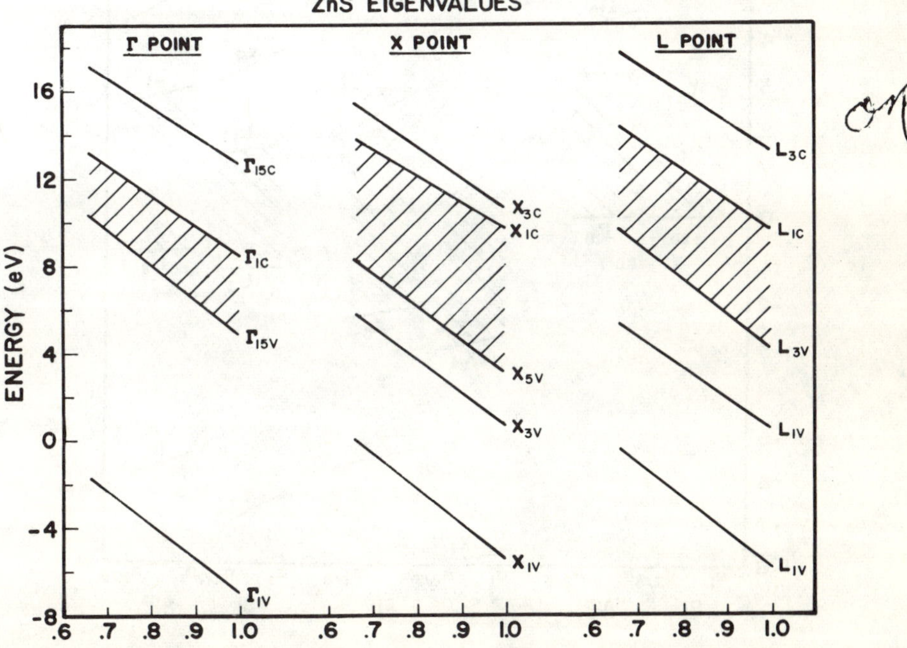

Fig 5. The variation of the Γ, X, and L eigenvalues of ZnS using X_α exchange term. α varies from Kohn-Sham value to Slater value

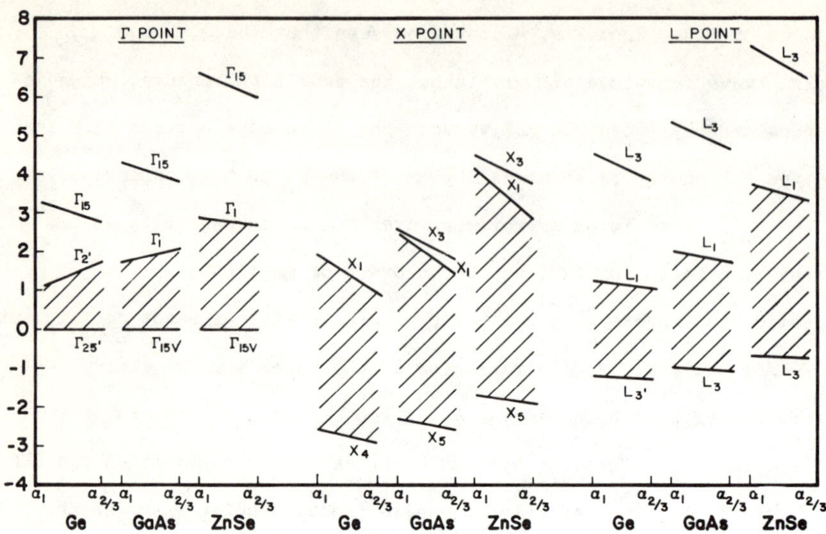

Fig 6a. The variation of the Γ, X, and L eigenvalues of Ge, GaAs, and ZnSe versus the exchange paramter α. α varies from Kohn-Sham value to Slater value.

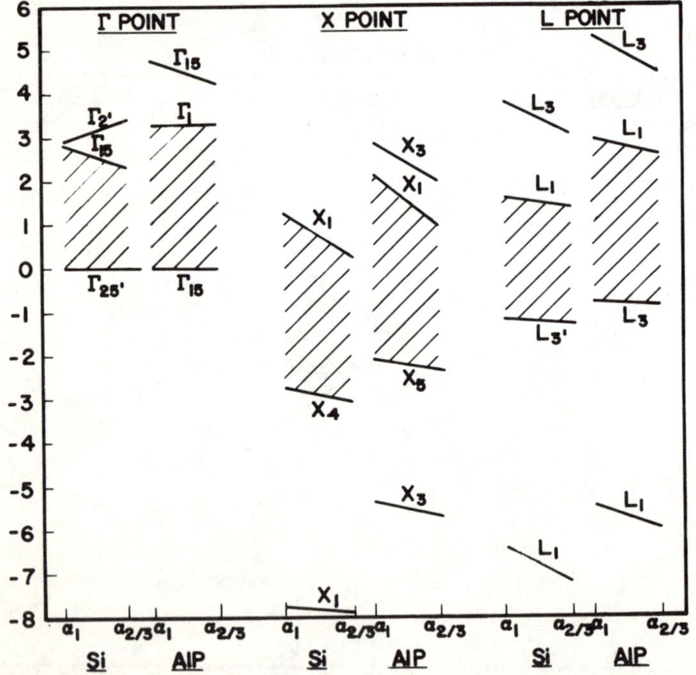

Fig 6b. The variation of the Γ, X, and L eigenvalues of Si and AlP versus the exchange parameter α. α varies from Kohn-Sham value to Slater value.

Table II. Figure 5 shows that again the dependence of eigenvalue upon

TABLE II - The SCOPW energy eigenvalues of ZnS obtained using the L, α_1 and $\alpha_{2/3}$ exchange approximations are given in the first three columns. In column 4 and 5 are given the α_1 and $\alpha_{2/3}$ binding energies calculated in the Liberman approximation. In column 6 are given the $\alpha_{2/3}$ binding energies calculated in the Slater approximation.

	E_L	E_{α_1}	$E_{\alpha_{2/3}}$	$B^L_{\alpha_1}$	$B^L_{\alpha_{2/3}}$	$B^S_{\alpha_{2/3}}$
Γ_{1v}		-7.00	-1.76	-7.12	-7.31	-8.26
Γ_{15v}	.88	4.75	10.36	8.97	8.65	2.58
Γ_{1c}	6.48	8.46	13.08	13.98	13.24	7.40
Γ_{15c}		12.63	17.08	19.01	18.89	11.91
X_{1v}		-5.48	.04	-4.28	-5.16	-7.93
X_{3v}		.77	5.80	2.61	2.52	-.70
X_{5v}	-1.93	3.12	8.26	6.07	6.06	1.89
X_{1c}	8.09	9.66	13.78	14.57	14.02	8.68
X_{3c}	9.26	10.59	15.38	16.59	16.18	9.44
L_{1v}	-5.87	-5.87	-.43	-4.96	-5.77	-8.25
L_{1v}		.49	5.27	1.93	1.99	-.45
L_{3v}	8.10	9.62	14.21	15.30	14.83	8.61
L_{3v}		13.27	17.72	19.81	19.50	12.39

exchange coefficient is linear, with the α_1 eigenvalues again the lowest. We find the same spreading characteristics. The $\alpha_{2/3}$ bands are closest together, with the α_1 bands next closest. The L bands are the most widely spread out. These statements are generally true, although exceptions can be seen in Figures 6a and 6b. The dependence of the individual bands upon the exchange coefficient depends upon the character of the bands (symmetry, p state content, etc.) rather than upon their energy as can be seen from the behavior of the Γ_{15c} and Γ'_{2c} states in Ge (Fig 6a) and Si (Fig 6b). The behavior of Γ, X, and L point bands for the two isoelectronic sequences Si, AlP and Ge, GaAs and ZnSe are shown to give a general idea of the behavior of the various bands with α for different compounds.

In the atomic case, the most widely spread energy levels, those of L and RHF, match experiment most closely. In the case of our SCOPW semiconductor calculations, the α_1 bands have just the correct amount of spread. This point is illustrated in Table III where experimental and

Table III - The direct band gaps (indirect where indicated) are given based on the free atomic and the SCOPW Models for different exchange approximations. The experimental value is also given.

Compound	FREE ATOMIC (eV)		SELF-CONSISTENT (eV)			
	Slater	Kohn-Sham	Slater	Kohn-Sham	Liberman	Exp
ZnS	4.30	3.68	3.71	2.72	5.60	3.84[a]
ZnSe	3.16	2.78	2.94	2.68	4.97	2.83[b]
GaAs	1.32	1.49	1.61	1.97	3.63	1.54[c]
Si(Δ_1^M-Γ_{25})			1.07			1.13
Ge(L_1-Γ_{25})			1.27	1.01		.76[d]
(Γ_2-Γ_{25})			1.18	1.73		.90[d]
CdS			2.50			2.58[e]

a J. L. Birman, H. Samelson, and A. Lampicki, G. T. & E Res. Developm. J. 1, (1961).
b. M. Aven, D. Marple, and B. Segall, J. Appl. Phys. Suppl. 32, 226 (1961).
c. M. Cardona, K. L. Shaklee and F. H. Pollak, Phys. Rev. 154, 696
d. G.G. MacFarlane, T. P. McLean, J. E. Quarrington and V. Roberts Proc. Phys. Soc. (London) 71, 863 (1958).
e. D. G. Thomas and J. J. Hopfield, Phys. Rev. 116, 573 (1959).

theoretical band gaps are given. It can be seen that in the case of almost every semiconductor for which we have SCOPW results, the α_1 band gap is within 0.2 eV of experiment, while the L gap is far too large and the $\alpha_{2/3}$ gap is usually too small. It can also be seen that for non-self-consistent crystal calculations based upon free atomic potentials the α_1 band gap is larger than when self-consistency is obtained. This fact explains why such calculations often require a smaller exchange coefficient than α_1 to compress the band structure down to α_1 self-consistent values. Clearly the only physically meaningful comparison is the comparison of self-consistent results with experiment. Few other direct transition energies are known experimentally to compare with the SCOPW results, but we have made comparisons of ε_2 curves which depend upon the bands throughout the Brillouin zone. There again, the α_1 results closely match experiment as to peak positions, while the $\alpha_{2/3}$ results are compressed at too small energies. The L results again are far too expanded. For some reason in the crystalline case, α_1 results are remarkably good.

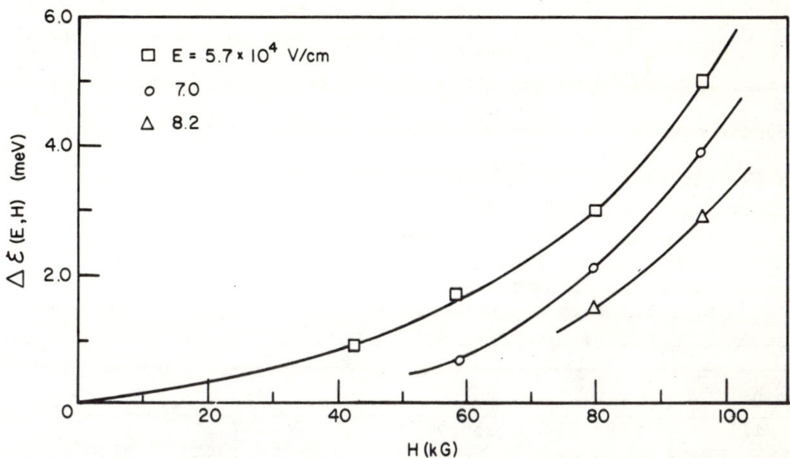

Fig 7. The variation of the Γ, X, and L binding energies of ZnS using X_α exchange term to calculate the orbitals and Liberman's approximation to calculate the binding energies. α varies from Kohn-Sham value to Slater value.

The dependence of ZnS binding energies on the exchange coefficient shown in figure 7, is not quite linear - some bowing is evident-and the variation is much smaller than the variation of the eigenvalues. All the binding energies, like the L eigenvalues, are much too widely spread out as compared to experiment. For example, the ZnS binding energy band gap is around 5.6 eV, whereas the experimental value is 3.84 eV.

The ZnS $\alpha_{2/3}$ binding energies, calculated in the α_1 approximation, are again too widely spread out, but the high energy spreading is much less than for the binding energies calculated in the L approximation. The spreading at the high energy end for both L eigenvalues and binding energies in the L approximation reflects the fact that the F(η) curve of Eq. 2-13 and figure 1 goes rapidly to zero beyond the Fermi energy (where

$\eta = 1$). The very rapid decrease tends to widely spread out the conduction bands in these crystalline calculations. We conclude that this behavior, which is not tested in atomic calculations, is unrealistic.

Another good test of wave functions, in addition to the calculation of total energies, is to calculate the Fourier transform of the electron charge density. The results for free atomic zinc and selenium packed in a crystal lattice (RHF) and SCOPW results for α_1, $\alpha_{2/3}$, and L exchange are compared with the experimental measurements of P. M. Raccah[14] in Table IV. At high k, ($k^2 = \frac{4\pi^2}{a^2}[h^2+h^2+1]$) the core state density is emphasized, while the small k values emphasize the valence contrubution.

TABLE IV ZnSe Form Factors. Experimental form factor are taken from P. M. Raccah. Tehoretical form factors are given based on relativistic Hartree-Fock and the SCOPW using the L, α_1 and $\alpha_{2/3}$ exchange approximations.

hkl	Exp	RHF	L	α_1	$\alpha_{2/3}$
111	158.55	155.80	156.17	157.68	156.22
200	14.86	11.22	11.44	11.72	11.15
220	189.80	189.79	189.05	191.14	189.48
311	129.10	125.90	125.03	126.40	125.50
222	11.63	10.33	9.82	9.61	9.50
400	162.31	162.13	161.39	163.57	162.51
331	109.55	109.47	109.39	111.13	110.34
420	12.09	11.59	11.29	10.85	11.08
422	140.56	143.30	142.82	145.37	144.43
440	128.76	128.86	128.29	131.05	130.19
531	88.37	88.29	87.81	89.74	89.18
620	117.46	117.28	116.57	119.37	116.78
533	80.77	80.73	80.26	82.25	80.46
622	12.28	12.26	12.17	12.15	12.11
444	107.69	107.78	107.04	109.80	107.37

Since the core states are very close to free atomic states, we again have an atomic-crystalline comparison. It is seen that for high k values the RHF results closely match experiment. According to P. M. Raccah, this is usually the case. The RHF results are very good at high k. The $\alpha_{2/3}$ and L results also closely match RHF and experiment, while the α_1 core is clearly several percent worse. At low k, where the valence wave functions are emphasized, the agreement with experiment is not impressive. All exchange approximations show an improvement over the free atomic RHF results because the SCOPW model allows the valence charge to spread out redistribute. But even the best results, produced by the α_1 approximation, are in poor agreement with experiment. The small deviation between the various approximations as compared to the discrepancy with experiment is consistent with the weak dependence of binding energies on exchange. The SCOPW wave functions seem to depend much more weakly upon the exchange constant than the eigenvalues do. But while α_1 eigenvalues are clearly superior to the others, the wave functions are only slightly better, and the α_1 binding energy band gap is slightly worse than the $\alpha_{2/3}$ gap.

III Comparison with Experiment, GaAs

In Table V the results of the SCOPW calculations using α_1 exchange

	Self-Consistent OPW (eV)	Adjusted[a.] OPW (eV)	Photoemission[b.] Exp. (eV)
Γ_{12c}	10.78	10.6	10.2 to 10.9
Γ_{1c}	9.11	9.0	9.0 to 9.1
Γ_{15c}	4.34	4.6	4.6 to 5.0
	1.61	1.54	1.35 to 1.5
Γ_{15v}	0.00	0.00	0.00
Γ_{1v}	-11.81	-12.4	
X_{3c}	2.62	2.5	2.40 to 2.45
X_{1c}	2.57	1.90	1.70 to 1.85
X_{5v}	-2.23	-2.3	- 2.3 to 2.6
X_{3v}	-6.29	-5.5	
X_{1v}	-9.48	-10.7	
Min. along Δ At 0.82 of way to X	2.43		
L_{3c}	5.36	5.3	5.0 to 5.3
L_{1c}	1.99	2.00	1.80 to 1.95
L_{3v}	-0.95	-0.9	-0.9 to -1.1
L_{1v}	-10.22	-11.1	-5.9 to 6.7
L_{1v}	-10.22	-11.1	

a. See Ref. 19, b. See Ref. 20

term are presented in the first column and the graph of the energy structure is given in Fig. 8. For comparison, in column two the adjusted OPW

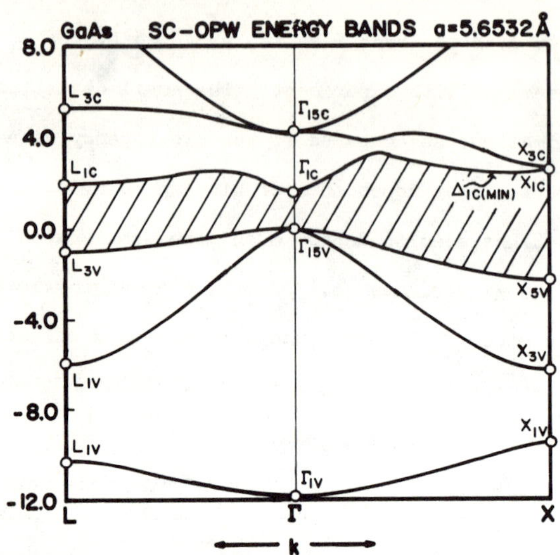

Fig 8. Energy value structure of GaAs.

calculations of Ref. 19 are given, and in column three the photoemission data of Ref. 20 are recorded. All three columns are in fairly close agreement. Most of the exceptions, which are underlined in the table, are related to the location of the X_{1c} or Δ_{1c} minimum in the conduction band. From the SCOPW calculations, the value of 4.34 eV is at least 0.26 eV too low when compared to the experimental value of 4.6 to 5.0 eV. This is the only important result which differs appreciably from the photo-emission data.

In order to compare the results with optical experiments, we calculate the imaginary part of the dielectric constant. The plot of calculated ε_2 versus energy is given by the solid curve in Fig. 9. Experimental results from Philipp and Ehrenrich[21] are represented by the dashed line on the same figure. The peak energies together with the experimental reflection results of Thomas, Wooley, and Ruberstein[22] are listed in Table

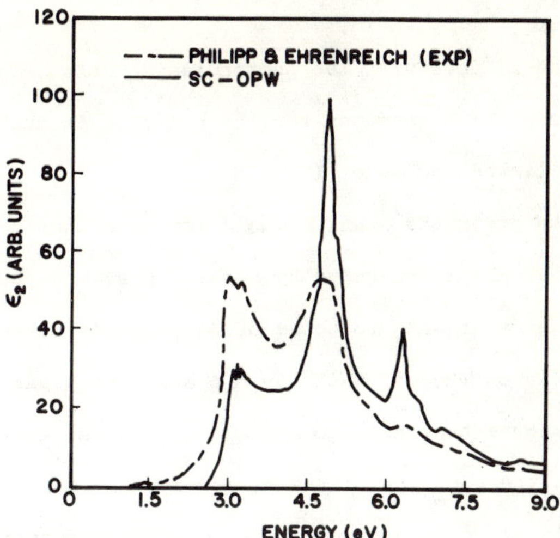

Fig 9. ε_2 plot of GaAs verus

VI. The calculated peak positions agree well with the experiments except that the theoretical E_1' peak appears 0.2 eV lower in energy than the measured value. Also note that the experimental E_o' peak, which is thought to originate from the $\Gamma_{15c} - \Gamma_{15v}$ transition, agrees better with the cal-

TABLE VI. The energy peaks of the imaginary part of the dielectric constant obtained from experiments of Philipps and Ehrenreich are listed in column one. In column two are given the reflectivity structure energies obtained by Thompson, Wooley and Ruberstein, and in column three are given the calculated energy peaks of the imaginary part of the dielectric constant. All values are in eV.

	ε_2(Exp)[a.]	Reflectivity (Exp)[b.]	ε_2(Cal.)
E_1	3.0	2.90	3.1
$E_1+\Delta$	3.2	3.13	
E_o'		4.45	4.43
E_2	4.7-4.8	5.04	4.8
E_1'	6.5	6.63	6.3
$E_1' + \Delta$		6.89	

a. See Ref. 21, See Ref. 22

culated value than it does with the photoemission data. The structure in the reflection data which accounts for the E_o' value is much weaker than other structures listed in Table VI.

The arguments presented[23] which would indicate that the X_{1c} (or Δ_{1c}) was lower than the calculated value by about 0.5 eV were based on relations of energy bands of GaAs to those of GaP, and on interpretations of pressure data. It was thought that the minimum energy gap in GaP was from the top of the valence band at Γ going to a minimum at X or along the Γ-X line. Looking at the band gap of the mixed crystal $GaAs_{(1-x)}P_x$ one finds only one change in slope as x changes from 0 to 1 somewhere above 0.5. Hence, the same SCOPW energy band programs were used to calculate GaP, and it was found that the lowest energy in the bottom conduction band was at the L point. If the small curving of the energy line of the mixed crystal is ignored, and straight lines are drawn between the Γ_{1c} position of GaAs and GaP and between the L_{1c} energies of the two compounds, the resulting indirect band gap for the mixed crystal will be that shown in Fig. 10. The one change in slope (at $x = 0.75$) is similar to the change found experimentally in Ref. 24. One word of caution has to be added. The X_{1c} minimum is only about 0.1 eV above the L_{1c} minimum. If X_{1c} were lower than L_{1c} (possible from the uncertainty of the calculation) the calculated graph would only change very near the GaP X_{1c} and L_{1c} location.

The calculated X_{1c} values are higher in energy for both compounds, GaAs and GaP, than the values determined from interpretation of experiment. In order to check that this is not an anomalous effect of the way the calculations are made, we calculated the band structure for a crystal in which a conduction minimum in the X direction is well established. It is fairly well established that Si has an off axis band gap with the maximum of the valence band at Γ, and the minimum in the conductuion band at Δ. Experimentally the energy difference is 1.13 eV[25] while the calculated

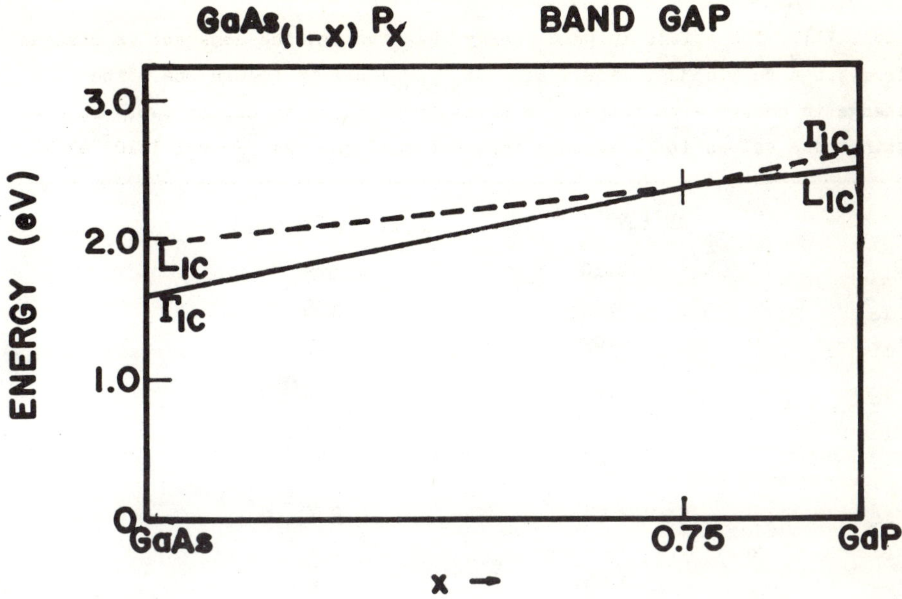

Fig. 10 The energy of the band gap (solid line) of $GaAs_{(1-x)}P_x$ as a function of concentration of As and P where x goes from 0 to 1.

SCOPW value is 1.07 eV. This is in very good agreement. In addition the calculated value is on the low energy side by 0.06 eV. Thus we conclude that the calculated X_{1c} values of GaAs and GaP are not an anomalous effect of this method of calculation.

By varying the lattice constant from 5.6532Å to 5.64Å, the effect of hydrostatic pressure was calculated. The results are given in Table VII. The first observation is that in column one all of the energy changes are positive with the minimum of the conduction band, Γ_{1c}, changing by approximately twice as much as each of the other energy changes. The values of the pressure coefficient given in the second column were obtained by assuming the band gap $(\Gamma_{1c}-\Gamma_{15v})$ pressure coefficient was $1.1x10^{-2}$eV/kb. the most important point is that the closure rate of the L_{1c} and Γ_{1c} conduction minimum is $-0.9x10^{-2}$eV/kb. The experimental value for the two lowest minimums was found to be $0.96x10^{-2}$ eV/kb to $0.97x10^{-2}$eV/kb by both

Table VII. The shifts in band energy when the lattice constant is changed from 5.64Å to 5.6532Å. ($\Delta E = E_{5.64}-E_{5.6532}$) are in column one. The change in energy with respect to pressure is given in column two. To obtain these values it is assumed that the band gap has $\frac{\Delta E}{\Delta P} = 1.1 \times 10^{-2}$ eV/kb.

	ΔE (eV)	ΔE/ΔP(X10^{-2}eV/kb)
Γ_{15c}	0.10	0.8
Γ_{1c}	0.23	1.8
Γ_{15v}	0.09	0.7
X_{3c}	0.09	0.7
X_{1c}	0.08	0.6
X_{3v}	0.06	0.5
L_{3c}	0.09	0.7
L_{1c}	0.12	0.9
L_{3v}	0.08	0.6

Balslev[26] and Hutson, Jayarman, and Coriell.[27] The vaule of Ref. 27 was obtained when they assumed the lattice constant was a linear function of pressure (as was assumed in this calculation). Also in Ref. 26, only two minima were found away from the zone center. One was at 0.34 above the minimum in the conduction band at Γ, and the other was at 0.78 eV above the primary minimum. These two values are in good agreement with the respective values of 0.38 eV and 0.82 eV eV found in this calculation.

Since the symmetry lines Γ-X and Γ-L of the SCOPW model have been calculated, one can obtain the effective mass along these lines. The values obtained are given in Table VIII. The calculated values appear higher than the experimental estimates. However, the ration of the Γ point effective mass to the L-point effective mass is about the same as the estimated ratio used in transport calculations.[23,28] One point should be noted M_L^* was only calculated along Λ and anisotropic effects were not considered. The effective mass at Γ was calculated for both the Δ and Λ lines and found to be isotropic at a lattice constant of 5.6532Å the

Table VIII. The Effective Masses of the three minimums in the condition bands are given for lattice constants of 5.6532Å and 5.64Å as well as estmates from experiments for the two lowest minimums. The values are given in a. u.

	Cal (a = 5.6532Å)	Exp.[a.]	Cal (a = 5.64Å)
M^*_Γ	0.11	0.07 to 0.08	0.14
M^*_L (along Λ)	1.78	1.2 to 1.4	1.78
$M_{0.82 \text{ of } X}$ (along Δ)	0.97		1.00

M^*_Γ along the Δ direction was 8% lighter than that along the Λ direction.

In these calculations relativistic effects have been neglected. In order to determine if relativity would change the relative minimum positions we examined estimates of relativistic corrections given by Herman.[29] It was found that the Γ_{1c} energy moves down relative to both of the L_{1c} X_{1c} energies. Furthermore, the L_{1c} energy moves down relative to the X_{1c} energy. Thus the relationship of the three minima do not change when relativistic corrections are added.

The results of the GaAs energy band calculation using SCOPW method with no adjustments agree quite well with photoemission experiments, hydrostatic pressure measurements, optical data, and investigations of $GaAs_{(1-x)}P_x$ crystals. One finds three conduction band minima. The lowest minimum occurs at the Γ-point. The next higher minimum is located at L, 0.38 eV above the Γ point minimum. And the third minimum is located on the symmetry line Δ, occuring 0.82 of the way from Γ to X at an energy value of 0.82 eV above the Γ minimum. The ratio of the effective masses of the Γ and L minima calculated along Λ agree well with estimates used in transport calculations.

IIId Correlation Effects

The question as to the magnitude of correlation effects has to be explained. In the absence of a full-fledged configuration interaction calculation or a Hylleras calculation, a very rough estimate can be obtained by following the procedure of Lundqvist and Ufford.[6] They used a correlation potential derived for a degenerate electron gas, interpolated between the high density and low density limits. Figure 11 (taken from Reference 6) shows two such correlation potentials. We chose the more rapidly varying potential of Carr and Maradudin[5] to maximize the effect of correlation and added the potential to the self-consistent RHF programs. The resulting shifts in the RHF eigenvalues, shown in Table IX, are reasonable if one examines Figure 11. The deepest electron states are in

Table IX. Correlation energy shifts calculated using Carr and Maradudin potential for Kr. The values are given in Ry.

	ΔE Carr
1s	0.18
2s	0.15
$2p^{1/2}$	0.15
$2p^{1/2}$	0.15
3s	0.13
$3p^{1/2}$	0.14
$3p^{3/2}$	0.13
$3d^{3/2}$	3.13
$3d^{5/2}$	0.13
4s	0.11
$4p^{1/2}$	0.11
$4p^{3/2}$	0.11

in regions of high electron density and undergo shifts of 0.15 Ry, while the most loosely bound electrons are in regions on low electron density and undergo smaller shifts. While we do not take the actual numbers too seriously, we believe that they give a good feel of the magnitude of the correlation corrections. It is interesting to note that the relaxed RHF 4p energies are slightly smaller than the experimental optical results which is reasonable since correlation should lower the total energy of the 36 electron atom more than the total energy of the 35 electron ion, thus slightly increasing the RHF ionixation energy. The actual calculated correlation correction of 0.11 Ry for the 4p states is not far from what

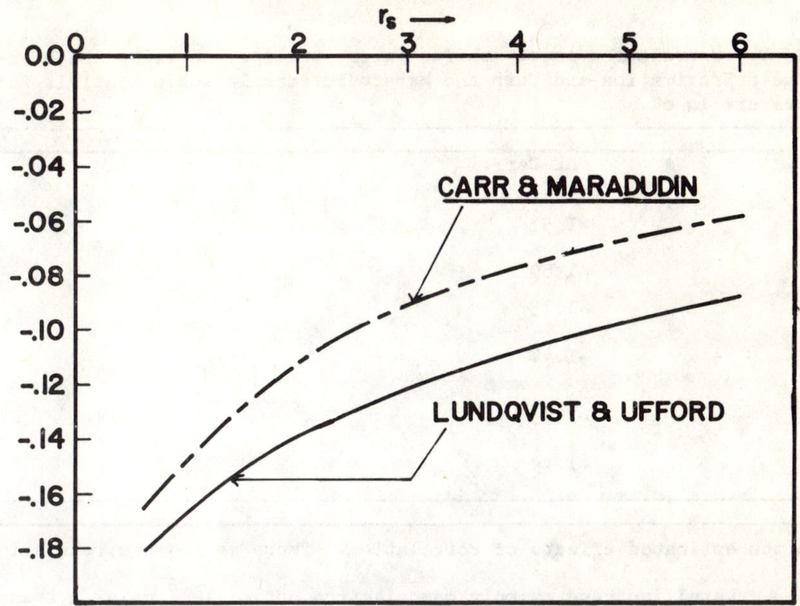

Fig 11. Static correlation potential verus r_s, the radius of a sphere containing one electron in a free electron gas. This figure was taken from Lundgvish and Ufford[6].

is actually required (0.09 Ry for the $4p^{3/2}$) and 0.06 Ry for the $4p^{1/2}$.

We thus conclude that in atomic calculations, RHF energies agree very closely with experiment for all states from the most tightly bound to the most loosely bound, and that the correlation corrections are very small, of the order of 0.1 Ry. We further conclude that in the atomic case, α_1 eigenvalues do not closely match experiment if core relaxation and the Lamb shift are properly taken into account.

To estimate the approximate size of correlation in crystal calculations, we added the correlation potential of Carr and Maradudin to our SCOPW programs, using the L exchange as the closest approximation to HF. The calculation for ZnS was carried to self-consistency. The resulting shifts (in eV) in the L eigenvalues are shown in Table X. It is concluded that if correlation can be roughly simulated in this manner, it will shift band energy differences by the order of 0.4 eV. The difference of energies produced by different exchange approximations is much larger

Table X. Correlation energy shifts in crystalline ZnS using the L exchange approximation and Carr and Maradudin correlation potential. The values are in eV.

	ΔE Carr.
Γ_{15v}	-1.72
Γ_{1cv}	-1.51
X_{5v}	-1.69
X_{1c}	-1.73
X_{3c}	-1.46
L_{3v}	-1.73
L_{1c}	-1.49

than the estimated effects of correlation. There is no core relaxation in the crystalline case as only one electron out of 10^{23} makes a transition to the excited state. We thus conclude that the criterion for success of any exchange approximation in a crystalline self-consistent calculation is that the resulting one-electron energy differences match experiment to within 0.4 eV.

IV. Conclusion

It is felt that a successful exchange approximation must work well in both atomic and crystalline systems. We found that the atomic RHF results agreed very closely with experiment once corrections were made for core relaxation and the Lamb shift. An order of magnitude examination of the correlation contribution indicated that correlation corrections were small relative to the energy deviations between the various exchange approximations in both the atomic and the crystalline cases. Consequently it is concluded that a successful exchange approximation in the crystalline case should give bands agreeing closely with experiment.

In our SCOPW calculations, α_1 eigenvalues (as shown in detail in section IIIc) give the best results, while $\alpha_{2/3}$ eigenvalues are too close together and L eigenvalues are too spread out. The various binding energies, calculated in the L approximation, are also too spread out. The various binding energies, calculated in the L approximation, are also too spread out. This spreading casts doubt upon the propriety of using the $\eta<1$ part of $F(\eta)$ as derived by the free electron theory. In spite of the agreement of the crystalline α_1 eigenvalues with experiment, the crystalline α_1 wave functions are not much better than the $\alpha_{2/3}$ and the L crystalline wave functions.

As can be seen by this work, the exchange approximation will have to be improved in band calculations before one can really judge correlation corrections. Also it should be pointed out that adjusting the band structure after it has been calculated without changing the model serves no useful purpose. The example of GaAs shows this quite clearly.

References

1. J. C. Slater, Phys. Rev. 81, 385 (1951).

2. W. Kohn and L. J. Sham, Phys. Rev. 140, A1133 (1966).

3. R. Gaspar, Acta Phys. Acad. Sci. Hung. 3, 263 (1954).

4. D. A. Liberman, Phys. Rev. 171, 1 (1966).

5. W. J. Carr, Jr. and A. A. Maradudin, Phys. Rev. 133, 371 (1964).

6. S. Lundqvist and C. W. Ufford, Phys. Rev. 139, A1 (1965).

7. T. Koopman, Physica 1, 104 (1939).

8. J. C. Slater, Quantum Theory of Atomic Structure Vol II, Ch 17 Appendix 22.

9. W. E. Rudge, Phys. Rev. (IN PRESS)

10. J. C. Slater, T. M. Wilson, and J. H. Wood, Phys. Rev. (IN PRESS)

11. R. N. Euwema, T. C. Collins, D. G. Shankland and J. S. DeWitt, Phys. Rev. 162, 710 (1967).

12. D. J. Stukel, R. N. Euwema, T. C. Collins, F. Herman, and R. L. Kortum, Phys. Rev. 179, 740 (1969).

13. T. O. Woodruff, in "Solid State Physics", ed. F. Seitz and D. Turnbull (Academic Press Inc., New York, 1957), Vol V, pp 367.

14. P. M. Raccah, R. N. Euwema, D. J. Stukel, and T. C. Collins, Phys. Rev. (To be published).

15. R. N. Euwema, D. J. Stukel, T. C. Collins, J. S. DeWitt and D. G. Shankland, Phys. Rev. 178, 1419 (1969).

16. M. Siegbahn, Nova Acta Regiee Socielates Scientiarium Upsatiensen Series IV, Vol 20 (1967).

17. C. E. Moore, Atomic Energy Levels, National Bureau of Standards Circular No. 467, Vol II (U. S. Government Printing Office, Washington, D.C. 1952)

18. I. P. Grant, Proc. Roy. Soc. A262, 555 (1961).

19. F. Herman and W. E. Spicer, Phys. Rev. 174, 906 (1968).

20. W. E. Spicer and R. C. Eden in "Proceeding of the 9th International Conf. on the Phys. of Semiconductors, Moscow, USSR", July 1968.

21. H. R. Philipp and H. E. Ehrenreich, Phys. Rev. 129, 1550 (1963).

22. A. G. Thompson, J. C. Woolley, and M. Rubenstein, Can. Jour. Phys. 44, 2927 (1966).

23. H. Ehrenrich, Phys. Rev. 120, 1951 (1960).

24. Otfried Madelung, Physics of III-V Compounds (John Wiley & Sons, Inc., New York, 1964) Ch6, p. 287.

25. F. Herman, R. L. Kortum, C. D. Kuglin, and R. A. Short in Quantum Theory of Atoms, Molecules, and Solid State: A Tribute to J. C. Slater, ed. P. O. Löwdin (Academic Press, New York, 1966).

26. I. Balslev, Phys. Rev. 173, 762 (1968).

27. A. R. Hutson, A. Jayaraman, and A. S. Coriell, Phys. Rev. 155, 786 (1966).

28. E. M. Conwell and M. Q. Vassell, Phys. Rev. 166, 797 (1968).

29. F. Herman (private communication).

V

EXCHANGE AND CORRELATION IN SIMPLE METALS

Stig Lundqvist

Chalmers University of Technology, Göteborg, Sweden

I. INTRODUCTION

This lecture was planned to be a complement to the lectures by Dr. T. Collins given at this symposium by continuing his discussion along a slightly different line. In particular we wish to discuss further the choice of the potential in one-electron theory including exchange and correlation effects. The work by Collins and coworkers [1] represents to my knowledge the most extensive comparative study of the Hartree-Fock theory and some simplified versions of this scheme in which the non-local Hartree-Fock potential is replaced by a local potential. Our discussion will be limited to the case of metals having free-electron-like structure, where one can use the results obtained for a uniform electron gas as illustrations.

We shall first review some basic facts about the Hartree-Fock theory and point out some of the shortcomings, and briefly comment on the use of various local approximations to the exchange potential. These questions will be discussed in section II.

In order to obtain a one-electron theory which does not suffer from the shortcomings of the Hartree-Fock theory one has to include the dynamical effects of the electron-electron interaction. The role of a static one-electron potential is taken over by a non-local, time-dependent (or energy dependent) quantity called the self-energy. The self-energy can be defined in an exact manner, but in practise one has to consider low-order approximations, which may form the basis of different self-consistent schemes. In particular, we would like to mention the dynamical screened potential discussed extensively by Hedin[2]. Numerical calculations over a range of metallic densities show that the effective exchange and correlation potential thus obtained is almost constant over the momentum range corresponding to the conduction band. This in turn implies that the potential can be well approximated by a local potential. The numerical values fall in between those given by the Slater and the Gáspár approximations. Further calculations[3] show that for momenta outside the Fermi sea, there is a characteristic variation in the potential. Therefore, one should be cautious with the use of the local approximation for excited bands and applications to low energy electron diffraction.

The results of the detailed calculations indicate a way how the exchange and correlation effects could be built into energy band calculations. It is possible to introduce a local approximation for the potential, which uses the electron gas results at the corresponding density. This can be formulated as a numerical prescription which can be fed into existing schemes for band calculations[4].

II. THE HARTREE AND HARTREE-FOCK APPROXIMATIONS

The simplest self-consistent field theory, the Hartree theory, assumes that the electrons move in the average electrostatic field of the lattice. The Hartree-Fock theory introduces important effects of the exclusion principle by adding to the average field the exchange potential. In momentum space and for a translationally invariant system the exchange potential is given by the formula

$$V_{ex}(\underline{k}) = -\int_{|\underline{k}'|<k_F} \frac{d^3k'}{(2\pi)^3} \frac{4\pi}{|\underline{k}-\underline{k}'|^2} =$$

$$= -\frac{e^2 k_F}{\pi}\left[1 + \frac{1-(k/k_F)^2}{2\,k/k_F}\, \ln\left|\frac{k+k_F}{k-k_F}\right|\right] \qquad \text{II:1}$$

The dispersion law, $E(\underline{k})$, of an electron is obtained by adding the kinetic energy, thus

$$E(\underline{k}) = \frac{\hbar^2 k^2}{2m} + V_{ex}(\underline{k}) \qquad \text{II:2}$$

The exchange potential varies continuously across the Fermi level, however there will be a logarithmic singularity in the derivative of the energy with respect to k, thus the $dE/dk \simeq \ln(k - k_F)$. Therefore the group velocity goes to infinity at the Fermi energy and hence the density of states vanishes at the Fermi level, which is a clearly inadmissable property for a metal.

Another shortcoming of the Hartree-Fock theory is with

regard to the width of the conduction band. From Eq. II:1 we find that the Fermi level is shifted down compared with the Hartree value by an amount $e^2 k_F / \pi$. The bottom of the band, however, is shifted down by twice that amount, i.e. $2 e^2 k_F / \pi$. Thus, the width of the conduction band in the Hartree-Fock exceeds that in the Hartree theory by $e^2 k_F / \pi$. Experimental results, e.g. from soft X-ray emission, indicate that the Hartree results are approximately correct and that the Hartree-Fock theory is at variance with experiments not only at the Fermi surface but also with regard to the overall structure of the conduction band.

The shortcomings of the Hartree-Fock theory just mentioned have of course been known since the early quantum theory of metals. The strange features of the exchange potential arise from the long range of the Coulomb potential. Indeed, replacing the Coulomb potential by a screened potential would for any reasonable value of the screening length remove the anomalies and reduce the effect of the exchange potential drastically. The experiments tell us both the density of states and the band widths are given reasonably well by the Hartree theory.

Thus, a dispersion law of the form (for the uniform model)

$$E(\underline{k}) = \frac{\hbar^2 k^2}{2m} + V_o \qquad \text{II:3}$$

where V_o is a constant would not be far off and anyway be far superior to the Hartree-Fock theory. This choice would correspond to a local approximation to the exchange and correlation potential.

Such approximations were introduced in the early fifties in an ad hoc manner by Gáspár (5) and by Slater (6). Gáspár's potential corresponds to taking the value at the local Fermi momentum in Eq. II:1 as representative, whereas Slater considered an average over the Fermi sea. From the arguments just given about the band widths it follows that the density dependence will be the same but that the numerical coefficient in Slater's potential is a factor 3/2 larger than the coefficient in Gáspár's potential.

It is known from a number of calculations of band structures that these approximate choices for the exchange potential give results which are on the whole in good agreement with experiments. In the following section we shall indicate how the more sophisticated treatment of the exchange and correlation problem, which has been given in particular by Hedin /2/ finally leads back to an effective potential for exchange and correlation, which only varies slightly with momentum over the conduction band, and thus corresponds approximately to a local potential. Moreover the numerical values are in between those given by Gáspár and Slater formulae.

III. AN EFFECTIVE EXCHANGE AND CORRELATION POTENTIAL FROM THE SELF-ENERGY

For brevity we have to refer to the original papers by Hedin [2] and by Lundqvist [3] for a detailed discussion of the self-energy. A very brief account of the key properties is given in the paper by this author on many-electron effects on optical properties in this volume. The salient feature is the introduction of the dynamical response of the medium, described by the wave number and frequency dependent dielectric function. Although the calculations are based on a linear approximation of the dielectric response, the treatment nevertheless goes far beyond the Hartree-Fock theory.

A summary of the most important results obtained by Hedin and by B.I. Lundqvist was given in the Proceedings of the Second Chania Conference 1968 [7]. For example the band widths and the density of states at the Fermi level agree with the Hartree values within a few percent. In particular, they calculated the dispersion law for quasi-particles and from this one can deduce an effective exchange and correlation potential. This potential is presented in a typical case in Figure 1 as a function of momentum together with the Hartree-Fock exchange potential and the approximations by Slater and by Gáspár. The latter are constants in momentum space, whereas the Hartree-Fock potential varies over a large interval and changes rapidly in the region of the Fermi momentum. The effective potential calculated from the self-energy shows only little dependence on momentum in the region up to the

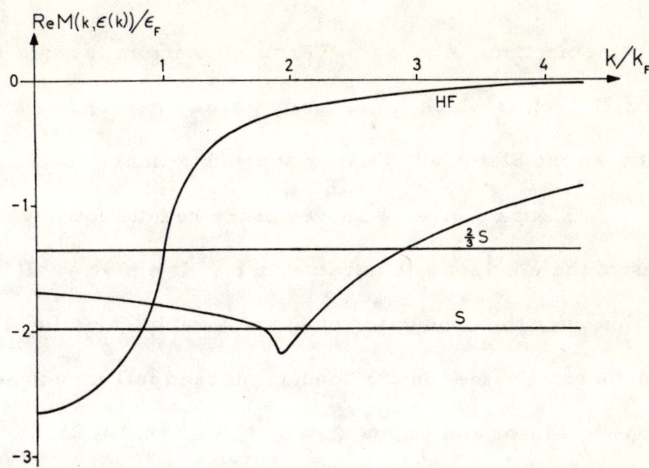

Fig. 1. Exchange correlation potentials for an electron gas after B.I. Lundqvist[3]. As a comparison the Hartree-Fock, the Slater and the Gáspár (2/3 Slater) approximations are also shown.

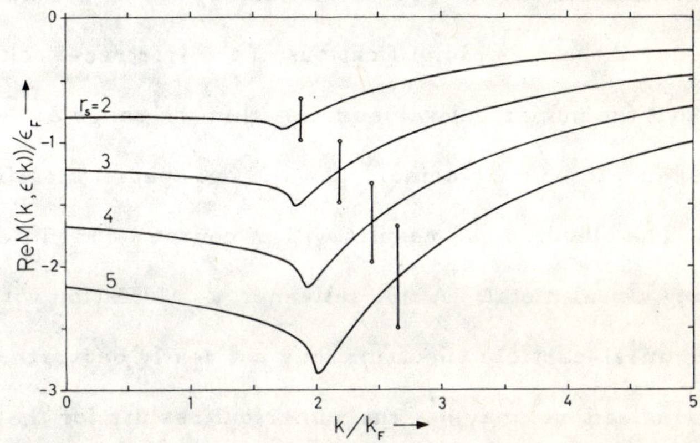

Fig. 2. Exchange correlation potentials for an electron gas over a range of values for the density parameter r_s. The rings on the vertical bars indicate the positions of the Slater and Gáspár approximations. After B.I. Lundqvist[3].

Fermi momentum. Above the Fermi momentum it has a rapid variation around $2k_F$ after which it gradually goes to zero. For $k < k_F$ it falls in between the Slater and Gáspár approximations.

Figure 2 gives a survey of the results for several different values of the electron gas parameter r_s. It shows at all densities a slow variation over the conduction band and a rapid variation around $2k_F$. The numerical values in the conduction band fall in between the approximations by Slater and Gáspár.

It should be noted that the potential obtained from the self-energy has its origin in dynamical effects of screening which go far beyond the Hartree-Fock theory. Therefore, the results will of course give no justification for the approximations by Slater and by Gáspár, who both introduced them as simplifications of the Hartree-Fock scheme. The fact that the numerical values fall within the range of these approximations seems to be pure coincidence without special significance.

The electron gas results will of course be modified when considering a real metal. A full self-energy calculation with determination of the quasi-particle spectrum may not really be worth the effort at present. Instead one may use the numerical results for the electron gas as a local density approximation. By using the spirit of the WKB-approximation one can determine a local wave-number, which isused to eliminate the explicit wave number in the self-energy. This results in a potential, which depends on the local density and of the energy. The results for the electron gas are of good use only if the density gradient is reasonably small. Therefore, we would like to apply this approach to the volume electrons only and to treat the care effects separately. The inclusion of exchange and correlation in such a local density approximation can easily be adapted to existing schemes for energy band calculations. A detailed account of the scheme indicated here will be published elsewhere[4].

ACKNOWLEDGEMENTS

This lecture is based heavily on work, published and unpublished, by Dr. L. Hedin and Dr. B.I. Lundqvist and I wish to thank them for the continuous discussion I had with them over a long time over this and related problems.

REFERENCES

1. T.C. Collins, R.N. Euwema, and D.J. Stukel, This volume.
2. L. Hedin, Phys. Rev. 139, A796 (1965).
3. B.I. Lundqvist, Phys. Stat. Sol. 32, 273 (1969).
4. L. Hedin, B.I. Lundqvist and S. Lundqvist, to be published.
5. R. Gáspár, J. Chem. Phys. 20, 1863 (1952); Acta. Phys. Hung. 2, 151 (1952).
6. J.C. Slater, Phys. Rev. 81, 385 (1951).
7. S. Lundqvist, in Electronic Structure in Solids, (ed. E.D. Haidemenakis), Plenum Press, New York 1969.

VI

MANY-BODY EFFECTS IN SPECTRA OF SIMPLE METALS

Stig Lundqvist

Chalmers University of Technology, Göteborg, Sweden

I. INTRODUCTION

Optical transitions in solids pose a number of interesting many-body problems. A number of possible new effects will show up when one goes beyond the one-electron approximation. One particular class of such effects is associated with the change in the one-electron spectrum itself. Because of the interactions the energy levels will no longer be infinitely sharp. The levels will broaden but there is in addition new characteristic effects appearing in the spectrum of conduction effects as well as that of core electrons. These effects are due to the strong coupling between particles and density fluctuations, i.e. plasmons and electron-hole excitations. Near the Fermi level there is in addition a strong coupling to phonons, giving rise to the well-known enhancement of the electron density of states.

As a result of the many-body interactions the density of states will differ appreciably from that in the one-electron approximation. A first step towards a many-body theory of optical effects would be to simply replace the one-electron density of states by the one including many-body effects in the theory of interband transitions. This procedure would be asymptotically exact in the case of x-ray photoemission and may also work fairly well in the case of x-ray absorption and emission except for some edge effects.

Some important effects are associated with the interaction between the electron and the hole left behind, the so called final state interactions. These effects may be quite strong in the uv absorption and photoemission spectra of metals. Of particular interest are the singular threshold effects occurring in soft x-ray absorption and emission spectra of metals. They are caused by the coupling of a deep hole with the particle-hole excitations of the conduction electrons. It is interesting to note that a singular edge exists also in the x-ray photo-emission spectrum of a deep hole, in which final state interactions can be neglected and the spectrum therefore reflects the dynamical spectrum of the hole itself.

We shall describe the nature of the one-electron spectrum of conduction electrons and deep holes in section II. In the subsequent section we shall discuss some experiments under the simple assumption that we can fold together the density of states as in the theory of interband transitions. In section IV we shall comment on the effects of final state interactions and discuss the threshold effects.

II. ENERGY SPECTRUM AND DENSITY OF STATES FOR INTERACTING ELECTRONS

Suppose that we have solved the one-electron problem in the average electrostatic potential seen by an electron and obtained the energy as a function of wave number, $\varepsilon = \varepsilon(\underline{k})$. The next step would be to include the dynamical effects of the interaction as well as exchange effects. Such effects cannot be described by an ordinary local potential. We consider for simplicity the case of a uniform or almost uniform distribution of conduction electrons, where specific effects of the periodicity are of minor importance. Such a generalized potential must be non-local in space and time. Its Fourier transform to momentum-energy space will accordingly show a dependence on both wave number and energy $\Sigma(\underline{k}, \varepsilon)$, where ξ is the energy variable. Σ is called the <u>self-energy</u> of the electron. Adding the self-energy to the average potential, we have to solve the equation

$$\varepsilon = \varepsilon(\underline{k}) + \Sigma(\underline{k}, \varepsilon) \qquad \text{(II:1)}$$

The self-energy includes all the interactions between the state considered and the system. This necessarily includes dissipative effects, which lead to the decay of the state. Consequently the self-energy must be a complex quantity, thus

$$\Sigma(\underline{k}, \varepsilon) = \Sigma_R(\underline{k}, \varepsilon) + i\,\Sigma_I(\underline{k}, \varepsilon)$$

The physics of a problem described by a complex potential is best represented in terms of the <u>spectral weight function</u> $A(\underline{k}, \varepsilon)$, defined by

$$A(\underline{k}, \varepsilon) = \frac{1}{\pi} \frac{|\Sigma_I(\underline{k}, \varepsilon)|}{[\varepsilon - \varepsilon(\underline{k}) - \Sigma_R(\underline{k}, \varepsilon)]^2 + [\Sigma_I(\underline{k}, \varepsilon)]^2} \qquad \text{(II:2)}$$

In the simple case that the imaginary part of Σ is independent of energy, the spectral function for fixed $\underline{k}$ will have a Lorentzian shape and Σ_I determines the width of the line. When the imaginary parts depends on energy there are no a priori restriction on the shape of the spectrum.

The spectral weight function is the general distribution function for electrons with regard to both wave number and energy and from $A(\underline{k}, \varepsilon)$ one finds the density of states $N(\varepsilon)$ and momentum distribution $n(\underline{k})$ by integration, thus

$$N(\varepsilon) = \frac{1}{4\pi^3} \int d\underline{k}\; A(\underline{k}, \varepsilon)\,;\quad n(\underline{k}) = \int_{-\infty}^{E_F} d\varepsilon\; A(\underline{k}, \varepsilon) \qquad \text{(II:3)}$$

where E_F is the Fermi energy.

Most of our present knowledge about the self-energy has been based on calculations, where vertex corrections have been neglected, i.e. using the formula

$$\Sigma(\underline{k}, \omega) = \frac{i}{(2\pi)^4} \int e^{i\omega'\delta}\, v(\underline{k}')\, \varepsilon_D^{-1}(\underline{k}', \omega')\, G_0(\underline{k} + \underline{k}', \omega + \omega')\, d\underline{k}' d\omega' \qquad \text{(II:4)}$$

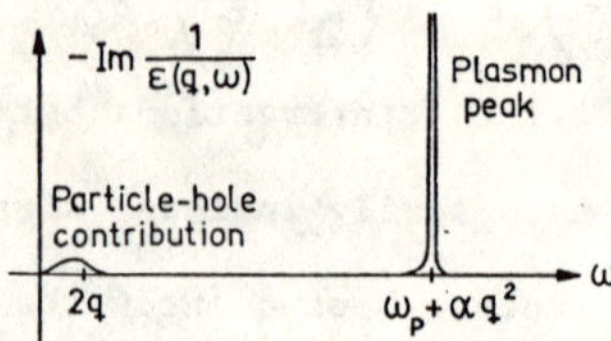

Fig. 1. Qualitative behavior of Im $\varepsilon^{-1}(q,\omega)$ for small q.

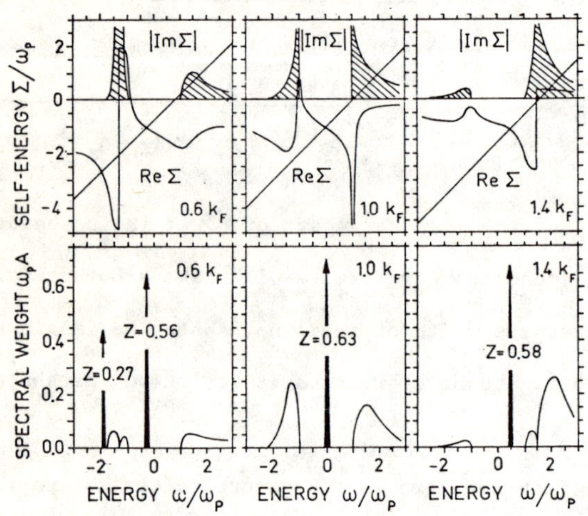

Fig. 2. The self-energy Σ and the corresponding spectral function A for $r_s=5$ as calculated by B.I. Lundqvist /2/. The crossings between the ReΣ -curves and the straight lines give the solutions to the Dyson equation. The numbers at the peaks indicate the strength of the lines.

where

$$G_0(\underline{k},\omega) = (\omega - \varepsilon(\underline{k}) \pm i\delta)^{-1}$$

and

$$v(\underline{k}) = \frac{4\pi e^2}{|\underline{k}|^2}$$

is the bare Coulomb potential. $\varepsilon_D(\underline{k},\omega)$ is the momentum and frequency-dependent dielectric function and δ is a positive infinitesimal. The Hartree-Fock theory corresponds to the choice $\varepsilon_D(\underline{k},\omega) \equiv 1$. The difference between Eq. II:4 and the Hartree-Fock theory shows up most clearly in the behaviour at small momenta, where the spectrum of $\varepsilon_D^{-1}(\underline{k},\omega)$ is dominated by the plasma resonance. The behaviour of Im $\varepsilon_D^{-1}(\underline{k},\omega)$ for small $|\underline{k}|$ is indicated in Fig. 1.

An approximation, which is most useful, is to take a plasmon-like sharp absorption for all $\underline{k}$ according to the formula

$$1/\varepsilon_D(\underline{k},\omega) = 1 + \frac{\omega_p^2}{\omega^2 - \omega_1^2(\underline{k})} \tag{II:5}$$

where ω_p is the classical plasma frequency and $\omega_1(\underline{k})$ the resonance frequency at wave number $\underline{k}$. The frequency $\omega_1(\underline{k})$ reduces to ω_p when $|\underline{k}| \to 0$ and is proportional to $|\underline{k}|^2$ for large $|\underline{k}|$. With this choice one can perform the frequency integration in Eq. II:4 and obtains

$$\Sigma(\underline{k},\omega) = -\frac{1}{(2\pi)^3}\int dq \frac{v(q)\, n_0(\underline{k}+q)}{\varepsilon_D[q, \varepsilon(\underline{k}+q)-\omega]} - \frac{\omega_p^2}{(2\pi)^3}\int \frac{v(q)}{2\omega(q)} \frac{dq}{\omega_1(q)+\omega-\varepsilon(\underline{k}+q)} \tag{II:6}$$

The first term in Eq. II:6 is a <u>screened exchange potential</u> and the second term describes the <u>Coulomb hole</u> around the electron.

Because of the plasma resonance in the dielectric function there will be a rapid variation in the real part of Σ at the frequence $\omega = \varepsilon(\underline{k}) + \omega_p$ for electrons and at $\omega = \varepsilon(\underline{k}) - \omega_p$ for holes. This behaviour of the electron gas seems to have been first noticed and

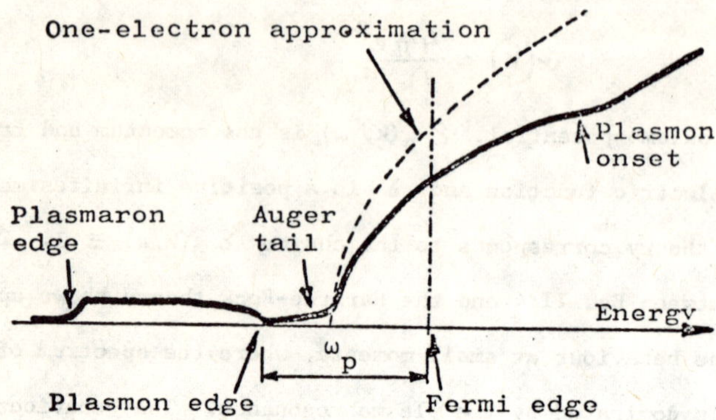

Fig. 3. Density of states for conduction electrons.

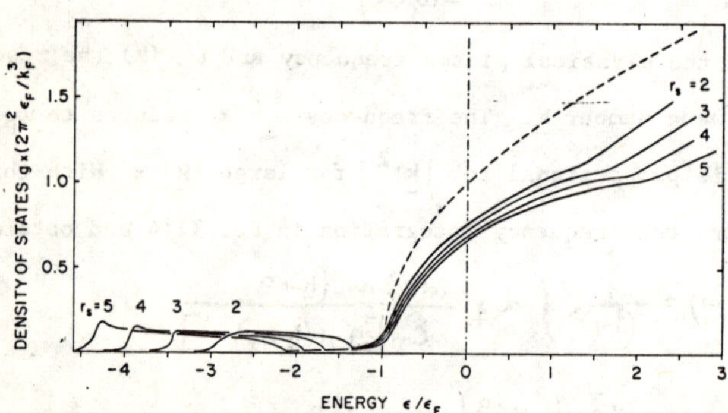

Fig. 4. The density of states for the values of the electron gas parameter r_s=2, 3, 4, 5. The dashed curve is the result of the one-electron theory, and the vertical broken line indicates the Fermi level.

discussed by Hedin et al /1/ and has been studied in great detail by Lundqvist /2,3/. Typical curves for Σ obtained with this dielectric function are shown in the upper half of Fig. 2. In the lower part of Fig. 2 we have given the results for $A(\underline{k},\omega)$ calculated from Eq. II:2.

For $k = k_F$ there is only one strong peak in the spectral function, corresponding to the usual quasi-particlé. For the other k-values in the figures, we find three solutions of the Dyson equation

$$\omega = \varepsilon(\underline{k}) - \Sigma_R(\underline{k},\omega)$$

One of them, however, falls at $\omega = \varepsilon(\underline{k}) \pm \omega_p$ where the damping is very strong, and therefore this solution is effectively suppressed. Of the two remaining solutions, one corresponds to the usual quasi-particle, i.e. a bare electron surrounded by a cloud of virtual plasmons and electron-hole excitations. For hole states, i.e. $k < k_F$, a new state appears which has an energy lower than that corresponding to a hole plus a plasmon, i.e. $\varepsilon(\underline{k}) - \omega_p$. This solution describes a resonant coupling between a hole and plasmons of different wavelengths and may be thought of as a hole coupled to a cloud of real plasmons. This coupled state, which is called a plasmaron, has a large oscillator strength in one-electron excitations. For electron states, $k > k_p$, there is no sharp state but a broad resonance with a sharp onset at $\omega = \varepsilon(\underline{k}) + \omega_p$.

The characteristic structure due to plasma effects will modify the density of states. This is demonstrated schematically in Fig. 3. The conduction band will retain approximately its parabolic shape. There will be an Auger tail below the band. At $\omega = \omega_p$ below the band there will be the onset of a secondary band due to the plasmaron states having an edge at the low frequency end. Above the Fermi level there is an increase in the density of states with a threshold at the frequency $\varepsilon_F + \omega_p$. Fig. 4 gives a survey of the density of state curves at four different values of the electron gas parameter r_s.

The preceding discussion has emphasized the strong effects of the electrón-plasmon on the spectrum of conduction electrons. Similar strong effects occur in the spectra of core electrons. For simplicity we limit the discussion to simple metals with small cores, so that the core electrons can be physically distinguished from the valence electrons and be well localized to a particular ion. The wave function of a core electron depends only weakly on the state of the outer electrons. The energy levels on the other hand are shifted by an appreciable amount, typically of the order of 5 - 10 eV, which is large on the scale of valence electron energies but is a small relative change in the energy of a core electron. The core shifts can be measured accurately by the method of x-ray photoemission spectroscopy as well as by x-ray absorption.

The shift of the quasi-particle energy of a core electron comes partly from changes in the average Coulomb field, partly from polarization effects. The Coulomb shift is due to the different valence charge distribution relative to that in a free atom. It generally results in a decrease of the binding energy. The polarization shift comes from the relaxation of the valence charge distribution around the hole created when we remove the electron. The valence electrons are drawn in towards the positive hole in the ion. This effect decreases the binding energy by half of the change in the Coulomb potential calculated at the core site, i.e. precisely the amount obtained if we calculate the self-energy of the hole using electrostatics. The shift in the core energy thus contains information about the valence electron distribution and polarizability, measured with the core electron as a probe. Theoretical calculations for simple metals (Li, Na, K. Al) by Hedin /4/ are in very good agreement with the experimentally observed peaks in the XPS spectra.

In analogy with the strong effects of the interaction between particles and plasmons previously discussed, there is a corresponding

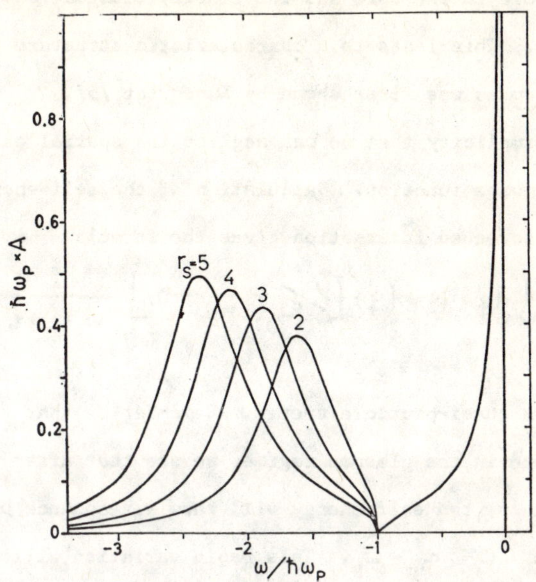

Fig. 5. Quasi-particle peak in the spectral function for a core electron and the associate plasmon satellite structure for different densities of the conduction electrons, measured by the electron gas parameter r_s. Calculated by B.I. Lundqvist /5/.

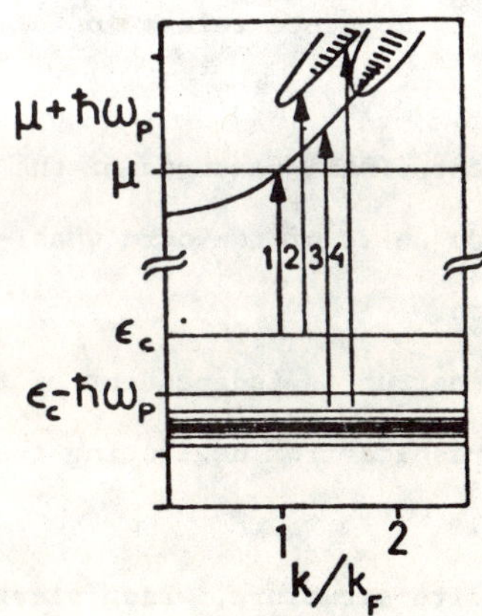

Fig. 6. Level structure in the one-electron spectrum of a metal after B.I. Lundqvist /5/.

coupling between a hole in the core and the density fluctuations of the conduction electrons. This leads to a characteristic structure in the core electron spectrum as was first shown by Lundqvist /5/.

We assume for simplicity that we can neglect the spatial extension of the core electron wave function. Calculation of the self-energy to lowest order in the screened interaction gives the formula

$$\Sigma(\varepsilon) = -\frac{i}{(2\pi)^4}\int dq\, d\omega\, v(q)\left[1/\varepsilon_D(q,\omega) - 1\right]\frac{1}{\omega - \varepsilon + \varepsilon_n + i\delta} \tag{II:7}$$

where ε_n is the core quasi-particle energy. Remembering that $\varepsilon_D^{-1}(\underline{q},\omega)$ has a strong resonance in the plasmon regime, we see that after integration over the frequency, the self-energy will show a resonance behaviour in the energy region $\varepsilon \simeq \varepsilon_n - \omega_p$. This rapid variation will give rise to two solutions of the Dyson equation. The second solution, however, gives a quite broad peak in the spectral function. The results for different densities of the electron gas are illustrated in Fig. 5. The spectrum is measured from the shifted quasiparticle energy, i.e. the zero of energy corresponds to complete relaxation of the electrons around the hole.

We summarize the characteristic features of the core spectrum

a. A large polarization shift of the core quasi-particle as previously discussed.

b. The shape of the spectrum is independent of the core level considered. This results from neglecting the actual size of the core wave function.

c. A pronounced satellite structure, which starts at $\omega = -\omega_p$, has a broad peak at $\omega = -(1.6 - 2.3) \times \omega_p$ for r_s in the range 2 - 5, and falls off at still more distant energies. The widths vary from $\frac{1}{2}\omega_p$ to $3/4\,\omega_p$, dependent on r_s.

d. An extended tail on the low-energy side of the quasi-particle peak. This tail is due to the coupling between the hole and screened electron-hole pair excitations, and corresponds to states of the whole system involving two holes, one in the core and one in the conduction band plus one excited electron above the Fermi sea.

e. There is an appreciable reduction of the spectral strength of the quasi-particle. Approximately half of the spectral strength corresponds to excitations close to the quasi-particle state and approximately the same strength corresponds to high excitations of the conduction electrons as described by the broad satellite structure.

The singular behaviour near the edge should be noted. When approaching the quasi-particle energy from below, the imaginary part of Σ decreases linearly. This corresponds to a $\omega \ln|\omega|$ -behaviour for the real part in the limit $\omega \rightarrow 0^-$. Consequently the spectral weight function diverges like $1/\omega^2 \ln^2\omega$ in this limit, giving an integrable contribution to the sum rule. This singularity at the threshold has the same basic origin as those occuring in the x-ray absorption and emission. We note that the spectral function has the form

$$A(\omega) = \sum_s |\langle N-1,s \,|\, N-1 \rangle|^2 \, \delta(\omega - \varepsilon_s) \qquad \text{(II:8)}$$

where $|N - 1\rangle$ is the ground state of the electron gas <u>without a core hole</u>, and $|N - 1,s\rangle$ is an excited state of the electron gas <u>in the presence of the core hole</u>, which acts as a static hole potential.

Thus the spectral function depends in an essential way on the overlap between the ground state of the unperturbed electron gas and the set of states of the electron gas with a core hole potential, which according to Anderson /6/ exhibits a singular behaviour.

We shall return to these questions in the following section, where we shall discuss how the structure in the core spectrum can be related to XPS and X-ray spectra.

III. QUALITATIVE DISCUSSION OF SOME EXPERIMENTS

We shall in this section use the results of the preceding section to some optical experiments. We shall consider a simple-minded extension of the theory of interband transitions where we replace the one-electron density of states by the density of states including interactions. Such an approach is, although intuitively appealing, not very consistent and does not represent a systematic approach, e.g. in the form of an expansion in terms of an interaction parameter. Nevertheless, it may give a qualitative idea what kind of effect that may occur. The refinement by including final state interactions will be commented in section IV.

a) X-ray photoemission (XPS)

The experiment, which is simplest related to the one-electron spectrum is X-ray photoemission, where the final state interactions can safely be neglected. In photoemission the final state of the system is written

$$|\,\text{final}\,\rangle = a^*_{\underline{\kappa}}\,|\,N-1, s\,\rangle \quad ,$$

where $a^*_{\underline{\kappa}}$ is the creation operator for the outgoing electron and $|\,N-1, s\,\rangle$ is a state of the system with a hole. For dipolar interaction the transition probability per unit time becomes

$$w \sim \sum_{\kappa,s} |\langle N\,|\,p\,a^*_{\kappa}\,|N-1, s\,\rangle|^2\,\delta\left(\varepsilon_{\kappa} - w - \varepsilon_s\right), \quad \text{(III:1)}$$

where $\underline{p}$ is the dipole operator and ε_s is the excitation energy of the system with a hole and w is the frequency of the light. For large enough energies we have that $a_{\underline{k}}\,|\,N\rangle = 0$ and hence we obtain

$$w \sim \sum_{\kappa,s} \Big|\sum_{\underline{k}} p_{\underline{\kappa}\,\underline{k}}\,\langle N-1, s|a_{\underline{k}}|N\rangle\Big|^2 \delta(\varepsilon_{\kappa} - w - \varepsilon_s) \quad \text{(III:2)}$$

$$\simeq \sum_{\kappa,\underline{k}} |\,p_{\kappa k}|^2\,A\left(\underline{k}, \varepsilon_{\underline{\kappa}} - w\right)$$

Hence, the energy distribution of the photoelectrons is given by

$$I(\varepsilon) \sim \varepsilon^{1/2} \sum_{\underline{k}} |P_{\alpha \underline{k}}|^2 A(\underline{k}, \varepsilon - \omega) \simeq \varepsilon^{1/2} P_{eff}^2 N(\varepsilon - \omega) \qquad \text{(III:3)}$$

where $N(\varepsilon - \omega)$ is the one-particle density of states. Thus the energy distribution of the photoelectrons will approximately give the density of states. The experimental results must, however, first be corrected for plasmon energy losses of the outgoing electron.

b) Soft X-ray emission

We consider only the simplest possible case where we can assume complete relaxation of the Fermi gas around the hole before the emission takes place. This limits the approximate validity to the light metallic elements such as Li, Be, Na, Mg, Al and K and excludes e.g. all transition metals.

In one-electron theory the intensity is given by

$$I(\omega) \sim \omega \sum_{k} |P_{ck}|^2 \delta(\varepsilon_x - \varepsilon_c - \omega) \simeq \omega P_{eff}^2 N(\omega) \ . \qquad \text{(III:4)}$$

If we neglect the relaxation of the valence electrons the golden rule gives

$$I(\omega) \sim \omega \sum_{k} |P_{ck}|^2 A(k, \omega - \varepsilon_c) \simeq \omega P_{eff}^2 N(\omega) \ . \qquad \text{(III:5)}$$

Thus, the density of states for conduction electrons should be observed /7/.

In this case, however, final state interactions may lead to edge effects at the threshold and the presence of the core hole may cause interference effect, which tend to reduce the intensity of the satellite band.

c) X-ray absorption

In the absence of a detailed theory we shall take a simple point of view and treat the new structure in the one-electron spectra as additional levels or groups of levels in a one-electron scheme. Such a discussion of plasmon effects in X-ray absorption in metals was given by Ferrell /8/.

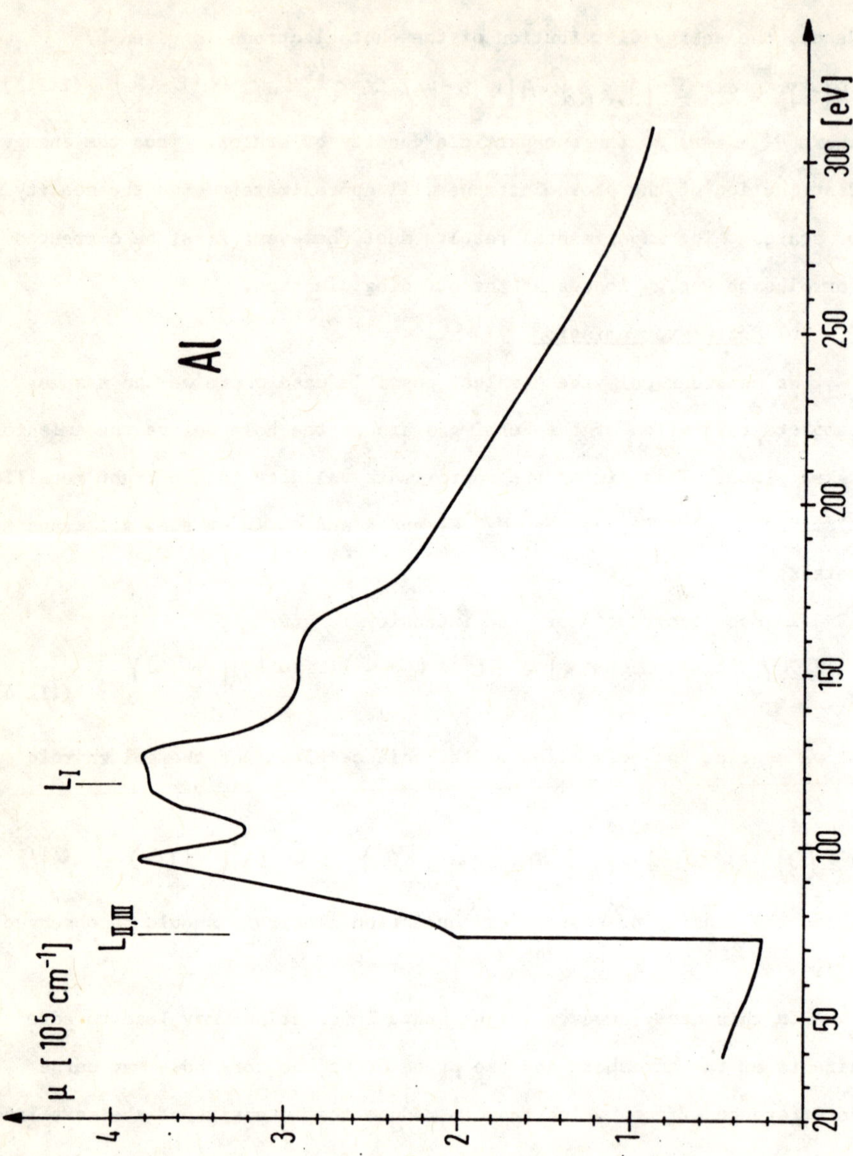

Fig. 7. Absorption coefficient of aluminium in the $L_{II, III}$ band from work by Haensel and coworkers /9/. The vertical bars indicate the onset of $L_{II, III}$ and L_I absorption.

Due to the presence of satellite structure in both core and conduction band spectra there should be a characteristic structure above the threshold as discussed by B.I. Lundqvist /5/. This is illustrated in Fig. 6.

There are four kinds of absorption processess (1) from core quasi particle to quasi particle in the conduction band, (2) from core quasi particle to unoccupied satellite, (3) from core satellite to quasi particle, and (4) from satellite to satellite. As the photon energy is increased, the transitions 1, 2, 3 and 4 contribute to the absorption in the mentioned order, starting at the absorption threshold energy, at $\hbar\omega_p$ above, at $\hbar\omega_p$ above, and at $2 \times \hbar\omega_p$, respectively.

Such a structure is qualitative agreement with the structure observed for several simple metals by Haensel and coworkers at DESY in Hamburg using synchrotron radiation. An example is given in Fig. 7, which gives the L shell absorption cross section in Al from ref /9/.

d) Photoemission and optical absorption in the ultraviolet

Photoemission experiments in the far ultraviolet range should reflect some structure due to the satellite band. However, it may be difficult to observe because it falls in the region of inelastically scattered electrons and the experimental situation is so far unsettled. The satellite structure should also be seen in the optical absorption but, again, the experiments give so far no clearcut evidence.

IV. BRIEF REMARKS ON FINAL STATE INTERACTIONS

The discussion given in these lectures have been based on a rather naive extension of the one-electron theory of interband transitions by using the density of states for an electron and a hole in an interaction system rather than that from an independent particle model. The optical experiments are however processes in which two quasi-particles are created in one event. The particles are usually generated very close together and may therefore interact strongly with each other before they get apart. This will imply that the spectral distribution may be quite different from that obtained under the assumption that the particle and the hole do not interact. Of the experiments mentioned in the preceding section only the X-ray photo-emission experiment gives a good justification for neglecting the interaction between the electron and the hole left behind.

In case of a sufficiently strong interaction between the electron and the hole a bound state, an exciton may be formed. A bound state requires that the state occurs in an energy gap of the system. There are also resonant decaying states if a quasi-bound state is imbedded in a continuum. There may also be non-dramatic effects of the electron-hole interaction, e.g. in the form of an overall distortion of the shape of the spectrum.

The inclusion of the final state interaction may interfere with the effects already included in the spectra of electrons and holes and may give rise to cancellation effects or possibly enhancements. A simple example is the X-ray line spectrum for transition between two core levels. In the approximation where both core states are of such small extension that we can neglect their extension, there will be a complete cancellation and no satellite effects will be observed at all. The physical reason for this is that both the initial and final core levels have the same interaction with the conduction band and therefore the conduction electrons notice no change during the transition. In real cases there will be an effect arising from the difference in the interaction with the conduction electrons in the two hole states.

Much attention has been given to the study of the edge effects, occuring in the X-ray absorption and emission spectra. These effects are related to an infrared divergence in metals of the electron-hole excitations close to the Fermi level and are connected with the orthogonality problem for an impurity imbedded in the electron gas discussed by Anderson /6/. It was pointed out by Mahan /10/ that the electron-hole interaction will give rise to anomalies near the X-ray transition edge of metals. Recently Nozières and coworkers have given a simple formula for this anomaly. A detailed discussion of these effects are beyond the scope of these lectures and we refer the reader to the original papers /11/.

ACKNOWLEDGEMENT

The author is indebted to Drs. L. Hedin, B.I. Lundqvist and J.W. Wilkins for innumerous discussions about the subject matter of these lectures.

REFERENCES

1. L. Hedin, B.I. Lundqvist and S. Lundqvist, Solid State Comm. 5, 237 (1967)

2. B.I. Lundqvist, Physik Kondensierten Materie 6, 206 (1967)

3. B.I. Lundqvist, Physik Kondensierten Materie 7, 117 (1968)

4. L. Hedin, Arkiv Fysik 30, 231 (1965)

5. B.I. Lundqvist, Physik Kondensierten Materie 9, 236 (1969)

6. P.W. Anderson, Phys. Rev. Letters 18, 1049 (1967)

7. L. Hedin, Soft X-Ray Band Spectra and the Electronic Structure of metals and Materials, ed. D.J. Fabian (Academic Press, London 1968)

8. R.A. Ferrell, Rev. Mod. Phys 28, 308 (1956)

9. R. Haensel, C. Kunz, T. Sasaki and B. Sonntag, J. Appl. Phys. (in press)

10. G.D. Mahan, Phys. Rev. 163, 612 (1967)

11. B. Roulet, J. Gavoret and P. Nozieres, Phys. Rev. 178, 1072 (1969)
    P. Nozières, J. Gavoret and B. Roulet, Phys. Rev. 178, 1084 (1969)
    P. Nozières and C. de Dominicis, Phys. Rev. 178, 1097 (1969)

VII

PHOTOEMISSION AND OPTICAL PROPERTIES OF METALS AND ALLOYS

by

P.O. Nilsson

Department of Physics

Chalmers University of Technology

Gothenburg, Sweden

Introduction.

This paper will be divided into two parts. In the first we shall briefly review the quantities measured in optical and photoemission experiments and their relation to the band structure. In the second part we shall review earlier investigations of optical transitions in the noble metals and present some new results of optical and photoemission experiments on Cu-Au and Ag-Au alloys.

I. THEORY.

The electronic states in a crystal can, within the one electron picture, be described by Bloch waves, $\psi_{n\bar{k}} = u_{n\bar{k}}(\bar{r})e^{i\bar{k}\cdot\bar{r}}$ and the corresponding energy functions, $E_n(\bar{k})$; the latter functions constituate the band structure of the crystal. When light of energy $\hbar\omega$ is incident upon the crystal the electrons may be excited from a state ψ_i to a state ψ_f , changing their energy by an amount $\hbar\omega = E_f - E_i$. To obtain the transition rate, the result of time dependent perturbarion theory can be used:

$$\frac{dW}{dt} = -\frac{4\pi}{\hbar}\int\frac{d\bar{k}}{4\pi^3}\,|\langle\psi_f|H_p|\psi_i\rangle|^2\,\delta(E_f-E_i-\hbar\omega) \quad \ldots (1)$$

$H_p = \frac{e}{m} A_o e^{i\bar{K}_p \cdot \bar{r}} \bar{e} \cdot \bar{p}$ is the perturbation operator due to the electromagnetic wave described by the vector potential $\bar{A} = A_0 \, \bar{e} \, e^{i(\bar{K}_p \cdot \bar{r} - \omega t)}$. Inserting the Bloch functions into Eq. (1) the matrix element is

$$\langle \psi_f | H_p | \psi_i \rangle = \int_{\text{unit cell}} \phi(\bar{r}, \bar{k}_i, \bar{k}_f) d\bar{r} \sum_j e^{i(\bar{k}_i - \bar{k}_f - \bar{K}_p) \cdot \bar{R}_j} \quad \ldots (2)$$

where $\bar{R}_j$ are the positions of the unit cells; ϕ is a periodic function: $\phi(\bar{r} + \bar{R}_j) = \phi(\bar{r})$. This implies that the matrix element is non-zero only when

$$\bar{k}_i = \bar{k}_f \quad \ldots (3)$$

because the photon wave vector $\bar{K}_p$ is small on the Brillouin zone scale. (The wave number of light in the UV region is of the order 10^5 cm^{-1} which should be compared with typical crystal momentum which is of the order of 10^8 cm^{-1}.) We now use the energy balance condition

$$\hbar \omega \frac{dW}{dt} = \frac{1}{2} \cdot \omega \varepsilon_o \varepsilon_2 \cdot \omega^2 A_o^2 \quad \ldots (4)$$

and obtain

$$\varepsilon_2 = \frac{8\hbar^2 \pi^2 e^2}{\varepsilon_o m^2 \omega^2} \int \frac{d\bar{k}}{4\pi^3} |M_{if}|^2 \delta(E_f - E_i - \hbar\omega) \quad \ldots (5)$$

where

$$M_{if} = \int d\bar{r} \, u_f(\bar{k}, \bar{r}) \, \bar{e} \cdot \nabla u_i(\bar{k}, \bar{r}) \quad \ldots (6)$$

By assuming M_{if} $\bar{k}$-independent the integral in Eq. (5) becomes

$$\varepsilon_2(\omega) = C \frac{|M_{if}(\omega)|^2}{\omega^2} \int \frac{1}{4\pi^3} \frac{dS_k}{|\nabla_k (E_f - E_i)|_{E_f - E_i = \hbar\omega}} \quad \ldots (7)$$

where

$$C = \frac{8\hbar^2 \pi e^2}{\varepsilon_o m^2}$$

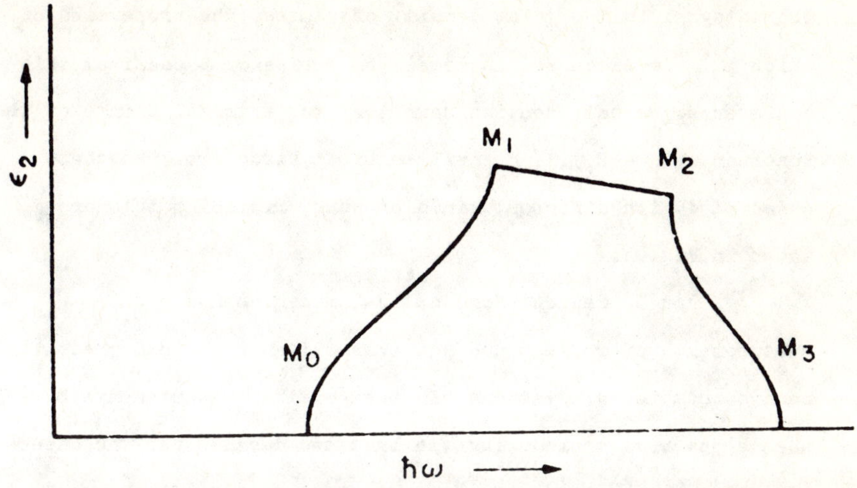

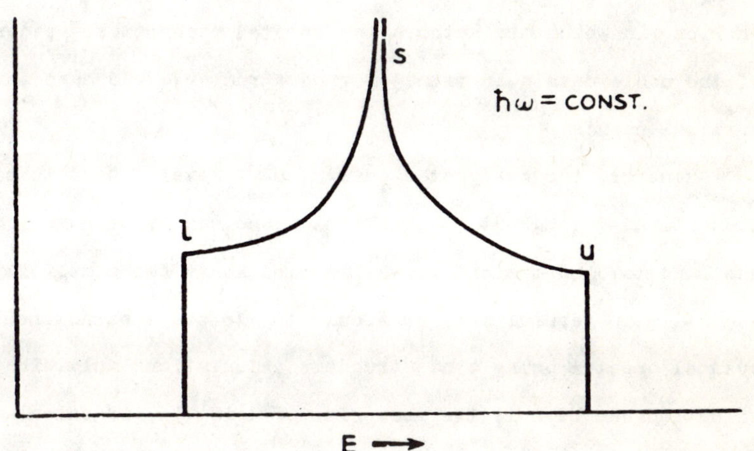

Fig. 1. Upper part: Structure in $\mathcal{E}_2$ due to critical points, i.e. extrema in the function $E_f - E_i$, which is min. at M_0, has a saddle point at M_1 and M_2 and is max. at M_3.
Lower part: Structure in the electron energy distribution due 2-dim. critical points. The optical and electron energy surface can have different types of tangency: min (l), saddlepoint (s) or max (u).

The integral is the joint density of states, the properties of which will be discussed in detail by Professor Bassani at this conference. We only mention here that for extremal points of the function $(E_f - E_i)$, ε_2 will exhibit structure. The shape associated with different kinds of such critical points are shown in Fig. 1.

ε_2 can be deduced from optical measurements such as measurements of reflectance and transmittance and compared with band structure calculations. In fact optical measurements have been a powerful tool during the last two decades for the understanding of the electronic properties of solids. With regard to metals one of the pioneer works was that of Ehrenreich and Philipp,[1] in which the authors interpreted the optical spectra of the noble metals in terms of band structures and associated critical points.

However, for one photon energy $\hbar\omega$, several band to band transitions throughout the Brillouin zone can contribute (only one band to band transition was assumed above for simplicity) and it is in general very difficult to interprete structure in optical spectra using band structure calculations only. The situation can be a little more clarified if ε_2 is calculated from the band structure and the different contributions are decomposed in this calculation. Calculations of this sort have recently been made[2] and we shall return to them later.

There are other experimental techniques which give information in addition to ε_2 including all optical measurements where a perturbation such as pressure, electric field or heat current, etc., is applied. In this way, using e.g. polarized

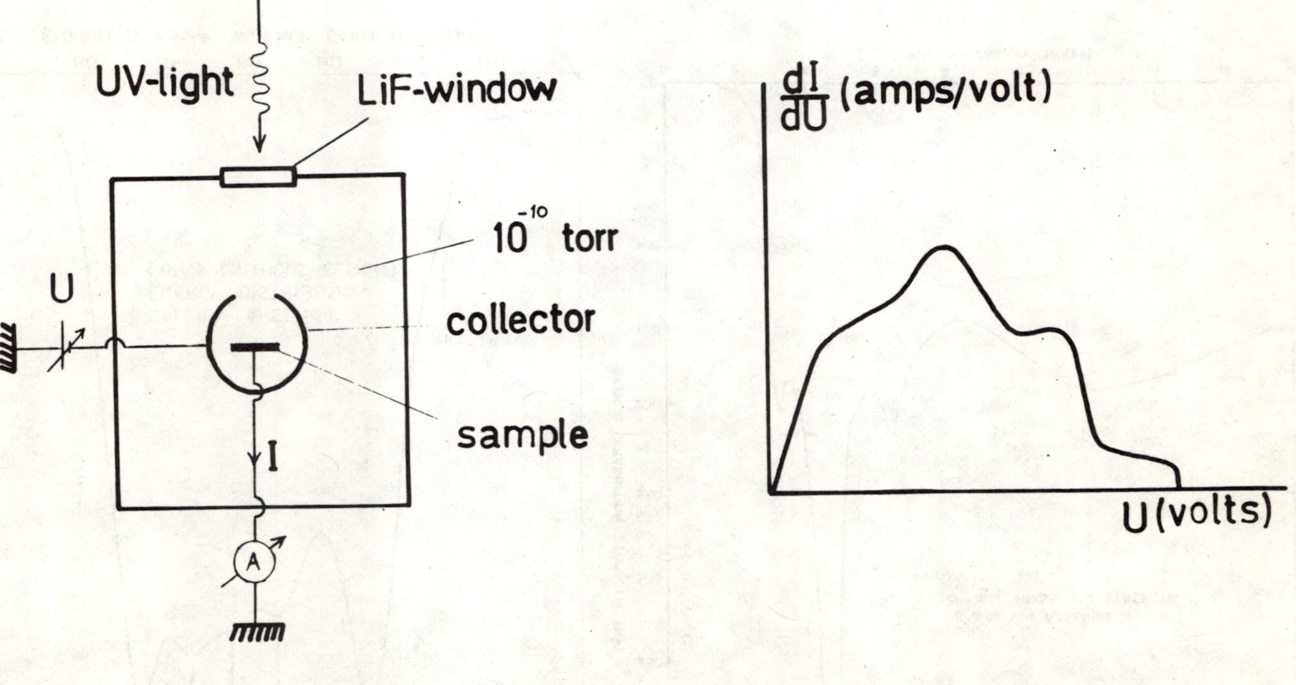

Fig. 2. Example of experiment set up to perform a photoemission experiment. The $\frac{dI}{du}$ versus U curve corresponds to the number of photoemitted electrons per eV as a function of their energy.

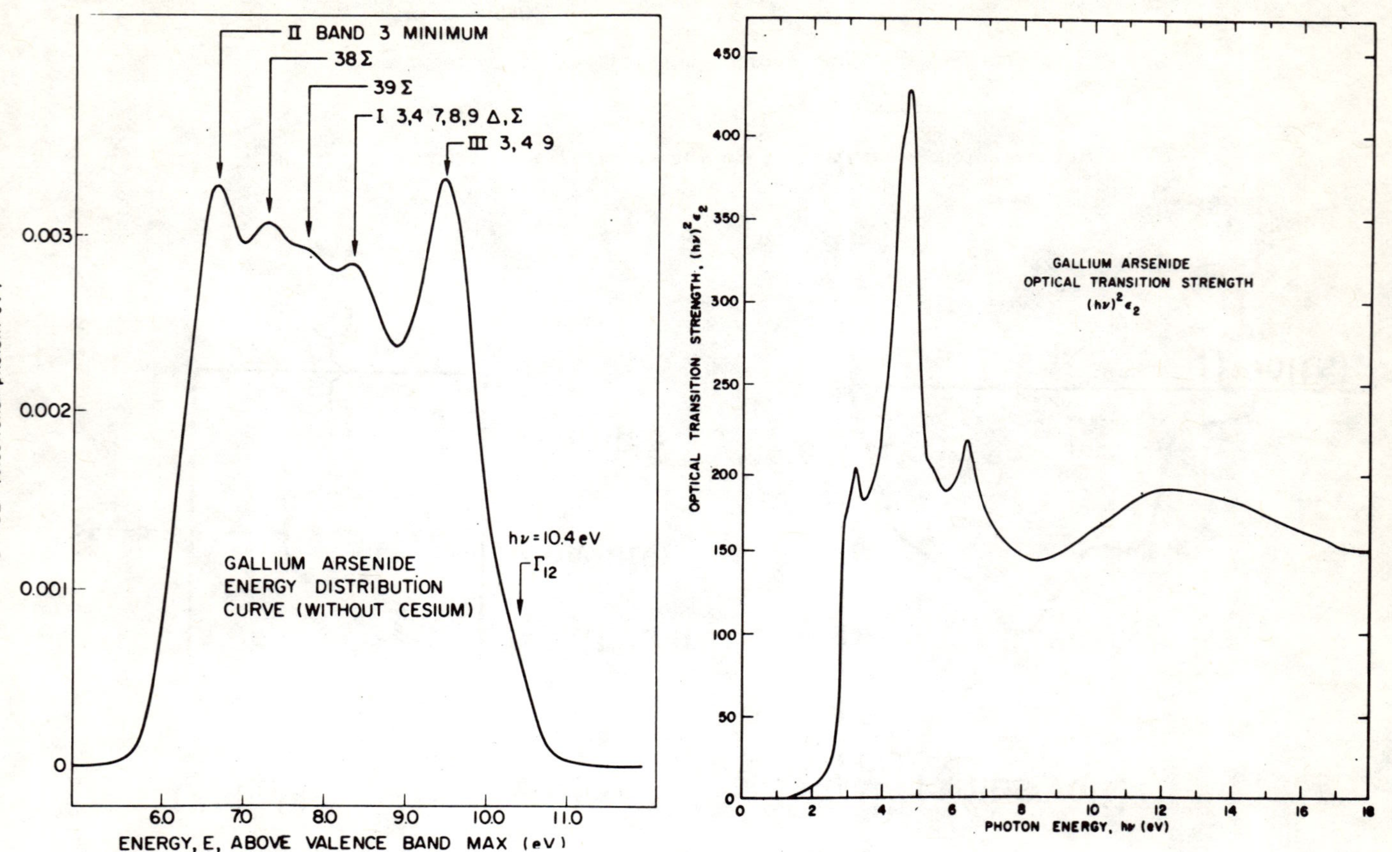

Fig. 3. Electron distribution curve of GaAs for $\hbar\omega$ = 10.4 eV.

light and a single crystal, the k-vector involved in the transition can be determined. Dr. Seraphin will discuss this in detail in his lectures.

Still another way of resolving the different contributions in an optical absorption spectrum is to perform photoemission experiments. Here the distributiona of the photo-emitted electrons are energy analyzed, shown schematically in Fig. 2. Fig. 3 shows an example of how a P.E. experiment can resolve the different optical transitions taking place for single photon energy $\hbar\omega$. In this experiment electrons having a final energy E are determined. The internal energy distribution in the crystal is thus

$$\tilde{N}(E,\hbar\omega) = \frac{C}{\varepsilon_2}\int |M(\omega,\bar{k})|^2 \delta(E_f - E_i - \hbar\omega)\,\delta(E-E_f) \quad \ldots (8)$$

Again, assuming the matrix element $\bar{k}$-independent we can transform the volume integral to a line integral:

$$\tilde{N}(E,\hbar\omega) = \frac{C}{\varepsilon_2}|M(\omega)|^2 \int \frac{dl}{|\nabla_k E_i \times \nabla_k E_f|_{\substack{E_f - E_i = \hbar\omega \\ E_f = E}}} \quad \ldots (9)$$

The integration is performed along the line in the Brillouin which is the intersection between the optical energy surface $E_i(\bar{k}) - E_f(\bar{k}) = \hbar\omega$ and the electron energy surface $E_f(\bar{k}) = E$. As seen in Eq. (9) structure in $\tilde{N}(E)$ is obtained when the gradients of $E_i(\bar{k})$ and $E_f(\bar{k})$ are parallel or anti-parallel to each other, i.e. when the optical and electron energy surfaces form a common tangent. The different kinds of these two-dimensional critical points are shown in Fig. 1. The locus of a critical point in $\bar{k}$-space when $\hbar\omega$ is varied is defined a critical line.[3] The function $E(\hbar\omega)$ along a critical line is of great interest

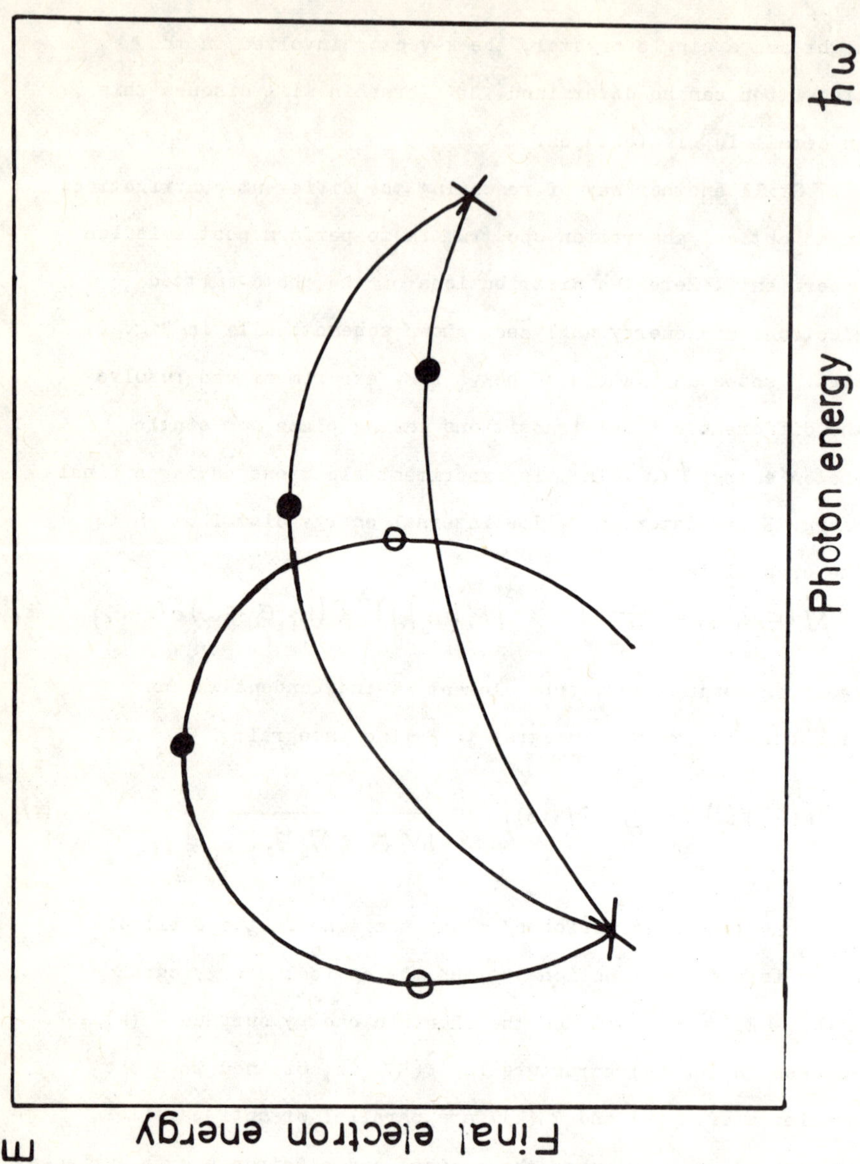

Fig. 4. Electron energy as function of photon energy along critical lines. o OCP, ● ECP, and X symmetry critical point (SCP).

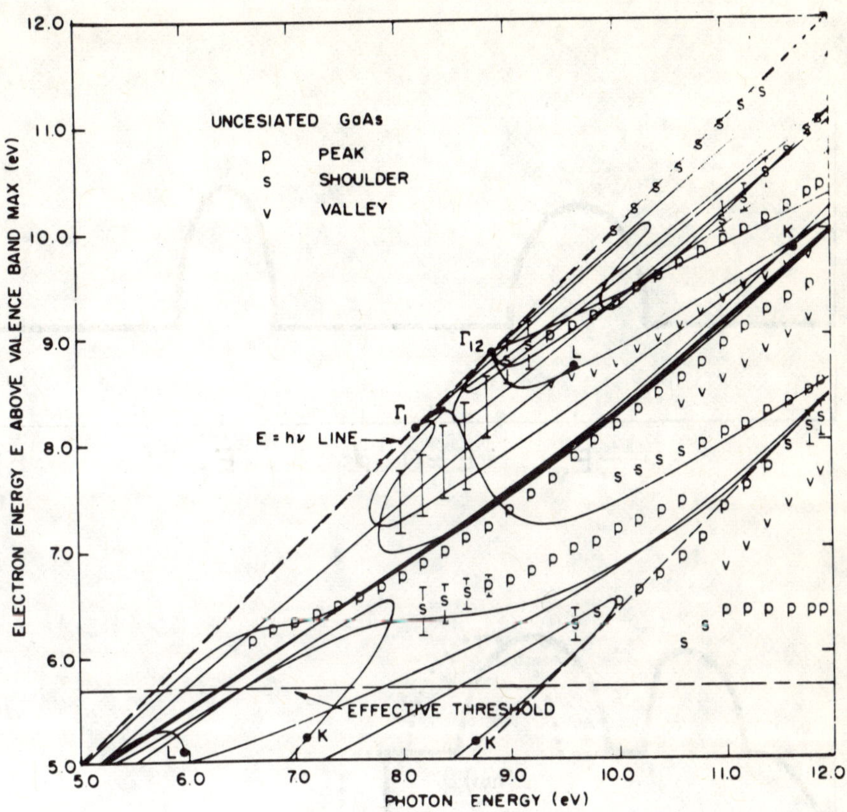

Fig. 5. Comparison between calculated E-$\hbar\omega$ plot and observed structure in photoemission data (from ref. 5).

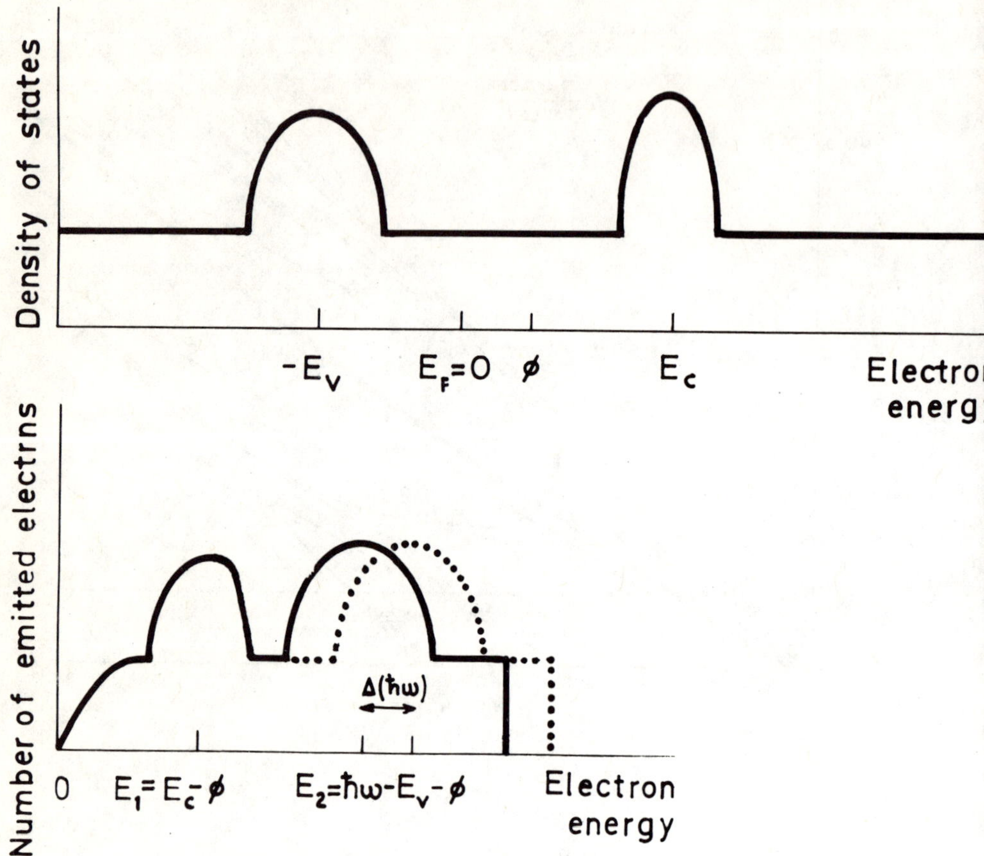

Fig. 6. Schematic density of states with one peak below and one above the Fermi level and corresponding EDC:s. If the so-called nondirect transitions dominate, peaks in the EDC, originating from the occupied density of states, move the same amount as the photon energy is changed (peak denoted E_2), while peaks from the unoccupied region are stable (E_1).

because it tells us how a piece of structure in $\tilde{N}(E,\omega)$ moves when changing the photon energy. An $E(\hbar\omega)$ plot is shown schematically in Fig. 4. Note that critical points of $E_f(\bar{k})$, called ECP, and critical points of $(E_f(\bar{k}) - E_i(\bar{k}))$, called OCP, imply extrema in $E(\hbar\omega)$ and $\hbar\omega(E)$ respectively. In the photoemission experiment the external electron energy distribution is measured. This is essentially the quantity $\tilde{N}(E)$ but distorted due to the electron-electron scattering, the surface barrier, the optical absorption etc.[4] However, these effects can be corrected for and the main effect consists of a smooth cut-off function at low electron energies (see later).

No complete analysis involving comparison between calculated $E(\hbar\omega)$ plots and experimental EDC:s have yet been reported. However, Eden[5] has made a comparison between EDC:s from GaAs and a $E(\hbar\omega)$ plot obtained from band structure along the principal symmetry lines. Part of the result is shown in Fig. 5 and may serve as an illustration of the power of the photoemission technique to resolve the details of a band structure. For some substances however, the photoemission results seem to be inconsistent with the above theory. When changing $\hbar\omega$ by $\Delta(\hbar\omega)$ the peaks and edges of the EDC:s in these cases are either moving $\Delta(\hbar\omega)$ or not moving at all, which corresponds to 45° or horizontal lines in the E vs $\hbar\omega$ diagram. These results may be interpreted so that the EDC:s are proportional to the product of the initial and final density of states (see Fig. 6):

$$\tilde{N}(E,\hbar\omega) \sim N_f(E)\, N_i(E-\hbar\omega) \qquad \ldots(10)$$

To understand why this occurs, Spicer[6] has suggested that $\bar{k}$-conservation in the usual one particle sense (Stated in Eq.(3)) is not an important selection rule in these cases. Imagine that the hole created is localized on one atom for a time which is comparable with the excitation time. This is likely to happen for states with small overlap, between neighbouring atom wave functions, e.g. the d-states in the noble metals. The appropriate wave functions are then no longer periodic and the eigenstates of the ground state, i.e. the Bloch waves, cannot be used to describe the exitation. Instead we are dealing with a many body excitation, involvning electronic relaxation around the hole. No complete theoretical treament of these so called nondirect transitions[x)] has yet appeared.

Let us, however, briefly see how the optical electronic transitions are generally described when the whole system of interavting electrons are taken into account. The many body wavefunction can be written as $\Psi_{\bar{k}_1 \ldots\ldots \bar{k}_N}(\bar{r}_1 \ldots\ldots \bar{r}_N, n)$. n is the index of the energy band and the $\bar{k}$:s are quantum numbers of the system. It should be noted that the k-vectors are not the momenta of the individual electrons. The wave function can be rewritten in "Bloch form":

$$\Psi_{\bar{K},I,n}(\bar{R},\bar{r}_{ik}) = \chi_{\bar{K},I,n}(\bar{R},\bar{r}_{ik})\, e^{i\bar{K}\cdot\bar{R}} \quad \ldots(11)$$

$\bar{R}$ is the center of gravity of the system, $\bar{r}_{ik} = \bar{r}_i - \bar{r}_k$, $\bar{K}$ is the total crystal momentum not equal to $\sum_{j=1}^{N} \bar{k}_j$ and I are continuous quantum numbers associated with the $\bar{r}_{ik}$:s. χ is a periodic function in $\bar{R}$. It can be shown that the selection rule for a dipole optical transition is now[7]

$$\bar{K}' = \bar{K} \quad \ldots(12)$$

x) Distinguish these first order transitions from the so-called indirect transitions. The latter are nonvertical transitions of second order, where the momentum is conserved via phonons.

where the prime refers to the excited state. The total momentum of the whole system is conserved and we have no selection rule for individual electrons. (Note e.g. that intraband transitions are allowed.) The formula for the dielectric constant is quite analogous to that in the one electron theory[7] (Eq. 7)

$$\varepsilon_2(\omega) = C\ \frac{|M_{nn'}(\omega)|^2}{\omega^2} \int \frac{1}{4\pi^3} \frac{dS_{\xi\xi'}}{\left|\nabla_{\xi\xi'}\left(E(\xi',n') - E(\xi,n)\right)\right|_{E'-E=\hbar\omega}} \quad \dots(12)$$

where $(\xi, n) = (\bar{k}_1, \dots \bar{k}_N, n)$ and where the matrix element $M_{n\xi, n'\xi'}$ is assumed not to depend on ξ. It is therefore not surprising that the single electron theory usually describes well what is observed in experiments. On the other hand the energy functions $E(\xi, n)$ and the matrix element $M(n\xi, n'\xi')$ do depend on the electron-electron interaction. To get a qualitative measure of the importance of these interactions it is necessary to use a model for the electron system. It is possible that such an explicit calculation will show that many-body effects, and thus the selection rule in Eq. 19, have to be taken into account in some cases when studying optical transitions. If this is the case the nondirect transitions observed in photoemission could be explained accordingly.

However, to explain the observed relationship (10) we may again use Eq. (5), but this time assume $\bar{k}_i \neq \bar{k}_f$

$$\varepsilon_2(\omega) = \frac{C}{\omega^2} |M(\omega)|^2 \frac{1}{4\pi^3} \int d\bar{k}_i N_i(\bar{k}_i) \int d\bar{k}_f N_f(\bar{k}_f)\, \delta(E_f - E_i - \hbar\omega) \quad \dots(13)$$

We now write $$N_f(E) = \frac{1}{4\pi^3} \int d\bar{k}_f N_f(E, \bar{k}_f) \quad \dots(14)$$

and analogically for $N_i(E)$.

We thus finally arrive at

$$\varepsilon_2 = \frac{32\hbar^2\pi^5 e^2}{\varepsilon_0 m^2} \frac{|M(\omega)|^2}{\omega^2} \int_{E_F}^{E_F+\hbar\omega} N_f(E) N_i(E-\hbar\omega)\, dE \qquad \ldots(15)$$

where E_F is the Fermi energy.

The internal electron energy distribution in this case is

$$\tilde{N}(E,\hbar\omega) = |M(\omega)|^2 N_f(E) N_i(E-\hbar\omega) \qquad \ldots(16)$$

In most cases it is not possible to determine $|M(\omega)|^2$ experimentally, so this quantity is incorporated in N_f and N_i which are then called "the optical density of states". We have thus arrived at Eq. (10).

Most of the optical and photoemission experiments designed to study the band structure of solids, have been carried out with photon energies in the visible and UV regions. There are, however, also some investigations in the x-ray region. The electron is then excited to very high energies, where it behaves as a free electron. This means that we need not be so concerned with the matrix elements. With regard to the x-ray photoemission the resolution is still moderate (appr. 1 eV) and worse than in the UV photoemission (app. 0.1 eV).

Finally it should be noted other experimental methods for studying the band structure of solids, inclose x-ray Compton scattering, soft x-ray emission and ion neutralization, but these techniques fall outside the scope of this review.

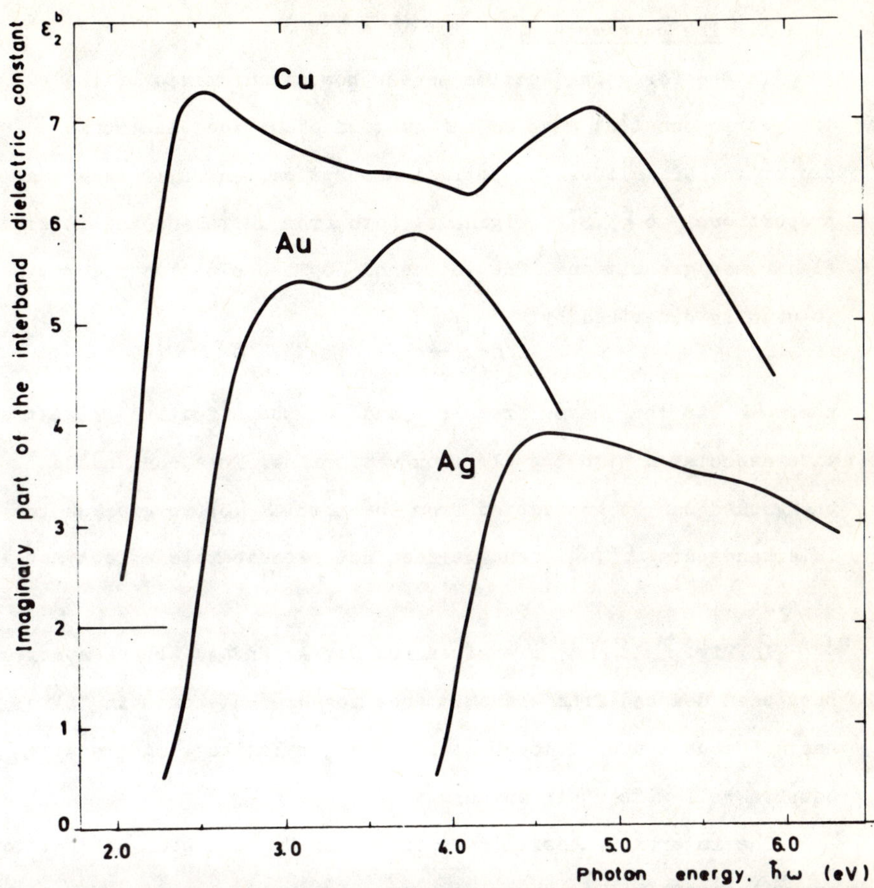

Fig. 7. The imaginary part of the dielectric constant for the noble metals Cu, Ag and Au. Note that the intraband part has been subtracted and that the Cu-curve is shifted two units upwards.

II. STUDIES OF THE NOBLE METALS AND THEIR ALLOYS

A. The pure metals.

In the foregoing section we saw how measurements of the dielectric constant give us information about the electronic properties of solids. The optical absorption, which is essentially proportional to $\varepsilon_2 \omega^2$, originates both from intraband and interband electronic transitions.[8] The intraband or free electron absorption is usually described by

$$\varepsilon_2^f = \frac{\omega_p^2}{\tau} \cdot \frac{1}{\omega^3} \qquad \ldots (17)$$

where ω_p is the plasma frequency and τ the effective relaxation time associated with the electron scattering. This continuous background can be subtracted from the total $\varepsilon_2(\omega)$ (to give us the interband part $\varepsilon_2^b(\omega)$) and we need not discuss this effect more here.

In Fig. 7 $\varepsilon_2^b(\omega)$ are given för Cu, Ag and Au. These spectra have been deduced from transmittance measurements on thin films and subsequent use of the Kramers-Kronig relations. The results compare well with other authors.[9]

The interband absorption spectrum of Cu have been subject to many discussions during recent years. In particular, the question of the relative contributions of direct and non-direct transitions is still very controversial; this is important because the same problem applies to many metals.

It is agreed that the absorption edge around 2 eV is caused by interband transitions from the top of the d-band to the Fermi level. Whether or not these transitions are direct, non-direct or exitonic in character is till an open question. The second peak at about 5 eV has been given several assignments, e.g. the direct transitions $X_5 - X_4'$, $L_2' - L_1$, $L_1 - L_2'$ and the

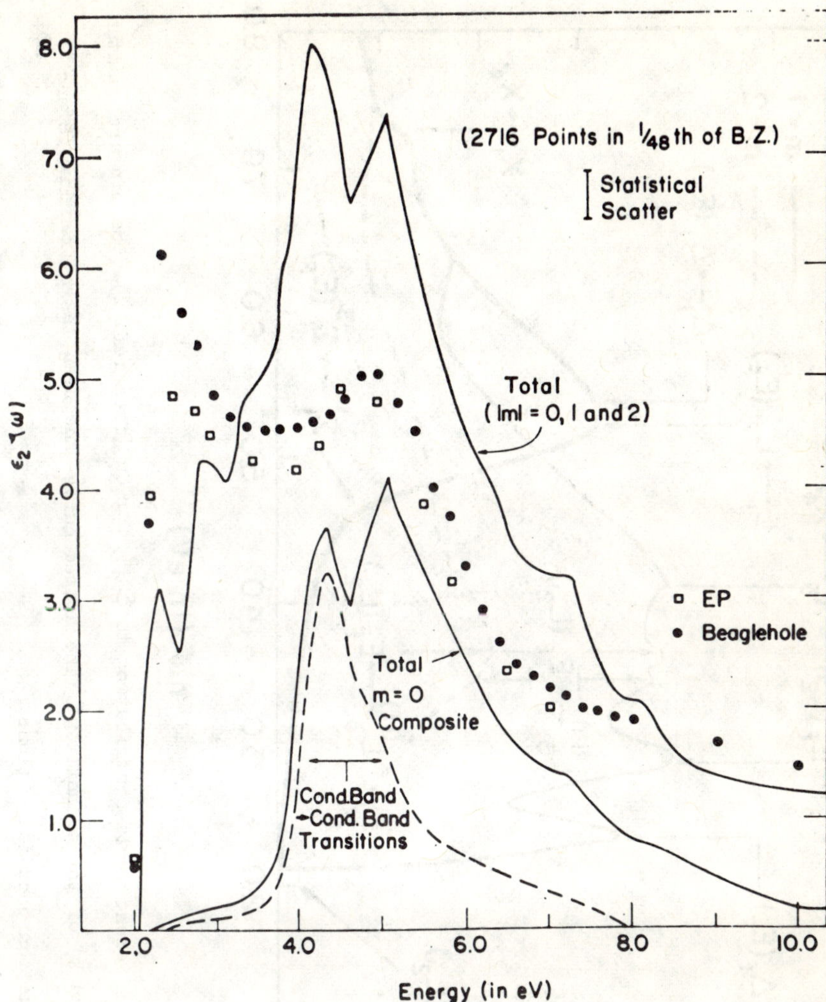

Fig. 8. $\varepsilon_2{}^b$ for Cu as calculated from band structure with assumption of direct transitions (from ref. 2) EP is experimental points from firs authors in ref. 9. In the data by Beaglehole the intraband part has not been subtracted.

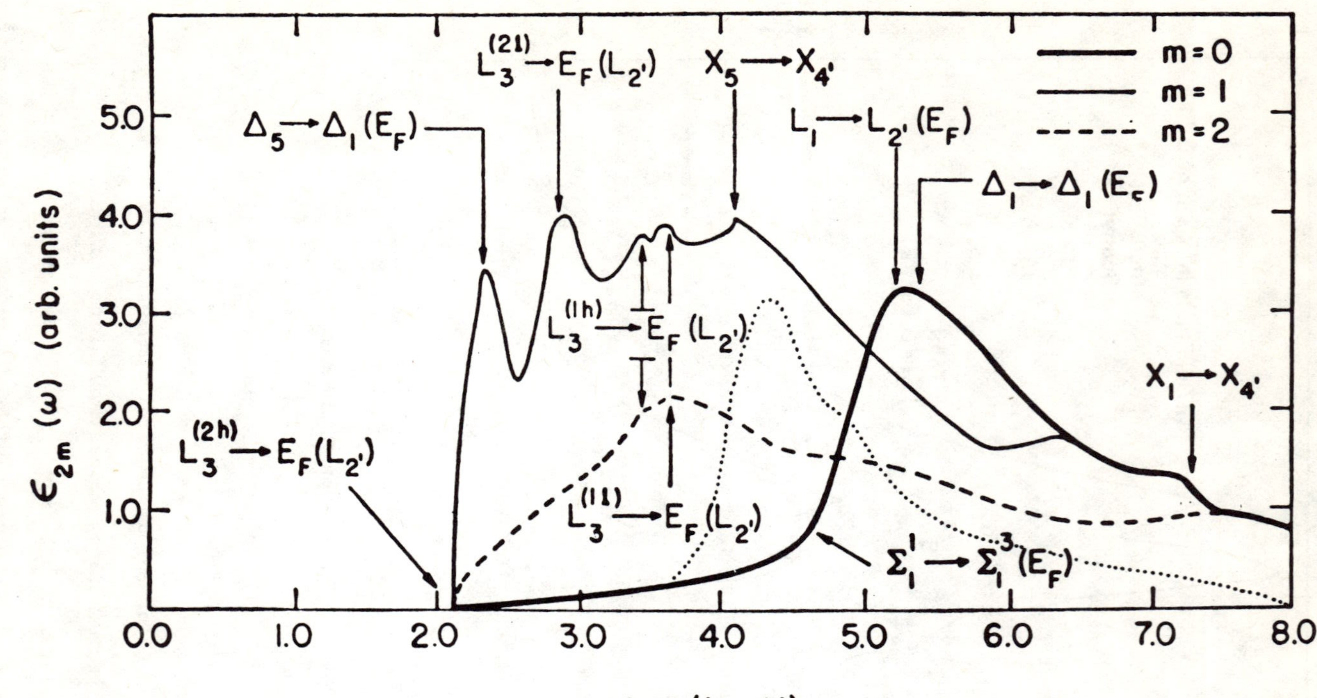

Fig. 9. Decomposition of the different parts contributing to ε_2^b for Cu (from ref. 2). The m:s are quantum numbers (the projection of L on $\tilde{k}$) of the transitions. The dotted curve has been taken from Fig. 8 and corresponds to conduction-conduction band transitions.

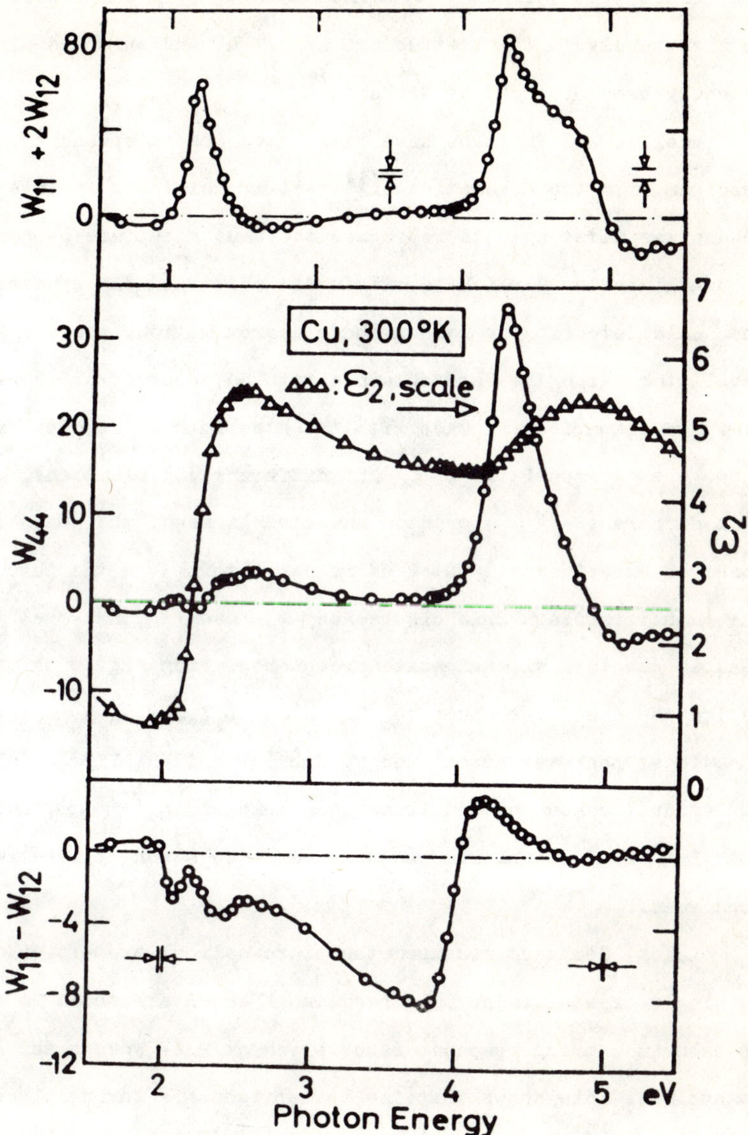

Fig. 10. The change in ε_2 due to different types of strain (from ref. 10). The upper curve results from hydrostatic pressure, the middle one from trigonal shear strain and the lower one from tetragonal shear strain.

non-direct transitions d-band-F.S. and d-band-X_4'. In our experiments we also resolve a faint structure at 3.7 eV and this has also recently been observed by Gerhardt[10].

Mueller and Phillips have calculated the interband $\varepsilon_2(\omega)$-spectrum with the assumption of direct transitions.[2] An interpolation scheme was first used to reproduce the band structure calculations by Segall and by Burdick to within 0.1 eV. The $\varepsilon_2^b(\omega)$-spectrum was then calculated in the random phase approximation, whereby the oscillator strengths were obtained with an accuracy of 20 %. In Fig. 8 a comparison is made with the experimental result. Fig. 9 shows the decomposition of $\varepsilon_2^b(\omega)$ by different contributions, arising from different quantum numbers. As seen, the sharp increase observed experimentally at 2 eV is not obtained in the calculations. The authors explain this discrepancy by assuming that the 2 eV peak is due to a virtual exiton resonance. From Fig. 9 we may assign the new 3.7 eV hump to $L_3^{lower} - L_2'$ and $X_5 - X_4'$ transitions. The 4.8 eV peak was associated by Mueller and Phillips mainly with the conduction-conduction transition denoted $L_2' - L_1$, because the $L_1 - L_2'$ transition was assumed to be strongly lifetime broadened.

Later, Gerhardt[10] measured the piezo-optical properties of Cu single crystals. The main results of which are shown in Fig. 10. As seen, no strong response is observed at 2 eV when shear strain is applied. This shows that the transitions are taking place in various parts of the Brillouin zone (B.Z.), which is consistent with transitions from the upper flat d-band to the Fermi surface. For trigonal strain, response is obtained at 4.3 eV. Due to symmetry reasons the corresponding transitions must take place close to the L-point and they were identified as $L_2' - L_1$ transitions, but admixture of $L_3^{lower} - L_2'$ and $L_1 - L_2'$ seemed also possible when inspecting the band structure diagram. Finally the response at 3.9 eV to tetragonal strain was associated with $X_5 - X_4'$ transitions. It should also be emphasized here that the piezo-optical experiments only give information around the B.Z. localized (i.e. direct) transitions. A consequence of this is that the relative strengths of the observed direct transitions, as compared to the non-direct ones, are uncertain.

In conclusion, the analysis of the ε_2 -spectrum för Cu, assuming direct transitions only, is interpreted mainly as follows: The 2 eV edge and its following peak is caused by transitions from the upper part of the d-band to the Fermi surface; the faint structure around 3.8 eV is due to $L_3^{lower} - L_2'$ and $X_5 - X_4'$ transitions; and the 4.8 eV peak originates from $L_2' - L_1$ and $L_1 - L_2'$ transitions.

Photoemission measurements on the other hand indicate that non-direct transitions contribute significantly to the optical absorption . Krolikowski derived the optical density of states of Cu from his photoemission measurements.[11] Using Eq. (13), i.e. assuming non-direct transitions, he then arrived at an ε_2-curve which very well reproduced the experimental ones. To obtain a peak at 4.8 eV, the peak in the density of states of the conducton band corresponding to X_4' was essential. This feature was taken from previous measurements by Berglund and Spicer[12] on cesiated specimens and gave a peak in ε_2 at 4.2 eV. Although this explanation may be correct, there is still another possibility. Again we calculate the ε_2 -spectrum from the calculated density of states assuming non-direct transitions. However, we pay little attention to the X_4' critical point since : (1) Gerhardt showed in

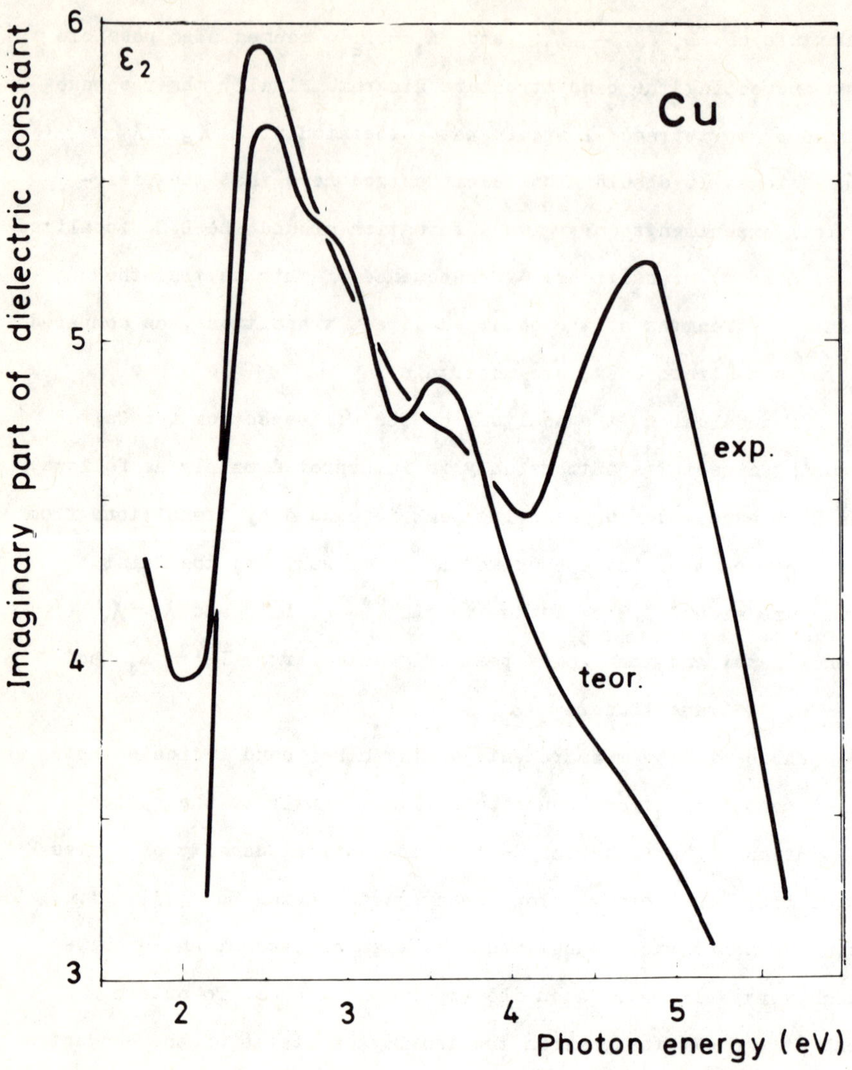

Fig. 11. ε_2 for Cu as obtained from experiment and from a calculation assuming nondirect transitions. A flat conduction band together with the d-density of states from ref. 15 was used.

his experiment (see above) that $X_5 - X_{4'}$ occurs already at 3.9 eV; (2) Mueller and Phillips showed that the $X_5 - X_{4'}$ transition should be very weak; (3) the strength of the density of states peak at $X_{4'}$ deduced from experiment is uncertain. The three peaks in the density of states of the d-band (see below) then correlate very well energywise with the features in ε_2. The discrepancy in position of the "5 eV peak" in ε_2 is thus removed. It should be noted that the experimental position of this peak differs from one author to another, possibly due to the fact that ε_2 is usually derived from reflectance measurements employing dispersion analysis. The position of the peak is then very sensitive to the shape of the reflectance curve around 5 eV. Our transmittance measurements, however, definitely give a position of 4.8 eV. As seen in Fig. 11 there is a big difference in amplitude between our experimental and calculated curve. In order to obtain the observed magnitude of the 4.8 eV transition one has to assume a large matrix element . There are evidently some contributions from direct transitions at 3.7 ($X_5 - X_{4'}$) and 3.8($L_2 - L_1$) eV, too, but they seem to be small.

The interpretation of the optical spectra of Ag and Au are even less clear because of fewer experiments and calculations. In Ag the absorption edge (Fig. 7) seems to be due to both the upper d-band to Fermi surface and to the $L_2' - L_1$ transitions according to Lynch,[13] who studied the optical properties of AgIn alloys. The second faint feature round 5 eV may reflect a transition from the density of states peak produced by L_3 and Γ_{25}' to the Fermi surface. This assignment is made from the available band structure calculations.

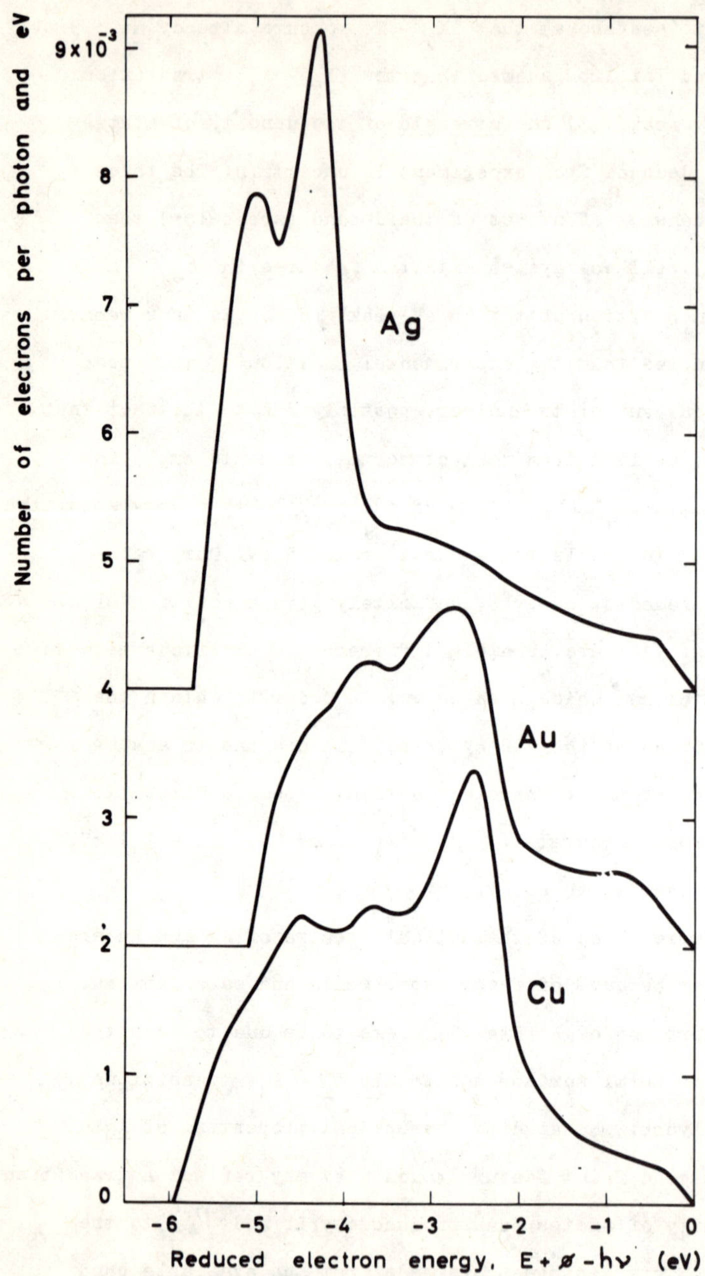

Fig. 12. Electron energy distributions as obtained from photoemission experiments on Cu, Ag and Au.

The optical absorption for Au also contains two peaks (Fig. 7). There is no published band structure for Au, but we know that the energy bands should resemble those of Cu and Ag if

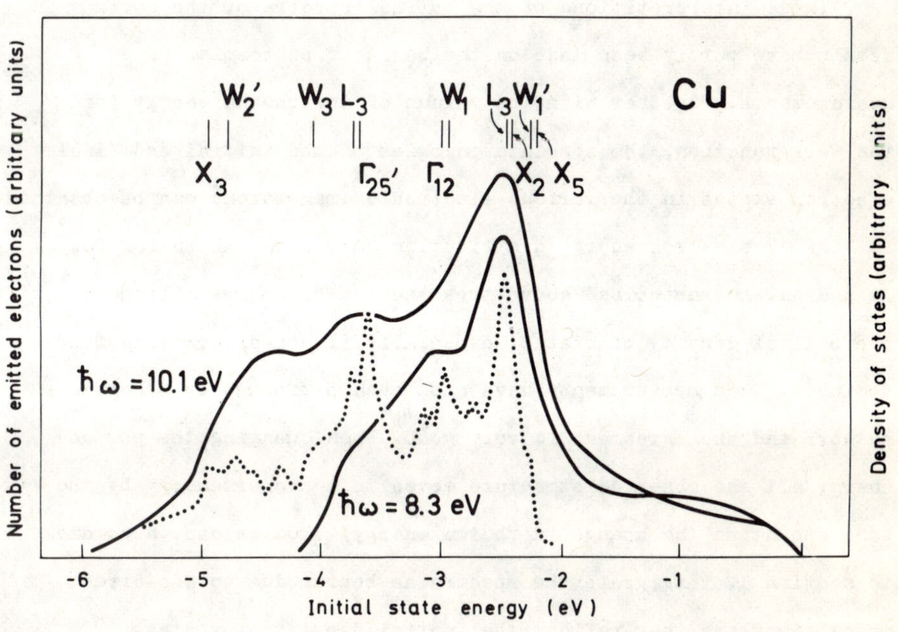

Fig. 13. Comparison between the measured electron energy distribution and the calculated [15] density of states for Cu.

not for the large spin-orbit coupling. The splitting due to this is of the order of 1 - 2 eV . From the cubic double group we find e.g. that the Γ_{12} level is unsplit (Γ_8), while the Γ_{25}' level splits into one two-fold (Γ_6) and one four-fold (Γ_8) level. The two observed peaks we now assign to non-direct transitions from the two density of states peaks to the Fermi surface in analogy with the interpretations of the Cu and Ag spectra. The density of states peaks may partly be produced by the upper Γ_8 and the Γ_6 critical points, respectively. Probably there are contributions from the L-point too.

These interpretations of the optical spectra of the noble metals have partly been made on the basis of photoemission measurements. Together with the values of the photon energy and the work function, the absolute energies of the initial and final electron states in the various electronic transitions can be obtained. In Fig. 12 electron energy distribution curves are shown for Cu, Ag and Au. As described above these bear a close resemblance to the optical density of states, especially if non-direct transitions dominate. Such measurements have also been performed by other authors and the agreement is very good.[11,14] When changing the photon energy all the observed structure moves to a higher energy by the same amount as the change in photon energy. Thus we can, according to section I, interprete the spectra as beeing due to non-direct transitions and they reflect the initial density of states.

Let us first study the Cu spectra. By the aid of Mueller's[15] density of states we now can identify the four features in the optical density of states as already discussed above. As seen in Fig. 13 the agreement is very good (error < 0.1 eV) if the calculated result is rigidly shifted. ~~eV.~~ The relative magnitudes as well as the shapes of the peaks do not agree so well. However,

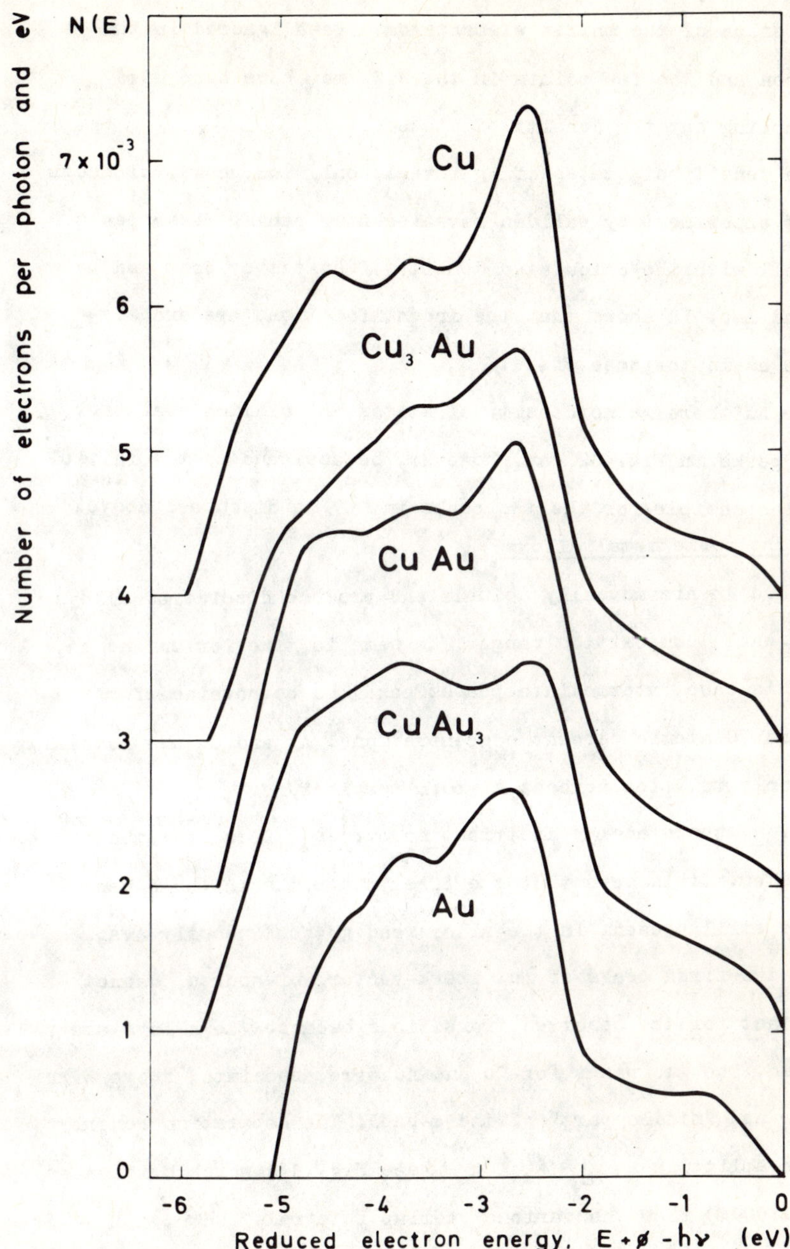

Fig. 14. Electron energy distribution curves for the Cu-Au system.

the variation of the matrix elements have been ignored in the comparison and too few points in the B.Z. may have been used when sampling out the density of states.

The density of states of Ag reveals only two peaks, although a recent experiment by Walldén[16] revealed four peaks. These peaks agree well with the calculated density of states by Snow and by Lewis and Lee. It shows that the indentifications are exactly the same as in the case of Cu.

For Au there is no density of states calculation avaiable. The two peaks in Fig. 12 can, however, be assigned to the initial states responsible for the two peaks in ε_2 as discussed above.

B. The noble metal alloys[x]

Ag and Au are mutually soluble and produce disordered alloys over the whole composition range. The same is true for Cu and Au, but in this case intermediate phases can also be obtained for 25, 50 and 70 atomic percent concentrations. Ag and Cu are mutually soluble only to a few percent at room temperature.

Fig. 14 shows energy distribution curves (EDC:s) for the Au-Cu system. It is seen that the first two peaks in the d-band of Cu are still present in the alloys and go continuously over into the two first peaks of Au. The $\bar{k}$-vector, of course, cannot be important for the observed transitions because the alloys are disordered. The two peaks for Cu and Au were associated above with the upper and "middle part" of the d-band. The separation roughly gives the splitting $\Gamma_{12} - \Gamma_{25}'$ (see Fig. 13) which originates mostly (90 %) from the various overlap integrals, $dd\sigma$, $dd\pi$ and $dd\delta$. As the splitting is of the same magnitude in Cu and Au, this indicates that the overlap integrals for Cu-Cu and Au-Au are for the same magnitude. It is therefore not unlikely

[x] This is a concentrated version of a paper to be published.

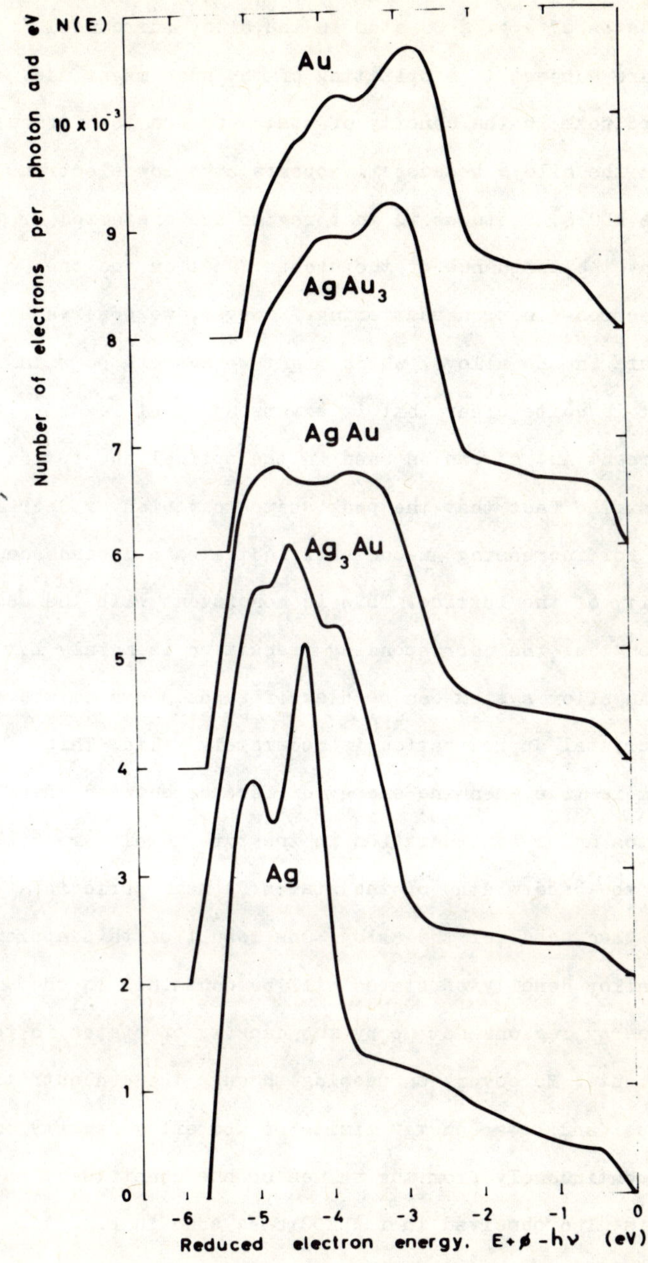

Fig. 15. Electron energy distribution curves for the Ag-Au system.

that the d-states of e.g. a Cu atom in the alloy surrounded by Au atoms are subject to a splitting of the same magnitude.

The third peak in the density of states for Cu is difficult to observe in the alloys because it appears at a low electron energy in the EDC:s. Features in that region are distorted or even masked by the influence of the escape function and the inelastic electron-electron scattering.[4] However, we observe a faint structure in the alloys, which might be associated with this peak. It is quite clear that it exists up to at least 20 atomic percent Au, as can be seen in the optical absorption in the alloys. The fact that the peak seems to vanish or decrease in magnitude for increasing Au-content, indicates a dependence on the periodicity of the lattice. This is consistent with the above interpretation that the corresponding transition is mainly $L_1 - L_2'$.

The Cu-Au alloy system can be classified as a system where the virtual crystal approximation[17] is moderately valid. This approximation is true when the energy difference between the electron states under consideration in the pure metals are much smaller than the band widths of the states. A mean periodic potential is then used for the alloy. One result of this approximation is that the alloy density of states will be contained in the union of the two energy regions set up by the density of states in the pure constituents. Moreover, on passing through the concentration range the upper and lower energy limits of the alloy density of states move continuously from the values of one constituent to the other. This is also observed in the EDC:s as seen in Fig. 14.

The Ag-Au alloy system behaves in a somewhat different manner, due to the fact that the Ag and Au bands are more separated than the Cu and Au bands. Fig. 15 shows EDC:s for the highest available photon energies. The Ag d-band for the Ag_3Au alloy seems to be roughly unaffected. The Au d-band is seen to lie just above the Ag band. Obviously the virtual crystal approximation, i.e. the use of a mean potential, is not valid in this case. We therefore have to use more refined methods to explain this result.

Velicky et. al.[18] have recently investigated the properties of the so-called coherent potential approximation by Soven[19] and Taylor[20]. Generally this theory gives a density of states which is roughly the weighed sums of the density of states for the constituents. It also shows that there exists a certain minimum band separation and a maximum concentration where the alloy bands are still split. Obviously the bands in Ag and Au are too close in energy to give split bands in the alloy (N.B., alloys with lower Au content have also been studied). It is important to remember that it is not possible to separate the contributions from Au and Ag states in the experimental data even if the bands are split. The Au states extend in energy down to the bottom of the Ag band although the spectral density is far larger for the Au band.

The Ag and Au states mix even more when increasing the Au content. For the Au_3Ag alloy no structure from the Ag atoms is observed, presumably because the Ag band is within the Au band.

The optical spectra of the Au-Cu alloys, which have also been measured, are entirely consistent with the EDC:s and the above interpretations. However, one observes that the absorption edge does not move linearly with composition. This shows that the virtual crystal approximation is not valid, not even for the Au-Cu system.

The details of the analysis of the alloy spectra will be

published elsewhere. We conclude with some general comments on the optical and photoemission spectra of alloys. For the pure metals the main features in the spectra can be explained by the band structure concept. The coherent potential approximation shows that for disordered alloys some kinds of energy bands remain. These are, however, broadened and have a varying spectral density i $\bar{k}$-space, but may still give rise to critial points. The alloy spectra contain structure which could be associated with these.

On the other hand, it is still not clear to what extent one measures the one electron state in an optical or photoemission experiment. The question is closely related to the unsolved problem of the non-direct transitions. If we do not observe the one electron band structure in the experiments we have to stick to the density of states concept. The features in the d-band spectra of the pure noble metal and alloy spectra are then due to manybody effects (or discussed in Section I) which are more localized to the excited atom than in the band structure interpretation. To complete the manybody interpretation more theoretical work is necessary.

References

1. H. Ehrenreich and H.R. Philipp, Phys. Rev. 128, 1622 (1962).

2. F.M. Mueller amd J.C. Phillips, Phys. Rev. 157, 600 (1967).

3. E.O. Kane, Phys. Rev. 175, 1039 (1968).

4. C.N. Berglund and W.E. Spicer, Phys. Rev. 136, A1030 (1964) and ref. 11.

5. R.C. Eden, Ph. D. Thesis, Stanford Electronics Laboratory 1967, Technical Report No. 5221-1

6. W.E. Spicer, Phys. Rev. 154, 385 (1967).

7. A.V. Sokolov, Optical Properties of Metals, London and Glasgow 1967, Chap. 9 and references

8. See e.g. H. Ehrenreich, Proc. Int. School of Physics, Course XXXIV New York and London 1966, edited by J. Tauc.

9. See e.g. for Cu: B.R. Cooper, H. Ehrenreich, and H.R. Philipp, Phys. Rev. 138, A 494 (1965) and D. Beaglehole, Proc. Phys. Soc. (London) 85, 1007 (1965); for Ag: H. Ehrenreich and H.R. Philipp, Phys. Rev. 128, 1622 (1962) and R.H. Huebner, E.T. Arakawa, R.A. McRae, and R.N. Hamm, J. Opt. Soc. Amer. 54, 1435 (1964); for Au: M. Otter, Z. Phys. 161, 163 (1961) and J.N. Hogdson, J. Phys. Chem. Solids 29, 2175 (1968).

10. U. Gerhardt, Phys. Rev. 172, 651 (1968).

11. W.F. Krolokowski, Ph. D. Thesis, Stanford Electronics Laboratory 1967, Technical Report No. 5218-1.

12. C.N. Berglund and W.E. Spicer, Phys. Rev. 136, A1044 (1964)

13. R.M. Morgan and D.W. Lynch, Phys. Rev. 172, 628 (1968).

14. D.E. Eastman and W.F. Krolokowski, Phys. Rev. Letters 21, 623 (1968).

15. F.M. Mueller, Phys. Rev. 153, 659 (1967).

16. L. Walldén, Phil. Mag. in print.

17. L. Nordheim, Ann. Physik 9, 607 (1931).

18. B. Velicky , S. Kirkpatrick and H. Ehrenreich, Phys. Rev. 175, 747 (1968).

19. P. Soven, Phys. Rev. 156, 809 (1967).

20. D.W. Taylor, Phys. Rev. 156, 1017 (1967).

VIII

"RELATIVISTIC" PHENOMENA IN SEMICONDUCTORS

W. Zawadzki

Institute of Electron Technology, Polish Academy of Sciences

Warsaw, Poland

INTRODUCTION

It has been the general practice almost from the beginning of the history of solid state investigations to treat electrons in the conduction band of metals and semiconductors as "free" carriers, i.e. to use the concepts developed for free particles to describe properties of carriers moving in a periodic potential of the lattice. This very powerful analogy reflects a more general and natural tendency to relate the unknown to the known, to apply in new situations the notions which have worked out well in the past. Such approach, in spite of its obvious limitations, very often provides an excellent starting point of scientific research. The purpose of this lecture is to extend the fruitful analogy between " free " electrons in semiconductors and free electrons in vacuum to the relativistic region and to indicate its interesting possibilities.

In the first section we specify our model of semiconductors , derive the energy-momentum relation, and introduce the relativistic analogy. In the remaining three sections we describe conduction electrons in semiconductors in absence of external fields, in a magnetic field, and in crossed electric and magnetic fields, respectively, emphasizing the analogy to free relativistic electrons in vacuum. In conclusion we mention some other phenomena and materials of interest.

1. TWO-BAND MODEL OF A SEMICONDUCTOR. RELATIVISTIC ANALOGY.

We begin by specifying our model of a semiconductor and derive the energy-momentum relation. In this we essentially follow a simplifical version of Kane's well known k.p procedure.[1]

The Schrödinger equation for an electron in periodic potential is

$$\left[\frac{\vec{p}^2}{2m_0} + V(\vec{r})\right]\Psi = \varepsilon\Psi \qquad (1.1)$$

where $V(\vec{r}+\vec{a}) = V(\vec{r})$, m_0 - the free electron mass, and $p_j = (\hbar/i)\,\partial/\partial x_j$ - the momentum operator. It is well known that this eigenvalue problem has solutions of Bloch type $\Psi_{nk} = u_{nk}(\vec{r})\exp(i\vec{k}\vec{r})$, where $u_{nk}(\vec{r})$ has the periodicity of the lattice. We shall, however, use the Kohn-Luttinger[2] representation $\chi_{nk} = u_n(\vec{r})\exp(i\vec{k}\vec{r})$, where the orthonormal set of $u_n(\vec{r})$ denotes the periodic parts of the Bloch functions taken at k= 0 (or any other band extremum). They satisfy the eigenvalue problem

$$\left[\frac{\vec{p}^2}{2m_0} + V(\vec{r})\right]u_n = \varepsilon_n u_n \qquad (1.2)$$

where ε_n is the energy at the bottom of the n'th energy band (at k = 0).

In absence of external fields we can look for the solution of the problem (1.1) in the following form

$$\Psi = \sum_{n'} c_{n'} u_{n'} \exp(i\vec{k}\vec{r}) \qquad (1.3)$$

where the sum is taken over different energy bands. We introduce expansion (1. 3) into Eq. (1.1), and notice that

$$\vec{p}^{\,2}[u_n \exp(i\vec{k}\vec{r})] = [(\vec{p}^{\,2}u_n) + 2\hbar\vec{k}(\vec{p}u_n) + \hbar^2k^2u_n]\exp(i\vec{k}\vec{r})$$

hence, taking into account Eq. (1. 2) , we get

$$\sum_{n'}[\varepsilon_{n'}u_{n'} + (\hbar\vec{k}/m_0)(\vec{p}u_{n'}) + (\hbar^2k^2/2m_0)u_{n'}]c_{n'} = \varepsilon\sum u_{n'}c_{n'} \tag{1.4}$$

Taking the scalar product of both sides with u_n the final set of equations is obtained in the form

$$\sum_{n'}\left[\left(\varepsilon_{n'} - \varepsilon + \frac{\hbar^2k^2}{2m_0}\right)\delta_{nn'} + \frac{\hbar\vec{k}}{m_0}\vec{p}_{nn'}\right]c_{n'} = 0 \qquad (n=1,2,\ldots) \tag{1.5}$$

where $\vec{p}_{nn'} = (u_n, \vec{p}u_{n'})$ is the matrix element of momentum operator. Eq. (1. 5) represents the initial problem (1.1) rewritten in the Kohn-Luttinger representation.

In the usual one-band effective mass approximation one eliminates the non diagonal elements of the Hamiltonian (1. 5) using a suitable canonical transformation. This is equivalent to the perturbation treatment with $\hbar\vec{k}\cdot\vec{p}_{nn'}/m_0$ terms as the perturbation. In the immediate vicinity of the band minimum (small k values) this leads to separation of different energy bands, the nondiagonal terms contributing to the effective mass value at the bottom of each band, according to the well known formula

$$\varepsilon_{nk} = \varepsilon_n + \frac{\hbar^2k^2}{2m_n^*} \qquad \frac{1}{m_n^*} = \frac{1}{m_0} + \frac{2}{m_0^2}\sum_{n'}\frac{p_{nn'}p_{n'n}}{\varepsilon_n - \varepsilon_{n'}} \tag{1.6}$$

As we have said above, this approximation is valid only for small k values, since it takes into account only the second order corrections to energy.

Now we shall specify our model. Namely, we assume that the conduction and valence bands are separated by the energy gap $\mathcal{E}_g$ its value being very small in comparison to energy distance to any other band. Furthermore we assume that matrix element p_{cv} between these two bands does not vanish. This means that the corresponding term in Eq. (1.6) is large, so that the interaction between the two bands in question cannot be treated as a small perturbation. This is, in fact, very similar to the perturbation problem for degenerate or nearly degenerate energy levels. It is well known that one should treat an interaction between such levels exactly, neglecting in the first approximation all other levels. Following this procedure we shall treat the interaction between the conduction and valence bands exactly, neglecting all other energy bands. The set (1.5) is then reduced to two equations. Choosing the zero energy at the bottom of the conduction band we obtain the following condition for the eigenenergies

$$\begin{vmatrix} -\mathcal{E}_g - \mathcal{E} + \frac{\hbar^2 k^2}{2m_0} & \frac{\hbar \vec{k}}{m_0} \vec{p}_{vc} \\ \frac{\hbar \vec{k}}{m_0} \vec{p}_{cv} & -\mathcal{E} + \frac{\hbar^2 k^2}{2m_0} \end{vmatrix} = 0 \qquad (1.7)$$

We want to neglect the free electron term $\hbar^2 k^2/2m_0$. It can be easily seen that this corresponds to neglecting the free electron contribution in Eq. (1.6), which in narrow gap semiconductor amounts to few percent. Thus, we finally obtain the energy-momentum relation for our model in the form

$$\mathcal{E}_{c,v} = -\frac{\mathcal{E}_g}{2} \pm \left[\left(\frac{\mathcal{E}_g}{2} \right)^2 + \mathcal{E}_g \frac{\hbar^2 k^2}{2m_0^*} \right]^{1/2} \quad (1.8)$$

with $1/m_0^* = 2 p_{cv} p_{vc} / m_0^2 \mathcal{E}_g$. Our effective mass at the bottom of the band is determined by the most important term in the formula (1.6) , in agreement with our assumptions.*

Now we want to find out to what extent our two-band model corresponds to the situation in real semiconducting materials. First, we notice that for $\hbar^2 k^2 / 2m_0 \ll \mathcal{E}_g$, i.e. for electron energies small compared to the energy gap, we can expand the square root and obtain $\mathcal{E}_c = \hbar^2 k^2 / 2m_0^*$, in agreement with the usual approximation. Thus for small electron energies our model is certainly valid, as long as the mass value is close to the real one, which means that the two interacting bands lie close enough to each other. For $\hbar^2 k^2 / 2m_0^* \gg \mathcal{E}_g$ we find from Eq. (1.8) that the energy varies linearly with quasi momentum $\hbar k$. On the other hand we know very well that at the edge of the Brillouin zone the $\mathcal{E}(k)$ curve should have an extremum. Thus in the range of energies the model

---

* Actually our model, assuming a spherical $\mathcal{E}(k)$ dependence from the interaction between just two energy bands, is an idealization. One can obtain relation (1.8) for the conduction and light - hole energy bands by considering the interaction between one s - like conduction band and three p - like valence bands degenerate at k = 0 . This would not lead to any substantial changes in our considerations.

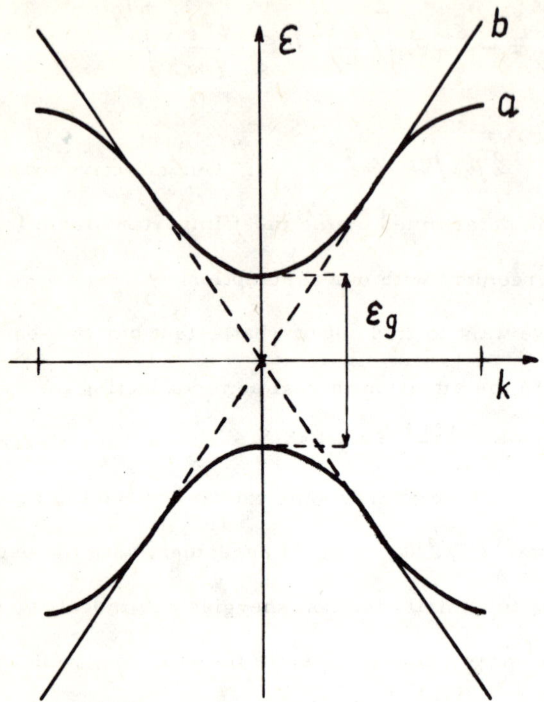

Fig. 1 Energy - wave vector dependence in the conduction and valence bands of a semiconductor (schematically) : a) exact relation, b) relation given by two-band model.

is not valid. Fig. 1 gives a schematical comparison of the $\varepsilon(k)$ dependence in real materials with that described by Eq. (1.8) . It can be seen that the two-band model is approximately valid up to the inflection point on the real curve. This is a substantial improvement in comparison to the one-band approximation, which describes only the parabolic part of the real curve. It is clear that in order to obtain the true $\varepsilon(k)$ relation one would have to take into account all energy bands in the set (1.5) . In the conduction band of InSb , having ε_g = 0.18 eV at 300°K,

Eq. (1.8) describes the conduction band up to the energies of $\varepsilon \simeq 0.45$ eV . The mass value is $m_o^* \simeq 0.013\ m_o$, so that the omission of the free electron term in Eq. (1.6) is well justified. The same may be said about all other narrow-gap materials.

Now we shall introduce the analogy mentioned in the title of this lecture. We notice that according to the relativistic mechanics the energy-momentum relation for free electrons in vacuum can be cast in the form analogous to Eq, (1.8), namely

$$\varepsilon = - m_o c^2 \pm \left[(m_o c^2)^2 + 2\, m_o c^2 \frac{p^2}{2m_o} \right]^{1/2} \qquad (1.9)$$

The analogy with our two-band formula is complete with the following correspondence:

$$m_o^* \longrightarrow m_o \qquad\qquad \varepsilon_g \longrightarrow 2m_o c^2 \qquad (1.10)$$

It is of interest to find out what value corresponds to the maximum " light velocity " in our model. We have

$$c = \left(\frac{2m_o c^2}{2m_o} \right)^{1/2} \longrightarrow \left(\frac{\varepsilon_g}{2m_{o*}} \right)^{1/2} = v_{max} \qquad (1.11)$$

Below we list the basic quantities and their approximate numerical values for the two situations

free electrons	electrons in semiconductors
m_o	$m_o^* \simeq 0.01\ m_o$
$2\ m_o\ c^2 \simeq 10^6$ eV	$\varepsilon_g \simeq 0.1$ eV
$c \simeq 3 \times 10^{10}$ cm/sec	$v_{max} \simeq 10^8$ cm/sec

It follows from the above table, and we want to emphasize it once again, that velocities of electrons in semiconductors are much lower than the real relativistic velocities. It is the formal similarity that makes our analogy possible, but the scale of values in both cases is vastly different.

In the above interpretation the expansion of the square root for small k values and the resulting quadratic dependence of energy on momentum corresponds to the " nonrelativistic limit " of small velocities.

Our analogy goes actually further. According to Dirac interpretation [3], one deals in vacuum with the " sea of electrons " having negative energies, and as long as no electrons miss in this sea, we do not observe them. However, if we excite an electron to the state with positive energy, we observe both the electron and the " hole " in the sea, i.e. the positron. This corresponds exactly to the picture of electrons and holes in the conduction and valence bands of a semiconductor. The masses of electrons and positrons are equal, and the same is true about electrons and holes in our two-band model.

2. ELECTRONS IN ABSENCE OF EXTERNAL FIELDS

If the electrons in a semiconductor are to behave like free relativistic particles, the first thing one would expect of them is the variation of their mass with velocity (or energy). The velocity of an electron in the crystal is

$$\vec{v} = \frac{1}{\hbar} \operatorname{grad}_k \varepsilon(k) \qquad (2.1)$$

Now we define the effective mass as a quantity relating the velocity to the momentum ,

$$m^* \vec{v} = \hbar \vec{k} \qquad (2.2)$$

Hence we obtain easily for a spherical band

$$\frac{1}{m^*} = \frac{1}{\hbar^2} \cdot \frac{1}{k} \quad \frac{d\varepsilon}{dk} \qquad (2.3)$$

It can be seen that if $\varepsilon(k)$ is not a quadratic function of k ., the effective mass depends on the energy. For the energy-momentum relation (1. 8) one easily obtains for electrons

$$m^* = m_o^* \left(1 + 2 \frac{\varepsilon}{\varepsilon_g} \right) \qquad (2.4)$$

One is normally used to another definition of the effective mass in solids, namely

$$M \vec{a} = \vec{F} \qquad \frac{1}{M} = \frac{1}{\hbar^2} \quad \frac{d^2}{dk^2} \qquad (2.5)$$

where $\vec{F}$ is the acting force and $\vec{a}$ the acceleration of an electron. We notice that the two definitions are equivalent in the parabolic region, given by Eq. (1. 6) where both masses are equal to the mass value at the bottom of the band m_o^* .

It should be emphasized that it is the momentum effective mass, as defined in Eq. (2.2) and (2. 3) , that appears as a basic quantity in all free - carrier properties of electrons in semiconductors. At first this feature seemed rather peculiar, but the relativistic analogy is again of help here. In the newtonian nonrelativistic mechanics we deal with constant mass of particles, relating both velocity to momentum and acceleration to force. However in the relativistic mechanics it is the mass which relates the velocity to the momentum: $m\vec{v} = \vec{p}$,

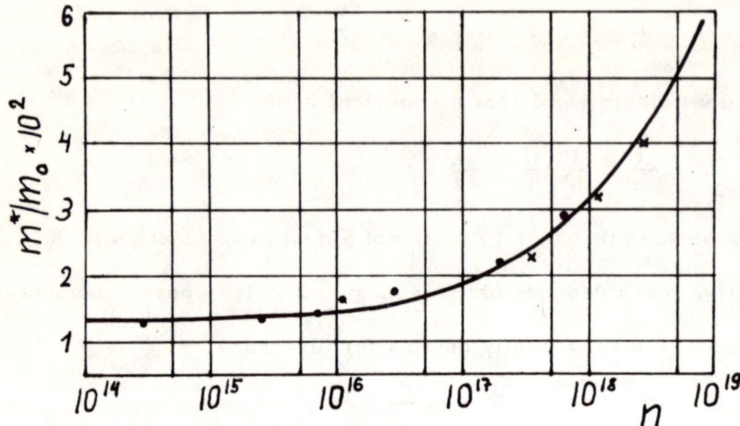

Fig. 2. Electron effective mass m^*/m_0 at the Fermi level versus electron concetration in the conduction band of InSb. The solid line is calculated according to the two-band model. The experimental points are from the optical reflectivity measurements of Spitzer and Fan, Phys. Rev. 106, 882 (1957) - crosses, and Faraday rotation data of Smith et al., Journ. Phys. Chem. Solids, 11, 131 (1959) - full dots. After Kolodziejczak (4).

that is the basic quantity in the theory. By an easy calculation this relativistic electron mass can be shown to be

$$m = m_o \left(1 - \frac{v^2}{c^2}\right)^{-1/2} = m_o \left(1 + \frac{\varepsilon}{m_o c^2}\right) \qquad (2.6)$$

where ε is given by Eq. (1.9) This is again in complete analogy to the energy dependence of effective mass, as given by Eq. (2.4). Thus the effective mass in semiconductors (2.3) reduces for small energies to the mass given in Eq. (2.5) in the same way as the relativistic mass m(v) of Eq. (2.6) reduces to the rest mass

m_o in the nonrelativistic limit of small velocities.

The energy dependence of effective mass in semiconductors is measured by studying free - carrier transport phenomena for strongly degenerate electron gas, when only the electrons having energies in the immediate vicinity of Fermi level participate in the transport of charge and heat. In such conditions we can relate directly the Fermi energy ζ to the electron concentration n in the band. For the conduction band described in Eq. (1.8) we have (Kolodziejczak [4])

$$\zeta = \frac{E_g}{2}\left(\sqrt{\Delta} - 1\right) \qquad m^*(\zeta) = m_o^* \sqrt{\Delta} \tag{2.7}$$

where

$$\Delta = 1 + 2\pi^2 \left(\frac{3}{\pi}\right)^{2/3} \frac{\hbar^2}{E_g m_o^*} n^{2/3}$$

Formula for the effective mass is obtained by substituting the expression for ζ into Eq. (2.4). Fig. 2 illustrates the energy dependence of electron effective mass in the conduction band of InSb , as determined from the free - carrier optical data. The solid curve is calculated from Eq. (2.8). It can be seen that the effective mass really exhibits the " relativistic " energy dependence. Similar relations are obtained for other narrow gap semiconductors.

Also the thermodynamic properties of electrons in a conduction band described by two-band Eq. (1.8), like the carrier concentration, specific heat, internal energy, etc. (Zawadzki and Kolodziejczak [5]) are formally identical with the properties of Jüttner' s relativistic gas of free fermions (see e.g. [6] and [7]). For example in the non-degenerate conditions obeying the Boltzmann statistics, the internal energy per carrier is not equal to the classical value of (3/2) $k_o T$

but at high temperatures it approaches 3 k_0T

3. ELECTRONS IN A MAGNETIC FIELD

We begin this section considering properties of relativistic electrons in vacuum in the presence of a magnetic field, and then proceed with our analogy describing the magnetic behavior of conduction electrons in semiconductors.

In order to find eigenenergies and eigenfunctions of free relativistic electrons in the presence of a uniform constant magnetic field H one has to solve the Dirac equation for this problem. This can be done analytically in closed form (8) (9). If a magnetic field is chosen along the z direction the resulting eigenenergies are given by the following formula

$$\mathcal{E}_{n\uparrow\downarrow} = -m_0c^2 + \left[(m_0c^2)^2 + 2m_0c^2\left(D_{np_z} \pm \tfrac{1}{2}\hbar\omega_c\right)\right]^{1/2} \quad (3.1)$$

where $D_{np_z} = \hbar\omega_c(n+\tfrac{1}{2}) + p_z^2/2m_0$. This resembles Eq. (1.9) with free electron energy $\vec{p}^2/2m_0$ replaced by D_{np_z}, where $\omega_c = eH/m_0c$ is the usual cyclotron frequency. This is due to the fact that the motion in the plane perpendicular to the direction of a magnetic field is quantized (diamagnetic or Landau quantization). There appears also the additional term $\pm\frac{1}{2}\hbar\omega_c$, associated with the spin magnetic moment of the electron, $\pm$ signs corresponding to its two possible orientations with regard to magnetic field. For magnetic energies small compared to the rest energy we can expand the square root and obtain

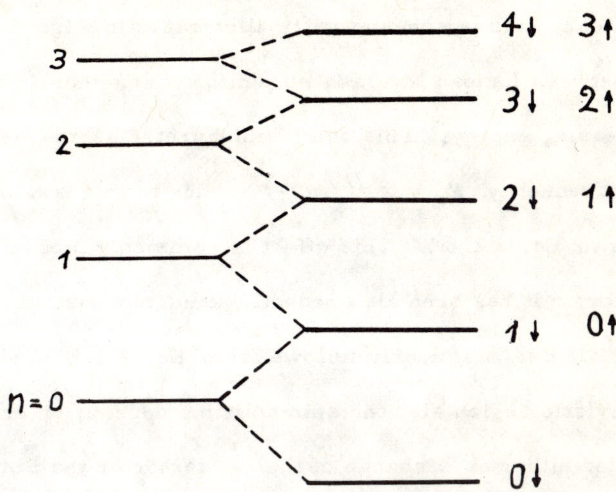

Fig. 3 . Energy levels of a free electron in a magnetic field according to the Dirac equation (schematically) . The Landau quantization is illustrated on the left, the final spin - split levels on the right.

$$\mathcal{E}_{n\uparrow\downarrow} = \hbar\omega_c\left(n+\tfrac{1}{2}\right) + \frac{p_z^2}{2m_0} \pm \frac{1}{2}\hbar\omega_c \qquad (3.2)$$

In the nonrelativistic limit, represented by Eq. (3.2), the spacing between Landau levels is uniform and equal to $\hbar\omega_c$. The spin (Pauli) splitting of each level is also equal to $\hbar\omega_c$. Introducing the spin splitting g factor,

$$\mathcal{E}_{\uparrow\downarrow} = \pm\frac{1}{2}\mu_B^0 g H \qquad (3.3)$$

where $\mu_B^0 = e\hbar/2m_0c$ is the Bohr magneton, we see that for free nonrelativistic electrons g = 2 .

It can be seen that the relativistic formula (3.1) introduces two features. This is schematically illustrated in Fig. 3 . First, the spacing between Landau levels is not uniform any more, but it decreases with increasing energy. This can be interpreted as the decrease of the cyclotron frequency $\omega_c = eH/m(\varepsilon)c$ due to increase of the mass, according to Eq. (2.6). This effect is commonly met in free particle accelerators. It has been also recently used to construct a free electron maser [10]. But, as clearly follows from Eq. (3.1) and Fig. 3, in the relativistic region also the spin splitting depends on energy. It can be regarded either as a change of the g factor or the Bohr magneton. It is easy to find the explicit energy dependence of the latter, assuming g = 2 . Using Eq. (3.1) we obtain after simple calculation

$$\mu_B = \lim_{H \to 0} \frac{\varepsilon_\uparrow - \varepsilon_\downarrow}{2H} = \frac{e\hbar}{2m_0c}\left(1 + \frac{\varepsilon}{m_0c^2}\right)^{-1} = \frac{e\hbar}{2m(\varepsilon)c} \qquad (3.4)$$

where $m(\varepsilon)$ is the usual relativistic electron mass, given in Eq. (2. 6).

Now we want to show that similar effects are observed in semiconductors with pronounced nonparabolicity of the conduction band. We shall not present the derivations, those can be found in original publications.

In III - V semiconducting compounds with narrow energy gap , like InSb, InAs and some others, the electron energy levels in the presence of a magnetic field can be to a good approximation represented by a formula analogous to Eq. (3.1) , namely (Lax et al. [11])

$$\varepsilon_{n\uparrow\downarrow} = -\frac{\varepsilon_g}{2} + \left[\left(\frac{\varepsilon_g}{2}\right)^2 + \varepsilon_g\left(D^*_{nk_z} \pm \frac{1}{2}\mu_B^0 g_0^* H\right)\right]^{1/2} \qquad (3.5)$$

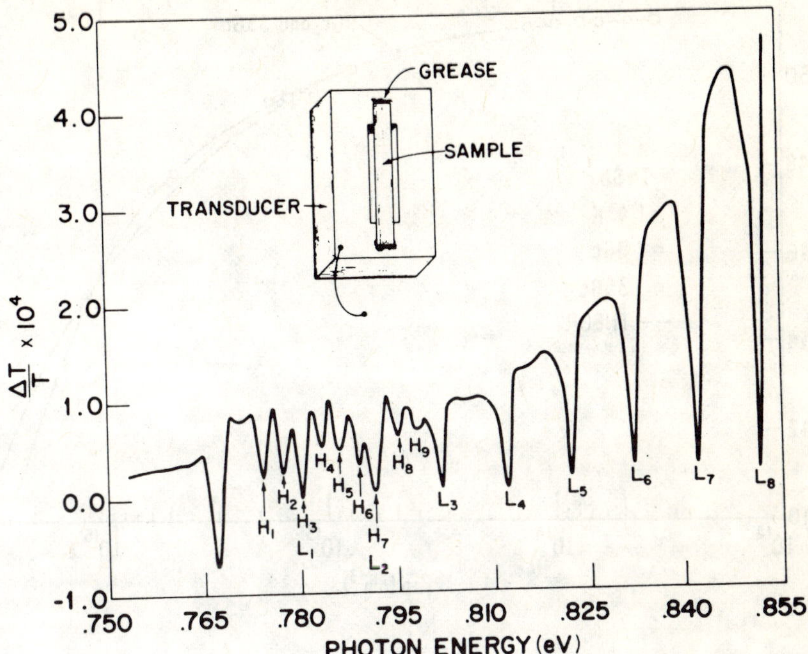

Fig. 4. Magneto-piezotransmission in germanium showing the indirect transitions from the top of the valence band to the Landau levels at the L point. The decreasing spacing of Landau levels due to band's nonparabolicity can be seen. After Aggarwal et al.[12]

where

$$D^*_{nk_2} = \hbar\omega_c^*(n+\tfrac{1}{2}) + \frac{\hbar^2 k_2^2}{2m_o^*} \qquad (3.6)$$

with $\omega_c^* = eH/m_o^*c$ being the effective cyclotron frequency, and g_o^* the effective spin splitting factor. The nonlinear spacing of Landau levels, both as a function of magnetic field intensity H and quantum number n, has been observed many times. Perhaps the most spectacular demonstration is that by Aggarwal et al [11] in germanium, presented in Fig. 4. L_n indicate transitions from the top of the

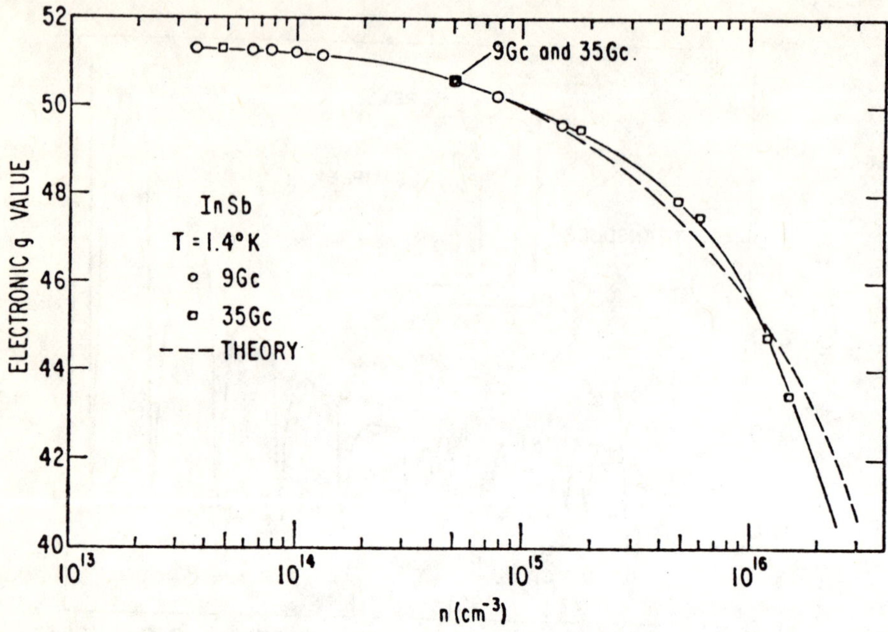

Fig. 5. Electron spin splitting g-factor at the Fermi level versus electron concentration in the conduction band of InSb . The dashed line is calculated according to formula (3.7). After Isaacson (16) .

valence band to the consecutive levels at the edge of the Brillouin zone (L point). The diminishing spacing between Landau levels is clearly seen. It turns out that for higher levels this decrease can be to a good approximation described by the increase of effective mass, according to Eq. (2.4).

The dependence of the effective g-factor on energy in III-V compounds was analyzed by Yafet (13). It turned out that his results for low magnetic fields can be expressed by the following formula (Zawadzki (14)),

$$g(\varepsilon) = 2\left[1 - \left(\frac{m_0}{m^*(\varepsilon)} - 1\right)\frac{\Delta}{3(\varepsilon_g + \varepsilon) + 2\Delta}\right] \qquad (3.7)$$

where Δ is the value of spin-orbit interaction at k = 0. For $\varepsilon = 0$ the above relation reduces to Roth's formula (15). It should be emphasized that although the initial values (for $\varepsilon = 0$) of the g-factor are anomalous (this follows both from the above formula and experiment), the energy dependence of the spin splitting is similar to that for free relativistic electrons, given by Eq. (3.4), since to the first approximation $g(\varepsilon) \sim 1/m^*(\varepsilon)$. The agreement between the theory and experiment carried out by Isaacson (16) in InSb is demonstrated in Fig. 5. The plot is made similarly to that in Fig. 2., i.e. for a given electron concentration n the Fermi energy and the effective mass is calculated using Eq. (2.7) and their values used to determine the g-factor according to Eq. (3.7). The energy dependence of electron g-factor has been observed also in other materials.

Wolff (17) applied formalism of the Dirac equation to derive interesting selection rules for transitions between Landau levels in bismuth.

4. ELECTRONS IN CROSSED ELECTRIC AND MAGNETIC FIELDS

The " relativistic " nature of electrons in semiconductors is perhaps most spectacularly demonstrated in a physical situation created by external crossed electric and magnetic fields. In the following considerations we shall omit the spin effects.

We begin using the one-band effective mass approximation. In other words we treat electrons in the conduction band of a semiconductor as free nonrelativistic electrons, with free-electron mass m_0 replaced by the effective mass m_0^*. For the purpose of the optical absorption in crossed fields this was done by Aronov [18]. The Schrödinger equation for the envelope function of a conduction electron in the presence of crossed fields is

$$\left[\frac{1}{2m_0^*}\left(\vec{p}+\frac{e}{c}\vec{A}\right)^2 + e\vec{E}\vec{r}\right]f(\vec{r}) = \varepsilon f(\vec{r}) \qquad (4.1)$$

The magnetic field is introduced by means of the vector potential $\vec{A}$. For a magnetic field in the z direction we choose $\vec{A} = [-Hy, 0, 0]$ so that $\vec{H} = \mathrm{curl}\,\vec{A} = [0, 0, H]$. We take the transverse electric field in the y direction, $\vec{E} = [0, E, 0]$. With the above choice of the vector and scalar potentials the coordinates in Eq. (4.1) can be separated. We look for the solution in the form

$$f(x,y,z) = (1/4\pi^2)\exp(ik_x x + ik_z z)\,\varphi(y) \qquad (4.2)$$

It is well known that the solutions of Eq. (4.1) without the electric field term given by the harmonic oscillator functions. It can be seen that the electric field, introducing a term linear in y, does not change the character of the solutions. It simply shifts the potential well, which corresponds to the shift of the oscillation center. Thus the solutions of our problem are given by

$$\varphi(y) = C_n \Phi_n\left(\frac{y - k_x L^2 + eEL^2/\hbar\omega_c}{L}\right) \qquad (4.3)$$

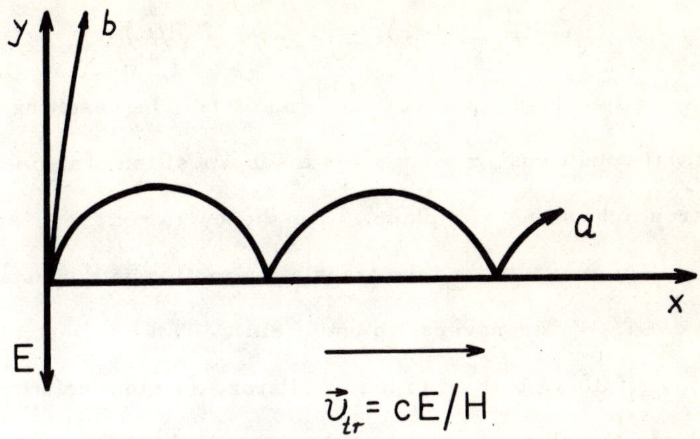

Fig. 6. Motion of a free relativistic electron in crossed electric and magnetic fields. (The magnetic field is perpendicular to the plane of the paper).
a) for $E / H \ll 1$ the motion is a superposition of the cyclotron motion with the angular velocity $\omega_c = eH / m_o c$ and the motion with constant velocity $v_{tr} = c E / H$, transverse to the direction of both fields.
b) for $E / H > 1$ the motion is essentially along the electric field direction, only slightly deflected by magnetic field.

where ϕ_n are the harmonic oscillator functions. The eigenergies are

$$\mathcal{E} = \hbar\omega_c^*(n+\tfrac{1}{2}) + \frac{\hbar^2 k_z^2}{2m_o^*} - eEk_x L^2 - \frac{m_o^* c^2}{2}\frac{E^2}{H^2} \qquad (4.4)$$

where $L = (\hbar c/eH)^{1/2}$ is called the magnetic radius. C_n is the normalization factor. For $E = 0$ Eqs. (4.3) and (4.4) reduce the eigenfunctions and eigenergies for the electron in a magnetic field alone.

One can interpret the above results by considering the classical motion of a free electron in crossed fields. The nonrelativistic equation of motion

$$m_0 \ddot{\vec{r}} = -e\vec{E} - (e/c)(\vec{v} \times \vec{H}) \qquad (4.5)$$

can be easily solved (see e.g. [19] p. 54). The resulting motion for the initial conditions $\dot{\vec{r}} = \vec{r} = 0$ is a superposition of the oscillatory cyclotron motion in xy plane, with the cyclotron frequency $\omega_c = eH / m_0 c$, and the translation motion with constant velocity $v_{tr} = cE / H$ transverse to both fields. This is illustrated in Fig. 6.

As follows both from the oscillatory quantum solutions of Eq. (4.3) and the classical picture, the described behavior is essentially of the magnetic-field type : the cyclotron motion is still the dominant feature and there is no net motion along the electric field.

Experiments on interband optical absorption and dispersion in crossed fields carried for low values E / H ratio confirm the above conclusions. As follows from Eq. (4.4), the main effect of transverse electric field is to pull Landau levels in the conduction band down by the amount $m_0^* c^2 E^2 / 2H^2$. Changing the sign of the mass, it can be seen that Landau levels in the valence band are pushed up by the electric field. This effect should be observable in interband optical transitions, which have resonant character whenever the photon energy is equal to the separation between pairs of Landau levels in both bands. This condition can be easily obtained from Eq. (4.4), neglecting the k_x and k_z terms, which do not affect positions of peaks,

$$\hbar\omega = \varepsilon_g + \hbar\omega_{c1}^*(n_1 + \tfrac{1}{2}) + \hbar\omega_{c2}^*(n_2 + \tfrac{1}{2}) - (m_{01}^* + m_{02}^*)c^2E^2/2H^2 \qquad (4.6)$$

where the subscripts refer to the conduction and valence bands.

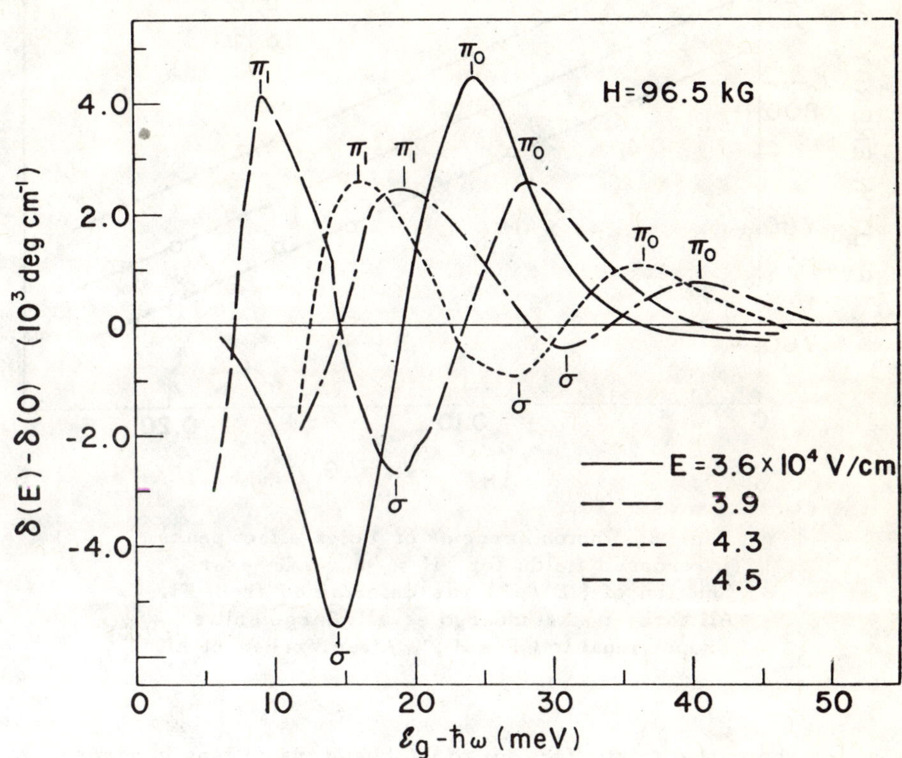

Fig. 7. Resonant Voigt effect in crossed fields for photon energies below the direct gap of germanium for H = 96, 5 kGs and various electric fields, up to E = 4, 5 x 10^4 V / cm. It illustrates magnetic-type behavior of electrons. The transitions shift to lower energies as E increases. After Vrehen et al. [20]

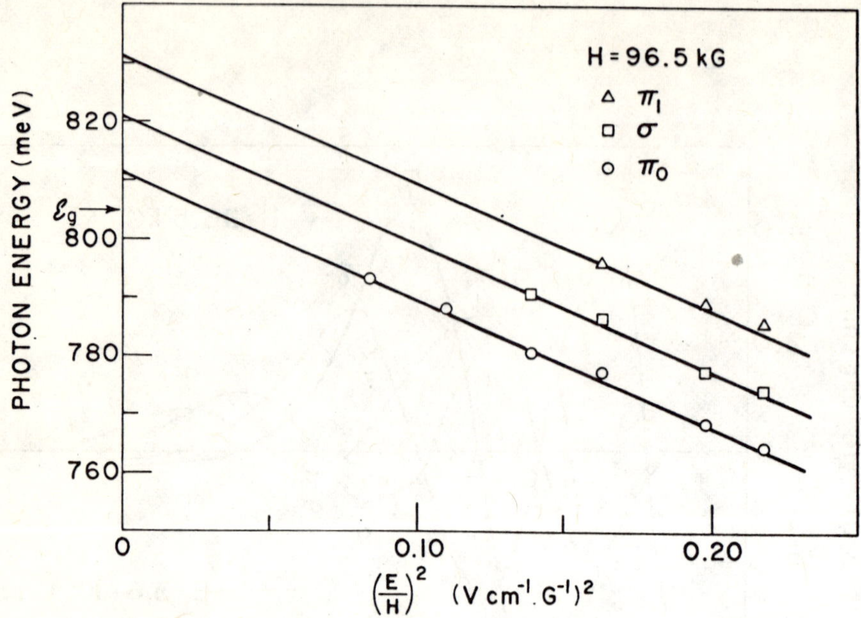

Fig. 8. Photon energies of Voigt-effect peaks in crossed fields for H = 96, 5 kGs as a function of $(E / H)^2$ as determined from Fig. 7. All three peaks undergo equally large shifts proportional to $(E / H)^2$. After Vrehen et al. [20]

Fig. 7. shows the Voigt effect due to interband transitions in germanium, measured for a fixed value of a magnetic field and different electric fields (Vrehen et al. [20]). The electric fields are strong enough to pull the interband transitions _into_ the gap. First, it is seen that the effect has a resonant behavior which is characteristic of quantized solutions, i.e. the magnetic-type motion. Second, the peaks really shift to lower energies for increasing electric fields. According to Eq. (4.6) the shift should be proportional to E^2 / H^2. Fig. 8 shows this to be in agreement with experiment.

Thus everything seems to work well, but it is not really so.

The trouble begins when we try to apply the theory for large values of E / H ratio, for example by making a transition $H \rightarrow 0$. According to solution (4.3) and Fig. 6 (a), as long as $H \neq 0$ the solutions are of magnetic type. In other words even a very small magnetic field seems to have dramatic influence on the character of the motion. This does not make much physical sense, since for large electric and small magnetic field the electric-type motion should occur. Also according to Eq. (4.6) for sufficiently large values of E / H ratio the Landau levels in the conduction and valence bands would cross each other.

On the other hand, the experiments performed for high electric and low magnetic field intensities (Reine et al [21]) clearly indicate that one deals with electric-field-type solutions in this range.

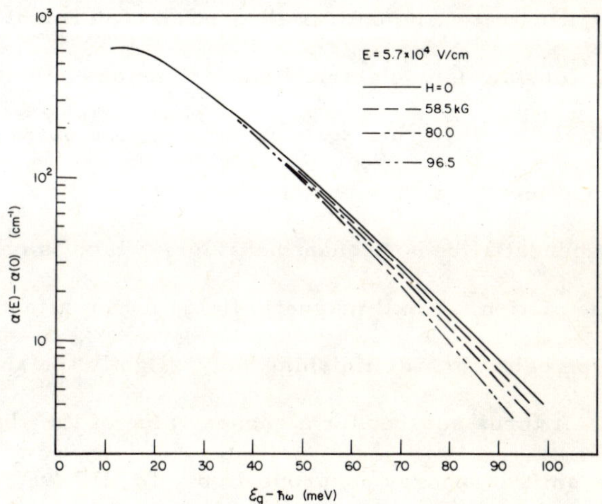

Fig. 9. Exponential absorption coefficient below the direct gap of germanium due to electric field $E = 5,7 \times 10^4$ V/cm for various values of transverse magnetic field. It illustrates electric-type behavior of electrons. The effect of the magnetic field is to decrease the photon-assisted tunneling current. After Reine et al. [21]

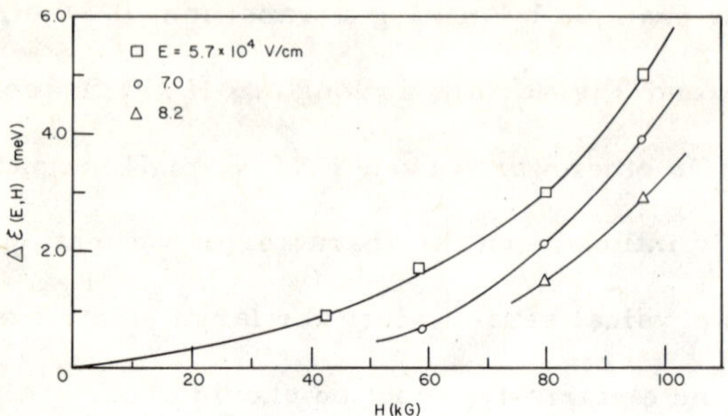

Fig. 10. The magnetic shift $\Delta\mathcal{E}$, as determined in part from Fig. 9, for a chosen value of the absorption coefficient $[\alpha(E) - \alpha(0) = 20\ cm^{-1}$ at a constant electric field. The shift is proportional to H^2 / E^2. After Reine et al.[21]

Fig. 9 illustrates the optical absorption of germanium in the presence of a constant high electric field for various values of transverse magnetic field. It is seen that for $H = 0$ one observes the usual Franz-Keldysh effect, i.e. the photon-assisted tunneling. The absorption has an exponential non-resonant behavior, which is characteristic of electric-type motion. Small magnetic fields do not affect dramatically the electric-type behavior, diminishing only slightly the absorption below the gap. It turns out that for a chosen value of the absorption coefficient the shift of energy is proportional to H^2 / E^2. The H^2 behavior is illustrated in Fig. 10 .

Thus, for low E / H ratios we deal experimentally with energy shift proportional to E^2 / H^2 and magnetic-type motion, and for low H / E ratios the shift behaves as H^2 / E^2 and the motion is of electric type. Both the character of motion and the energy shift behavior indicate that the theory should treat both fields symmetrically, whereas the initial approach, as summarized in Eqs (4.1), (4.3) and (4.4), does not possess this feature.

The relativistic analogy again helps to find the key to our problem. It was used by Lax(22) to predict correctly the H^2 / E^2 behavior in the tunneling regime. To begin with, we notice that, according to the nonrelativistic motion of free electrons in crossed fields, the troublesome large values of E / H correspond to large transverse velocities $v_{tr} = cE/H$ (see Fig. 6 (a)). It becomes obvious that for sufficiently large values of E / H this description can not be valid, since it would lead to possibility of producing $v_{tr} > c$, which can not be. The remedy is simple : one may not use the nonrelativistic equation of motion (4. 5) for E/H comparable to unity or larger, and it is necessary to use the relativistic equation

$$\frac{d}{dt}\left[m_0\left(1-\frac{v^2}{c^2}\right)^{-1/2}\vec{v}\right] = -e\vec{E} - \frac{e}{c}(\vec{v}\times\vec{H}) \qquad (4.7)$$

This problem is treated in standard electrodynamics textbooks (ref. (19) p. 59). For small value of E / H this treatment reduces to the nonrelativistic limit, and the motion is of the type described in Fig. 6 (a). It turns out that as long as $cE / H < c$ it remains essentially the magnetic-type motion. However for $cE / H > c$ it becomes the electric-type motion : there are no oscillations and there is net motion and acceleration in the direction of electric field. *

---

* It should be mentioned that according to solutions of relativistic equation (4.7) $v_{tr} = cE/H$ only for $v_{tr} \ll c$, so that putting $cE/H > c$ does not lead to electron velocities larger than c.

This is illustrated in Fig. 6 (b). In other words, according to the relativistic mechanics there is a transition from magnetic- to electric-type motion as the electric field is increased, which is the only reasonable physical result.

In order to obtain this result quantum mechanically it is necessary to solve the Dirac equation for a free electron in crossed fields . However, since we are interested here in general features of the motion and not in details of the spin effects, it is enough to consider the Klein-Gordon equation for a spinless particle. The stationary Klein-Gordon equation for our problem reads

$$\left[\left(\vec{p}+\frac{e}{c}\vec{A}\right)^2-\left(\frac{\varepsilon}{c}+\frac{eV}{c}\right)^2+m_0^2c^2\right]\psi=0 \qquad (4.8)$$

Taking the vector and scalar potentials as in Eq. (4.1), and separating the variables according to Eq. (4.2), we obtain

$$\left[-\frac{\hbar^2}{2m_0}\frac{\partial^2}{\partial y^2}+\alpha y+\frac{m_0}{2}\left(\frac{e^2H^2}{m_0^2c^2}-\frac{e^2E^2}{m_0^2c^2}\right)y^2\right]\varphi(y)=\lambda\,\varphi(y) \qquad (4.9)$$

where α and λ are constants. The closer examination of Eq. (4.9) is very instructive. For $E = 0$ it becomes an eigenvalue problem for the harmonic oscillator. In this case we obtain the usual magnetic Landau levels, quantized in terms of $\hbar\omega_c$, in other words we deal with magnetic-type motion. As long as $E < H$, i.e. as long as the magnetic term in the parentheses is larger than the electric one, we still have a parabolic potential well, so that the eigenenergies are quantized and the motion remains of magnetic type. However for $E > H$ the coefficient in front of y^2 term becomes

negative, the motion is not bound any more, there is no quantization, and we deal with electric-type solutions (Weber functions).
For the free electrons the transition between the two cases occurs for $cE/H = c$, in agreement with the classical relativistic result.

Going back to our relativistic analogy one should expect similar behaviour of electrons in semiconductors. According to the correspondence indicated in Eq. (1.11) the transition between the two types of motion should occur for

$$c\frac{E}{H} = v_{max} = \left(\frac{\varepsilon_g}{2m_o^*}\right)^{1/2} \tag{4.10}$$

The above intuitive argument is confirmed by the rigorous treatment with the use of the two-band model (Zawadzki and Lax[23]), which we shall now present in brief. We start with the Hamiltonian for an electron in a periodic potential in the presence of uniform electric and magnetic fields

$$\mathcal{H} = \frac{1}{2m_o}\left(\vec{p} + \frac{e}{c}\vec{A}\right)^2 + e\vec{E}\vec{r} + V(\vec{r}) \tag{4.11}$$

Calculating the eigenvalue problem with the use of the Kohn-Luttinger functions and returning to the coordinate representation, the set of equations for the envelope functions is obtained in the following form

$$\sum_{n'}\left[\left(\frac{1}{2m_o}P^2 + e\vec{E}\vec{r} + \varepsilon_n - \varepsilon\right)\delta_{nn'} + \frac{\vec{p}_{nn'}}{m_o}\vec{P}\right]f_{n'}(\vec{r}) = 0 \tag{4.12}$$

Here $\vec{P} = \vec{p} + (e/c)\vec{A}$ is the kinetic momentum. For $H = 0$ and $E = 0$ the above set reduces to Eq. (1.5).

In principle it is now possible to decouple the bands in Eq. (4.12) by a suitable canonical transformation and arrive at the one-band effective mass approximation (4.1), which was our starting point. However, as demonstrated by Zak and Zawadzki[24], the decoupling procedure is valid only if, among other restrictions, the following condition is satisfied

$$\frac{m_o^* c^2 E^2}{2H^2} \ll \mathcal{E}_g \qquad (4.13)$$

This explains why for low values of E / H the one-band effective mass approximation is valid and agree with experiments.

Since we want to go beyond the restriction (4.13) we shall again consider the two-band model. Neglecting as before the free electron term we obtain the set of two equations

$$\begin{bmatrix} e\vec{E}\vec{r} - \mathcal{E} & \frac{1}{m_o}\vec{p}_{cv}\vec{P} \\ \frac{1}{m_o}\vec{p}_{vc}\vec{P} & e\vec{E}\vec{r} - \mathcal{E}_g - \mathcal{E} \end{bmatrix} \begin{bmatrix} f_1 \\ f_2 \end{bmatrix} = 0 \qquad (4.14)$$

With the previous choice of $\vec{E}$ and $\vec{H}$ this set can be solved by substitution.

The final equation is

$$\Big[-\frac{\hbar^2}{2m_o^*}\frac{\partial^2}{\partial y^2} - \alpha y + \frac{m_o^*}{2}\Big(\frac{e^2H^2}{m_o^{*2}c^2} - \frac{2e^2E^2}{m_o^*\mathcal{E}_g}\Big)y^2 + + \frac{3}{8}\frac{\hbar^2}{m_o^*}\Big(\frac{eE}{eEy - \mathcal{E}_g - \mathcal{E}}\Big)^2\Big]\varphi(y) = \lambda\varphi(y) \qquad (4.15)$$

where $\alpha = \hbar\omega_c^* k_x - eE(\mathcal{E}_g + 2\mathcal{E})/\mathcal{E}_g$ and $\lambda = \mathcal{E}(\mathcal{E}+\mathcal{E}_g)/\mathcal{E}_g - \hbar^2(k_x^2 + k_z^2)/2m_o^*$

The solution in terms of $\varphi(y)$ is

$$f_1(\vec{r}) = exp(ik_x x + ik_z z)|eEy - \varepsilon_g - \varepsilon|^{1/2}\varphi(y)$$

Eq. (4.14) is very similar to Eq. (4.9). The additional last term on the left-hand side is usually very small.* It is seen that we deal with two kinds of motion, according to the sign of the coefficient in front of y^2 term. The transition between magnetic- and electric-type motion occurs exactly for field intensities predicted by Eq. (4.10). It can also be observed that the criterion (4.13) means the electric field term $2e^2 E^2 / m_o^* \varepsilon_g$ to be negligible in comparison with the magnetic term $(e H / m_o^* c)^2$. Thus in this case solutions of one-band equation can be reduced to the solutions of one-band equation (4.1).

Fig. 11 shows an experimental observation of the transition from electric- to magnetic- type of behavior as the magnetic field is increased in the crossed field configuration. Again the optical absorption of Ge is plotted as a function of photon energies below the direct gap. For H = 0 the Franz-Keldysh exponential tail is observed, which is an indication of the electric-field-type of behavior. As the magnetic field is increased, we first observe the decrease of

----

* This term results from the fact that the two-band set (4.13) bears similarities to the Dirac equation rather than to the Klein-Gordon equation. With spin included it can be brought exactly to the form of the Dirac equation, and solved by strictly relativistic methods (Aronov and Picus (25)).

absorption below the gap, and then, for magnetic fields sufficiently high, the well-pronounced resonance in absorption below the gap. This is characteristic of magnetic-field-type quantized behavior. The experimental value of E / H ratio at the transition point agrees rather well with the theoretical prediction for the actual values of $\mathcal{E}_g$ and m_0^* in Ge.

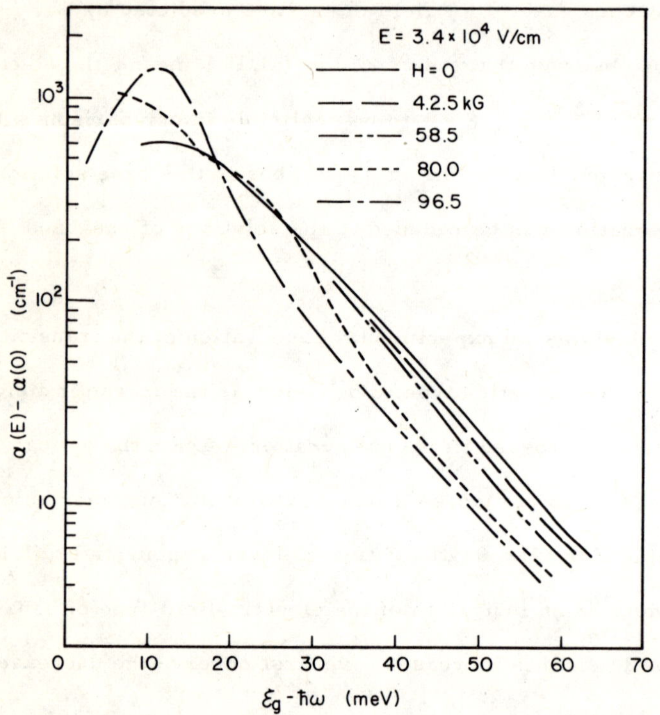

Fig. 11. Absorption coefficient below the direct gap of germanium due to an electric field E in the presence of a transverse magnetic field H . The continuous transition from electric-field-type interband photon-assisted tunneling (at zero magnetic field) to magnetic-field-type resonant interband absorption (at high magnetic fields) can be seen. After Reine et al.[21]

In conclusion we emphasize that the relativistic and the two-band approach of Eqs (4.9) and (4.14) treats both electric and magnetic fields symmetrically and predict correct behavior in both limiting cases of electric and magnetic types of motion [25] [26].

5. OTHER PHENOMENA

The analogy between the free relativistic electrons and electrons in narrow gap semiconductors is not restricted to the physical situations described above. For example this analogy is valid also for bound states. The celebrated hydrogen model for impurity atoms in semiconductors uses,. of course, the analogy to behavior of real hydrogen atoms in vacuum, as described by nonrelativistic quantum mechanics. However, for energy levels of impurity electrons lying deep in the energy gap it is again necessary to use the two-band model, which immediately leads to an analogue of the Dirac equation for the Coulomb potential (Keldysh[27], Yafet [28]). One can also mention here the analogy between the exciton-a bound state of electron and hole interacting via the Coulomb potential in a solid, and the positronium atom - the bound state of electron and positron coupled by the same interaction in vacuum.

Finally we want to indicate some very interesting possibilities offered by mixed crystals of semiconducting materials. By varying the chemical composition of a solid solution it is possible to change continuously the band parameters of the material. In particular, the mixed system of $Cd_x Hg_{1-x} Te$ allows to prepare the material with vanishing energy separation between interacting energy bands: $\varepsilon_g \simeq 0$ In this case the energy-momentum relation (1.8) becomes

$$\varepsilon = \frac{\hbar |p_{cv}|}{m_0} k \tag{5.1}$$

The linear dispersion relation has been observed experimentally in $Cd_{0.1}Hg_{0.9}Te$ (Galazka and Sosnowski[23]. In relativistic mechanics of free electrons the linear dependence of energy on momentum corresponds to the ultra-relativistic limit of velocities. In other words the above system allows to study some " ultrarelativistic " properties of electrons whose absolute energies are still well within the range of newtonian mechanics.

REFERENCES

1. E. O. Kane, J. Phys. Chem. Solids, 1, 249 (1957).

2. J. M. Luttinger and W. Kohn, Phys. Rev. 97 , 869 (1955).

3. P. A. M. Dirac, Proc. Roy. Soc. (London), A 126, 360 (1930).

4. J. Kolodziejczak, Acta Phys. Polonica, 20, 289 (1961)

5. W. Zawadzki and J. Kolodziejczak, Phys. Stat. Solidi, 6, 409(1964).

6. H. Arzelies, " Thermodynamique Relativiste et Quantique ", Gauthier-Villars, Paris 1968.

7. J. L. Synge, " The Relativistic Gas ", Noth Holland Publ. Comp., Amsterdam 1957.

8. I. I. Rabi, Zeits. f. Physik, 49, 507 (1928).

9. M. H. Johnson and B. A. Lippmann, Phys. Rev. 76, 828 (1949).

10. J. L. Hirshfield and J. M. Wachtel, Phys. Rev. Lett., 12, 533 (1964).

11. B. Lax, J. G. Mavroides, H. J. Zeiger, and R. J. Keyes, Phys. Rev. 122, 31 (1961).

12. R. L. Aggarwal, M. D. Zuteck, and B. Lax, Phys. Rev. Letters, 19, 236 (1967).

13. Y. Yafet, in Solid State Physics, Vol. 14 (F. Seitz and D. Turnbull, eds.), Academic Press, New York.

14. W. Zawadzki, Phys. Letters, 4 , 190 (1963).

15. Laura M. Roth, B. Lax, and S. Zwerdling, Phys. Rev. 114, 90(1959).

16. R. A. Isaacson, Phys. Rev. 169, 312 (1968).

17. P.A. Wolff, J. Phys. Chem. Solids, 25, 1057 (1964).

18. A. G. Aronov, Fiz. Tverd. Tela, 5, 552 (1963)/English transl.:

Soviet Phys.-Solid State, 5, 402 (1963).

19. L. D. Landau and E. M. Lifshitz, " Classical Theory of Fields ", Addison-Wesley, Inc., Reading, Massachusetts 1959.

20. Q. H. F. Vrehen, W. Zawadzki, and M. Reine, Phys. Rev. 158, p. 702 (1967).

21. M. Reine, Q. H. F. Vrehen, and B. Lax, Phys. Rev. 163, 726 (1967).

22. B. Lax, Proc. 7th Intern. Conf. Semicond., Dunod, Paris 1964, p. 253.

23. W. Zawadzki and B. Lax, Phys. Rev. Letters, 16, 1001 (1966).

24. J. Zak and W. Zawadzki, Phys. Rev. 145, 536 (1966).

25. A. G. Aronov and G. E. Picus, Zh. Eksperim, i Teor. Fiz. 51, 281 (1966); ibid. 51, 505 (1966). / English transalation. Soviet Phys. - JETP 24, 188 (1967); ibid. 24, 339 (1967).

26. Margaret H. Weiler, W. Zawadzki, and B. Lax, Phys. Rev. 163, 733 (1967).

27. L. V. Keldysh, Soviet Physics - JETP, 18, 253 (1964).

28. Y. Yafet, Bull, Amer. Phys. Soc. 11, 200 (1966).

29. R. R. Galazka and L. Sosnowski, Phys. Stat. Solidi, 20, 113 (1967).

IX

MODULATED REFLECTANCE IN THE STUDY OF BAND STRUCTURES

B. O. Seraphin
Michelson Laboratory, China Lake, California 93555

I. THE FAMILY OF MODULATED REFLECTANCE TECHNIQUES

The various modulated reflectance techniques developed during the last five years increasingly gain importance as a diagnostic tool in an exploration of the energy band structure of solids. All techniques modulate the reflectance of a sample through a periodic change of an external parameter such as an electric field, pressure, or temperature. Modulation of the reflected beam is detected synchronously and phase-sensitively, separated from the unmodulated background, and amplified.

Various modulation parameters define a whole family of modulated reflectance techniques such as electroreflectance, piezoreflectance, and thermoreflectance. The sample can be "prestressed" by one parameter and modulated by a second, establishing combinations of the three basic techniques. Modulation by rotation is possible, and a magnetic field can be added to all techniques.

Under similar conditions, modulated reflectance techniques better define the spectral contrast of structure than do static reflectance techniques. A rather featureless reflectance curve is replaced by a modulated trace rich in structure compressed into narrow regions of photon energy.

A response so sharply localized in energy suggests a correlation to discontinuities in parameters of the band structure. Consequently, band structure analysis in particular has taken an interest in the results of modulated reflectance techniques. Most of what follows will therefore be reviewed from this point of view. Other promising aspects, such as the apparent potential of electroreflectance in surface physics for example, take second place.

In a considerable effort during the last decade, energy band calculations have been used to establish the key features of the electronic structure of solids, to provide a basis for the qualitative interpretation of numerous experimental results, and to guide further investigations.[1]

As a matter of economy rather than principle, the most successful of these calculations rely on some input from properly interpreted experimental results. Although we can solve the Schroedinger equation for such complicated systems as crystals, the definition of a physically realistic crystal potential is complicated by exchange, correlation, and relativistic effects. Recent improvements in the computational tools permit us to treat these effects more rigorously, but the return is hardly worth the increase in effort. We can improve accuracy more efficiently by adding reliably interpreted experimental features to the procedure. A rough classification of the available methods can actually be obtained by considering the first-principle effort that enters the initial stage of computation before some parameter of the computational scheme is adjusted to experiment.[2,3]

The accuracy of such empirically adjusted calculations depends (to a varying degree) on the proper interpretation of experimental results. The collation of existing band models and new experimental evidence, which has been amazingly successful in the past, must be continued. The better the general features of a band structure are known, the more successful further analysis becomes.

Information from a number of experimental methods is available for the identification of the lowest band extrema. Away from these fundamental edges, however, the interpretation of reflectance measurements was the main source of experimental information. A bridge to band structure parameters is built by assigning structure in the reflectance spectrum to critical points - analytical singularities in the interband density-of-states function which determines the imaginary part ε_2 of the dielectric function.

As the field moves into a phase of further refinement, at least two limitations of the assignment procedure evolve. The first relates to the sole responsibility of critical points for structure in the static reflectance curve.[4] After all, the ε_2 curve derived from such a reflectance measurement represents the sum of all electronic transitions at a given photon energy, originating in large and widely separated areas of the Brillouin zone. Coarse structure in ε_2 reflects the spectral profile of this sum rather than contributions from a localized area near critical points. Their contributions probably appear as fine structure only, which may or may not be near the broad peaks with which they must not be exclusively associated.

This aspect strongly influences the interpretation of optical spectra. No longer can ε_2 be expected to assume the line shape predicted at critical points; the number of critical points is irrelevant to the number of coarse structural features in the experiment. The predominance of "peaks" in ε_2 rather than the predicted thresholds no longer calls for the unlikely arrangement of two critical points back-to-back at nearly the same energy. Instead, ε_2 reflects the energy contour of a summation over extended rather than localized regions in the Brillouin zone. Instead of adjusting high-symmetry transitions to the coarse structure in the experimental spectrum, a calculated band structure must reproduce ε_2 by adding up all transitions possible at a given photon energy, irrespective of their origin in momentum space.[5]

The importance of the critical-point concept as a bridge between experiment and theory is untouched by this development. True, we can no longer associate structure in ε_2 with critical points - but the need to know their transition energy, their class, and their location in the Brillouin zone still exists. Band structure calculations depend on the input from experiment, and critical points serve as the basis for the adjustment procedure. The type of experiment from which this input is to be gained, however, must change, and we must look for a mechanism that separates the massive background of noncritical transitions from the much smaller contribution of a critical point.

The second limitation of the assignment procedure results from the fact that the experimentalist leaves the theorist considerably short of information required for the full identification of a critical point. No direct procedure is available for the assignment of photon energies, at which structure is observed in static reflectance, to key transitions to which the parameters of the calculation are adjusted. Circumstantial evidence guides the theorist in unfolding the one-dimensional sequence of transition energies into the three dimensions of k-space. The more he knows about a band structure, the less ambiguous the assignment becomes. However, this interplay introduces the danger of a cyclic progression of errors. Agreement between theory and experiment may confirm the original assignment rather than reflect accuracy in the calculation of the band structure.[2,3]

In view of this situation, the development of modulated reflectance techniques was timely. We are acutely aware of the necessity to clearly identify, using criteria directly derived from experiment, the transitions to which semiempirical calculations are being fitted. We further realize that structure in ε_2 cannot be directly associated with critical points and that we must turn to a mechanism, such as the modulation of the reflectance, that apparently lifts a critical-point contribution from the noncritical background.

The separation of critical points from noncritical background in modulated reflectance, versus the observation of both in the static version, is first suggested by entirely different line shapes. On a qualitative basis, the modulated response can be approximated by the differential of the predicted ε_2 line shapes at critical points, whereas this approximation calls for considerable imagination in the case of static reflectance. The spectral width of the modulated response is typically 20 to 50 times smaller than the width of reflectance structure. And - last but by far not least - no response is obtained over large regions of the spectrum, suggesting that only limited areas of the Brillouin zone are affected by the modulation.

The separation of critical points and background is probably accomplished by utilizing the analytic singularities of the joint density-of-states function at critical points. A periodic change in the band structure generated by the modulation parameter is most effectively seen at these singularities. Tuning the detection to the phase and frequency of the modulation amplifies the synchronously modulated contribution from the slope discontinuity of the joint density-of-states function, rejecting the unmodulated background from areas where this function is smooth.

The high spectral resolution resulting from this separation must not be overemphasized, however. It is of rather minor significance that structure can now be located on the spectral scale to 0.001 eV rather than to 0.1 eV. A far greater promise rests with the increase in diagnostic potential inherently provided by the isolation of the critical-point contribution. The higher sensitivity to external perturbation, provided by the modulation, will probably lead to the identification of critical points without assistance from existing band structure calculations. Static reflectance, originating in extended and widely separated areas of the Brillouin zone, responds more reluctantly to diagnostic perturbation and only to the extent that the response characteristics of widely scattered contributions do not average out in destructive interference. A vectorial modulation parameter will automatically perturb the optical isotropy of a cubic crystal. Anisotropies are observed as functions of the angle between this modulation vector and the axes of the crystal or the polarization of the incident light, which can be explained without further assumption by one, and only one, location of the correlated critical point.

A variety of external parameters can be used for the modulation of the band structure. The way of "forming the derivative" on the joint density-of-states function differs from technique to technique. No attempt has so far been made to delineate the similarities and differences of these techniques, nor do we have more than a qualitative understanding of their basic mechanisms.

The potential of the various techniques with respect to an analysis of the band structure varies according to the character of the modulation parameter, vectorial or scalar, and its impact on the band structure or the optical transition probability.

The vectorial character of the modulation in electroreflectance and piezoreflectance lowers the symmetry of the sample crystal.[6,7] In a cubic crystal, for example, the dielectric tensor and, consequently, all optical properties are isotropic. Under the action of the electric field (or the uniaxial strain), the cubic symmetry is destroyed, a preferred direction is established, and the tensorial character of the dielectric function becomes important. Orientation and polarization effects are observed that can be interpreted in terms of the k-space symmetry of the critical point.[8,9] The symmetry information to be obtained through matrix element effects is particularly important for critical points off the center of the Brillouin zone. The necessary tensor analysis is familiar from piezo-optical studies.[10]

In both techniques the symmetry-breaking effect of the uniaxial modulation introduces tensorial anisotropies of the response that enter through the matrix element of the optical transition probability. In electroreflectance the electric field adds anisotropies of nontensorial character that enter through the interband part of the transition probability. A strong sensitivity of this interband part to the direction of the electric field results, which increases the diagnostic value of electroreflectance. The basic mechanism centers the action of the electric field at the critical points, leaving noncritical parts of the band structure unaffected. This results in extremely sharp spectra.[11-15]

The basic mechanism of piezoreflectance is quite different, however, and could account for broad structure cut out from a large background, as usually observed.[16-19] The wavelength of the strain field is long compared to the wavelength of the light or the atomic dimensions of the crystal. This results in a shift of the band structure as a whole, by a different amount or direction for different subbands, so that degeneracies are lifted. In a sense, the piezo-optical effect can therefore be described as taking the derivative of ε_2, properly superimposing the various derivatives at degenerate points. This effect operates throughout the Brillouin zone, and critical points are lifted from the background mainly because they occur at slope discontinuities of the density-of-states function. This establishes a remarkable contrast to electroreflectance. In piezoreflectance, critical points dominate in regions of large density of state, i.e., large interband reduced mass. Electroreflectance, breaking the translational symmetry of the crystal along the field direction, centers changes in the density of states at the critical points proper, in a manner that makes critical points with small reduced interband mass emerge more clearly.[20]

The effect of a temperature modulation on the band structure as a whole places thermoreflectance and piezoreflectance in the same class. The scalar character of the modulation parameter, however, gives thermoreflectance a special position with respect to the vectorial techniques that can expect guidance in the identification of a critical point through anisotropies of the response.

Thermoreflectance lacks this potential. A simplicity of the basic mechanism seems to compensate for this handicap, however.[21-24] In principle, the number of possible effects of temperature modulation on the band structure is large, particularly in semiconductors. A few examples of successful analysis demonstrate, however, that one or the other of these effects seems to dominate.[25] Good agreement between experiment and theory is obtained by assuming that the temperature change simply shifts the spectral position of an interband edge. In other cases, an assumed change in the electron-phonon interaction reproduces the observed line shapes. This in turn makes thermoreflectance applicable to a study of electron-phonon interactions such as indirect, phonon-assisted transitions.[26]

Magneto-optics has made considerable contributions to our knowledge of the parameters of the lowest extrema of valence and conduction band. In principle, magnetoreflectance should be applicable at photon energies above the fundamental edge where absorption measurements fail. Scattering is strong for these photon energies, however, and the resulting lifetime broadening wipes out the quantum effects produced by the magnetic field. Magnetoreflectance studies in the past have been restricted to materials with small, zero, or inverted energy gaps such as InSb, Bi, graphite, and HgTe.

Modulation of the Landau-level structure improves sensitivity and resolution sufficiently to observe an oscillatory pattern in magnetoreflectance that extends far beyond the fundamental absorption edge. If, in addition to the static magnetic field, the sample is subjected to modulation by an electric field, stress, or temperature, the reflectance responds in a spectrum that is characteristic for transitions between the Landau ladders of valence and the conduction band.[27-33]

Modulated reflectance has considerably expanded the range of magneto-optical experiments beyond the photon energies of the fundamental edge and has produced results in a number of new materials. We can expect modulated magnetoreflectance to extend the established potential of magneto-optical studies to a wider spectrum of energies and phenomena.

We have tried to sketch in this introduction the position of modulated reflectance with respect to band structure analysis. In the following part we will show on the example of electroreflectance how the information available to band structure analysis from optical studies can be upgraded in precisely the two aspects on which the reexamination of the assignment procedures centers. The fundamentals of a symmetry analysis of electroreflectance spectra will be described that rests on the separation of critical from noncritical contributions and utilizes the symmetry-breaking effect of the electric field. Although presently in an early stage only, directional experiments rotating the field direction with respect to the crystal frame will eventually give to the analysis of higher interband edges a degree of certainty that presently, mainly through cyclotron resonance and magneto-optics, can be achieved for the lowest band extrema only.

II. SYMMETRY ANALYSIS OF ELECTROREFLECTANCE SPECTRA

a. Scope, Intention, and Problems of the Analysis

We have pointed out in the first section of this lecture, how semi-empirical band models crucially depend on an accurate assignment of structure in optical spectra to critical points in the band structure, as characterized by transition energy, symmetry class and location in the Brillouin zone.[2,3] Of these three criteria of identification, only the first can directly be read out from static reflectance spectra. Consequently, band structure analysis faces the problem of unfolding the observed one-dimensional sequence of transition energies into the three dimensions of the Brillouin zone, and of assigning a symmetry class. In this process, the theorist is presently guided mainly by circumstantial evidence and a close cooperation with existing band models. The more he already knows, the more reliable the assignment becomes.

As the source of experimental information in this assignment procedure, modulated reflectance is superior to static reflectance for mainly two reasons. First, the modulated response correlates to localized rather than extended regions of the Brillouin zone.[4] Second, modulation by an electric field or a stress establishes a preferred direction and lowers the symmetry of the sample crystal; consequently, anisotropies of the reflectance response are observed as the modulation vector rotates relative to the crystal frame.[34-37]

Phillips[6] first recognized the diagnostic value of these anisotropies in an analysis of the band structure. Based on a generalized Franz-Keldysh theory independently developed by Aspnes and coworkers,[38-39] Phillips outlines the diagnostic potential of an assignment procedure based on modulated reflectance spectra. The symmetry-breaking effect of the modulating electric field should result in anisotropies of the electroreflectance response on essentially two different levels. Nontensorial anisotropies should enter through the interband part of the optical transition probability (the electro-optic effect), complemented by tensorial anisotropies expected to arise from the matrix element. Taking account of the two types of anisotropies, an interpretation of properly conducted electroreflectance experiments should be consistent only with certain locations and classes of critical points.

The proposed analysis received a strong impulse when a systematic study revealed distinct angular anisotropies of the electroreflectance response.[40,41] The results demonstrate that characteristic signatures can be extracted from proper directional experiments, that are indicative of location and symmetry features of a critical point correlated to electroreflectance structure.[42]

In the past, various attempts have been made to retrieve information beyond the mere spectral location of structure in electroreflectance spectra. They can all be described as interpretations of the spectral line shape. Two approaches in particular are typical for these efforts. One consists of a one-to-one correlation of the experimental spectrum with a matching member of a family of spectral functions $\Delta R/R(\omega)$ derived from an assumed mechanism such as the Franz-Keldysh effect[43-45] or exciton decay.[46] A second approach uses a systematic variation of the electric potential across the external terminals of the system in order to produce a similar variation of the effective modulating field.[47]

Both approaches are based on the expectation that a correlation between electrical and optical modulation can be established and that quantitative interpretation of the line shape in terms of a certain value of the electric field F is meaningful. The results are not convincing in the sense that they preclude alternate assignments. Severe handicaps enter through the drastic dependence of the line shapes upon surface conditions,[48,49] the nonlinearities of the basic mechanism[38,39] and the inhomogeneities of the modulating field in present experimental configurations.[50-52] At the present state of experiment and theory it is fair to state that line shape interpretations as one-to-one correlations of experimental and calculated spectra have failed to link up unambiguously.

The symmetry analysis developed by Bottka and coworkers,[42] executing Phillips original suggestion,[6] attacks the problem from a different angle. Spectra recorded along trajectories of various orientation of the electric field with respect to the crystal frame are interpreted in terms reminiscent of the analysis of cyclotron resonance spectra. Anisotropies of the cyclotron resonance frequency with respect to the direction of the magnetic field are consistent with a specific location of the correlated band minimum in the Brillouin zone, giving the appropriate effective mass tensor components. In essence, the diagnostic value of cyclotron resonance is based on the directional sampling of the effective mass by the magnetic field.

In an analogous manner, the electro-optical effect, aside from any particular mechanism, responds to the loss of translational symmetry of the crystal along the field direction. The correlation of electro-optical effect and interband effective mass is probably very complex and by far not as straightforward as in cyclotron resonance. Irrespective of the actual mechanism, however, its symmetry elements must appear in an angular scan and permit the interpretation of angular anisotropies in terms of k-space location and topographical features of the interband energy surface at critical points.

b. Anisotropies of the Electro-Optical Effect

In view of the difficulties of identifying structure of electroreflectance from line shape interpretation, we will now investigate the possibility of extracting "signatures" from a spectrum that are characteristic of the symmetry features of the correlated critical point. We will treat the problem on two successive levels. We will first consider the effect of an electric field on a singular critical point, either located at the center of the Brillouin zone, or in an off-center position, one of the many symmetry-equivalent branches. Except for a few basic assumptions, we will not prescribe a particular mechanism for the electro-optical effect on such a singular critical point. Anisotropies of the response are expected that are nontensorial in character and are particularly pronounced at saddle-points of the interband energy surface.

Summation of the effect over the equivalent valleys considerably reduces the influence of the orientation of the electric field, causing a loss in angular contrast.[53]

This is not of critical concern, however, since such a summation refers to the electro-optical effect "in the dark." As we sample the perturbed state of the dielectric tensor by reflecting polarized light off the crystal, we must consider the polarization dependent oscillator strength of the various valleys. This results in tensorial anisotropies of the optical sampling procedure that are superimposed upon the nontensorial anisotropies existent in the dark without sampling. Between these two levels of diagnostic anisotropies, nontensorial in the electro-optical effect and tensorial in the optical sampling, we expect to retrieve symmetry information on the location and character of a critical point.

We first discuss the electro-optical effect at a single, nondegenerate critical point located at the end-point of a vector $\bar{k}_i$ in the Brillouin zone. If $\bar{k}_i = 0$, the critical point occurs at the center of the Brillouin zone and no further summation is required. If $\bar{k}_i \neq 0$, the special point under consideration is a member of a symmetry-degenerate set of equivalent points located at the ends of a vector star produced by the lattice symmetry. The total electro-optical effect at such an off-center multivalley critical point consists of the sum of all the partial effects at each equivalent branch, one of which we first discuss separately.

There is no need to specify a particular mechanism for this electro-optical effect D_i. Experiment and the basic nature of the effect suggest the following three assumptions, on which the symmetry analysis will be built:

1. Among other parameters, the effect depends upon magnitude and direction of the electric field $\bar{F}$, as well as upon the reduced interband effective mass μ_i in field direction

$$\frac{1}{\mu_i} = \frac{1}{m^*_{ei}} + \frac{1}{m^*_{hi}} = \frac{1}{\hbar^2}\frac{\partial^2 E_{n'}}{\partial k_i{}^2} - \frac{1}{\hbar^2}\frac{\partial^2 E_n}{\partial k_i{}^2} \quad ; \quad i = x,y,z \quad , \tag{1}$$

where m^*_e and m^*_h are the effective masses of electrons and holes, respectively. Mass values of either sign are admitted in different directions, with a negative mass corresponding to a maximum of the band separation $(E_{n'} - E_n)$ at the critical point.[38,39]

In its first part, this assumption reflects experimental fact. Its second part rests on the perturbation of translational symmetry by the electric field along its direction. It further postulates that the topography of the interband energy surface, as described by the interband mass, can be explored by directional experiments.

2. Again following experimental fact, it is assumed that the effect cannot distinguish between equal fields of opposite direction.

3. At a saddle point, the effect disappears along $1/\mu_i = 0$. If this line or surface separates sectors of opposite sign of the interband mass, it is assumed that the electro-optical effect in the two sectors is described by two different branches of a functional relation.

This assumption, probably disputable, is suggested by a minimum of the band separation along a direction of positive mass, and a maximum along a negative-mass direction - features entering the electro-optical effect on a fundamental level.

These three assumptions permit us to draw a few conclusions based on symmetry arguments. We will define our terms as we proceed.

We describe the field direction by two angles Θ and Φ, which are counted from the z- and x-axes, respectively, of a Carthesian coordinate system attached in its origin to the tip of the $\bar{k}_i$-vector with the z-axis parallel to $\bar{k}_i$. We can then write the first assumption

$$D_i = D_i \left[\bar{F}(\Theta,\Phi) \quad ; \quad \mu_i(\Theta,\Phi) \quad ; \quad \ldots\ldots \right] \tag{2}$$

If $\bar{k}_i$ is aligned along a direction of high symmetry such as Δ or Λ, cylinder symmetry around $\bar{k}_i$ can be assumed and Eq. (2) depends upon Θ only.

Directions of lower symmetry such as Σ do not necessarily have rotational symmetry around $\bar{k}_i$ and dependence upon both Θ and Φ must be retained.

The second assumption requires

$$D_i(\bar{F},\mu_i) = D_i(-\bar{F},\mu_i) \quad , \tag{3}$$

and it follows

$$\left(\frac{\partial D_i}{\partial \Theta}\right)_{\Theta = 0,\frac{\Pi}{2},\Pi} = 0 \tag{4}$$

for all values of Φ.

Parabolic topography at the critical point dictates the same sign for the interband mass in all directions.

At a saddle point, however, opposite signs are assumed in sections that are separated by "cones" on which $1/\mu_i = 0$.

Accepting the third assumption, we expect two different branches of an electro-optical function in these sections of opposite sign. We can define

$$D_i(\Theta = 0,\Pi;\Phi) = D_{iL}$$

$$D_i(\Theta = \tfrac{\Pi}{2};\Phi_1) = D_{iT_1} \quad ; \quad D_i(\Theta = \tfrac{\Pi}{2};\Phi_2) = D_{iT_2} \tag{5}$$

which are assumed with the electric field aligned either along $\bar{k}_i$ or in two special directions in a plane perpendicular to $\bar{k}_i$. For a parabolic critical point, all three functions can belong to the same branch. At a saddle point of rotational symmetry, $D_{iT_1} = D_{iT_2} = D_{iT}$.

We can similarly define μ_{iL} and μ_{iT_1}, μ_{iT_2}. Three components of the interband-mass tensor are required along directions of lower symmetry such as Σ.

TRAJECTORY I

Critical point along	Equivalence for field along ⟨100⟩	Equivalence for field along ⟨110⟩	Oscillator strengths $n_i(\xi)$, $p_i(\xi)$
Δ	$D_1 = D_2 = D_T$ $D_3 = D_L$ $D_i' = 0$	$D_1 = D_T$ $D_2 = D_3$ $D_1' = 0$ $D_2' = -D_3'$	$p_1 = 0$ $p_2 = p_3 = 1$ $n_1 = 1$ $n_2 = 1 - n_3 = \sin^2\xi$
Λ	$D_1 = D_2 = D_3 = D_4$ $D_1' = D_2' = -D_3' = -D_4'$	$D_1 = D_2$ $D_3 = D_4 = D_T$ $D_i' = 0$	$p_1 = p_2 = p_3 = p_4 = \frac{2}{3}$ $n_1 = n_2 = \frac{2}{3}(1 + \cos\xi\cdot\sin\xi)$ $n_3 = n_4 = \frac{2}{3}(1 - \cos\xi\cdot\sin\xi)$
Σ	$D_1 = D_2 = D_{T_1}$ $D_3 = D_4 = D_5 = D_6$ $D_1' = D_2' = D_3' = D_4' = 0$ $D_5' = -D_6'$	$D_1 = D_2 = D_3 = D_4$ $D_5 = D_L$ $D_6 = D_{T_2}$ $D_1' = D_2' = -D_3' = -D_4'$ $D_5' = D_6' = 0$	$p_1 = p_2 = p_3 = p_4 = \frac{1}{2}$ $p_5 = p_6 = 1$ $n_1 = n_2 = \frac{1}{2}(1 + \sin^2\xi)$ $n_3 = n_4 = \frac{1}{2}(1 + \cos^2\xi)$ $n_5 = \frac{1}{2}(1 + 2\sin\xi\cdot\cos\xi)$ $n_6 = \frac{1}{2}(1 - 2\sin\xi\cdot\cos\xi)$

Table I :

Elements of the electro-optical response Eq. (9) as a function of the field direction along trajectory I (Fig. 1), as described by the trajectory coordinate ξ. For critical points along the principal directions Δ, Λ, and Σ, the center columns list equivalence relations for the D_i and their derivatives $\partial D_i/\partial\xi$ at the given directions of the electric field. The oscillator strengths $n_i(\xi)$ and $p_i(\xi)$ for the two directions of polarization $\bar{e}_\perp$ and $\bar{e}_\parallel$ are listed on the right as a function of ξ.

TABLE II

TRAJECTORY II

Critical point along	Equivalence for field along ⟨100⟩	⟨111⟩	⟨110⟩	Oscillator strengths $n_i(\xi)$, $p_i(\xi)$
Δ	$D_1 = D_2 = D_T$ $D_3 = D_L$ $D_i' = 0$	$D_1 = D_2 = D_3$ $D_1' = D_2' = -D_3'$	$D_1 = D_2$ $D_3 = D_T$ $D_i' = 0$	$p_1 = p_2 = \frac{1}{2}$ $p_3 = 1$ $n_1 = n_2 = \frac{1}{2}(1 + \sin^2\xi)$ $n_3 = \cos^2\xi$
Λ	$D_1 = D_2 = D_3 = D_4$ $D_1' = D_3'$ $D_2' = -D_4'$	$D_1 = D_3 = D_4$ $D_2 = D_L$ $D_2' = 0$ $D_1' = D_3' = -D_4'$	$D_1 = D_3 = D_T$ $D_2 = D_4$ $D_1' = D_3' = 0$ $D_2' = -D_4'$	$p_1 = p_3 = \frac{1}{3}$ $p_2 = p_4 = 1$ $n_1 = n_3 = \cos^2\xi + \frac{2}{3}\sin^2\xi$ $n_2 = \frac{1}{3}(1 + \sin^2\xi) + \frac{2\sqrt{2}}{3}\cos\xi\cdot\sin\xi$ $n_3 = \frac{1}{3}(1 + \sin^2\xi) - \frac{2\sqrt{2}}{3}\cos\xi\cdot\sin\xi$
Σ	$D_1 = D_2 = D_{T_1}$ $D_3 = D_4 = D_5 = D_6$ $D_1' = D_2' = 0$ $D_4' = D_5' = -D_3' = -D_6'$	$D_1 = D_3 = D_6 = D_T$* $D_2 = D_4 = D_5$ $D_3' = D_6' = 0$ * $T_1 < T < T_2$	$D_1 = D_{T_2}$ $D_2 = D_L$ $D_3 = D_4 = D_5 = D_6$ $D_1' = D_2' = 0$ $D_3' = D_6' = -D_4' = -D_5'$	$p_1 = 0$; $p_2 = 1$ $p_3 = p_4 = p_5 = p_6 = \frac{3}{4}$ $n_1 = 1$; $n_2 = \sin^2\xi$ $n_3 = n_6 = \frac{1}{4}(2+\cos^2\xi-2\sqrt{2}\cdot\cos\xi\cdot\sin\xi)$ $n_4 = n_5 = \frac{1}{4}(2+\cos^2\xi+2\sqrt{2}\cdot\cos\xi\cdot\sin\xi)$

Table II :

Elements of the electro-optical response Eq. (9) as a function of the field direction along trajectory II (Fig. 1), as described by the trajectory coordinate ξ. For critical points along the principal directions Δ, Λ, and Σ, the center columns list equivalence relations for the D_i and their derivatives $\partial D_i/\partial\xi$ at the given directions of the electric field. The oscillator strengths $n_i(\xi)$ and $p_i(\xi)$ for the two directions of polarization $\bar{e}_\perp$ and $\bar{e}_\parallel$ are listed on the right as a function of ξ.

TABLE III

TRAJECTORY III

Critical point along	Equivalence for field along ⟨100⟩	⟨111⟩	⟨110⟩	Oscillator strengths $n_i(\xi)$, $p_i(\xi)$
Δ	$D_1 = D_2 = D_T$ $D_3 = D_L$ $D_i' = 0$	$D_1 = D_2 = D_3$ $D_1' = D_2' = -D_3'$	$D_1 = D_2$ $D_3 = D_T$ $D_i' = 0$	$p_1 = p_2 = \frac{1}{2}(1 + \cos^2\xi)$ $p_3 = \sin^2\xi$ $n_1 = n_2 = \frac{1}{2}(1 + \sin^2\xi)$ $n_3 = \cos^2\xi$
Λ	$D_1 = D_2 = D_3 = D_4$ $D_1' = D_3'$ $D_2' = -D_4'$	$D_1 = D_3 = D_4$ $D_2 = D_L$ $D_2' = 0$ $D_1' = D_3' = -D_4'$	$D_1 = D_3 = D_T$ $D_2 = D_4$ $D_1' = D_3' = 0$ $D_2' = -D_4'$	$p_1 = p_3 = \sin^2\xi + \frac{2}{3}\cos^2\xi$ $p_2 = \frac{1}{3}(1 + \cos^2\xi - 2\sqrt{2}\cdot\cos\xi\cdot\sin\xi)$ $p_4 = \frac{1}{3}(1 + \cos^2\xi + 2\sqrt{2}\cdot\cos\xi\cdot\sin\xi)$ $n_1 = n_3 = \cos^2\xi + \frac{2}{3}\sin^2\xi$ $n_2 = \frac{1}{3}(1 + \sin^2\xi + 2\sqrt{2}\cdot\cos\xi\cdot\sin\xi)$ $n_4 = \frac{1}{3}(1 + \sin^2\xi - 2\sqrt{2}\cdot\cos\xi\cdot\sin\xi)$
Σ •	$D_1 = L_2 = D_{T_1}$ $D_3 = D_4 = D_5 = D_6$ $D_1' = D_2' = 0$ $D_4' = D_5' = -D_3' = -D_6'$	$D_1 = D_3 = D_6 = D_T$ * $D_2 = D_4 = D_5$ $D_3' = D_6' = 0$ * $T_1 < T < T_2$	$D_1 = D_{T_2}$ $D_2 = D_L$ $D_3 = D_4 = D_5 = D_6$ $D_1' = D_2' = 0$ $D_3' = D_6' = -D_4' = -D_5'$	$p_1 = 1;\quad p_2 = \cos^2\xi$ $p_3 = p_6 = \frac{1}{4}(2+\sin^2\xi+2\sqrt{2}\cdot\cos\xi\cdot\sin\xi)$ $p_4 = p_5 = \frac{1}{4}(2+\sin^2\xi-2\sqrt{2}\cdot\cos\xi\cdot\sin\xi)$ $n_1 = 1;\quad n_2 = \sin^2\xi$ $n_3 = n_6 = \frac{1}{4}(2+\cos^2\xi-2\sqrt{2}\cdot\cos\xi\cdot\sin\xi)$ $n_4 = n_5 = \frac{1}{4}(2+\cos^2\xi+2\sqrt{2}\cdot\cos\xi\cdot\sin\xi)$

Table III:

Elements of the electro-optical response Eq. (9) as a function of the field direction along trajectory III (Fig. 1), as described by the trajectory coordinate ξ. For critical points along the principal directions Δ, Λ, and Σ, the center columns list equivalence relations for the D_i and their derivatives $\partial D_i/\partial\xi$ at the given directions of the electric field. The oscillator strengths $n_i(\xi)$ and $p_i(\xi)$ for the two directions of polarization $\bar{e}_\perp$ and $\bar{e}_\parallel$ are listed on the right as a function of ξ.

For high-symmetry directions, we can write the ratio $\mu_{iT}/\mu_{iL} = c$. Its magnitude at a saddle point characterizes the topography of the interband energy surface by determining the angle under which the cone $1/\mu_i = 0$ opens around the principal direction $\bar{k}_i$.

This cone angle, determined by the ratio c of transverse to longitudinal interband mass introduces a major degree of freedom into the analysis. Different angular line shapes of the electro-optical response obtain for the two cases $c > 1$ and $c < 1$.

Band models sometimes provide guidelines for this ratio along directions of high symmetry. An analysis ab initio must forego these suggestions, however, and consider both possibilities separately.

At a multivalley critical point located off the center of the Brillouin zone, the total response is the sum of the partial effects at the separate branches, aligned differently with respect to the electric field. Bottka and Rössler have developed the angular algebra of this superposition[53] and have calculated the spectral line shapes for the principal locations of critical points.

We will bypass this stage, where the sample is still in the dark, and immediately proceed to the optical sampling of a multivalley star for which an electric field has broken the cubic symmetry of the dielectric function. We will investigate the effect of an electric field on off-center critical points along the three principal directions, Δ, Λ, and Σ. If the field direction coincides with one of these three principal directions, symmetry establishes relations of equivalence among the various D_i's, that are listed in Tables I, II and III. They simply result from the equality of the electro-optical effect for equal angles between $\bar{F}$ and $\bar{k}_i$, the two principal directions in a plane perpendicular to $\bar{k}_i$ for a Σ point.

These equivalence relations simply reflect the symmetry of the cubic lattice. Although they are an essential part of the complete symmetry analysis, alone they do not reveal more than the structural features of the sample crystal. Only in a combination with the superimposed anisotropies of the sampling process do they deliver information on the location of a critical point.

In listing the equivalent branches of critical points along the principal directions Δ, Λ, and Σ of a cubic lattice, we have used the same subscript for points at $\bar{k}_i$ and $-\bar{k}_i$, and only points on one side of the (001) plane are noted. For the Σ direction, two transverse components D_{iT_1} and D_{iT_2} are listed. Throughout the following discussion, we use the notation:

	1	[100]	
Δ	2	[010]	
	3	[001]	
	1	[111]	
Λ	2	[$\bar{1}$11]	
	3	[$\bar{1}\bar{1}$1]	
	4	[1$\bar{1}$1]	(6)
	1	[110]	
	2	[1$\bar{1}$0]	
Σ	3	[101]	
	4	[$\bar{1}$01]	
	5	[011]	
	6	[0$\bar{1}$1]	

c. Selection Rules

The field-perturbed dielectric function described by the D_i is now probed by reflecting polarized light off the sample crystal. This results in tensorial anisotropies that are superimposed on the nontensorial anisotropies of the electro-optical effect.

Depending upon specific selection rules, the incident photon is either accepted or rejected by a pair of initial and final states near a critical point. Among other parameters such as the character of this initial and final state, the selection rule depends upon the direction of the photon polarization with respect to the principal direction of the critical point. Consequently, at a multivalley point the photon can be accepted by some and rejected by other equivalent branches, depending upon the relative alignment of the particular star vector with the direction of polarization. In the unperturbed state, the various directions average out to a polarization-independent isotropy of the optical properties in a cubic crystal. Under the directional perturbation of the electric field, however, the various branches are no longer equivalent. Depending upon their orientation with respect to field and polarization, they are sampled with different weights by the incident photon, resulting in anisotropies of the reflectance response.

The effect D_i at the i-th branch therefore contributes to the total effect a fraction $f_i D_i$, where f_i represents an oscillator strength

$$f_i \sim \cos^2\theta \tag{7}$$

depending upon the angle θ between an allowed direction and the polarization vector $\bar{e}$. We register Eq. (7) as the first assumption on the sampling process.

Relating this allowed direction to the star vector $\bar{k}_i$ requires a second and more restrictive assumption. In writing

$$f_i \sim 1 - (\bar{e}\cdot\bar{k}_i)^2 \tag{8}$$

we identify the allowed direction with the direction perpendicular to the star vector $\bar{k}_i$. This second assumption holds for transitions between states of p-symmetry and states of s-symmetry, a case of common occurrence in semiconductors.

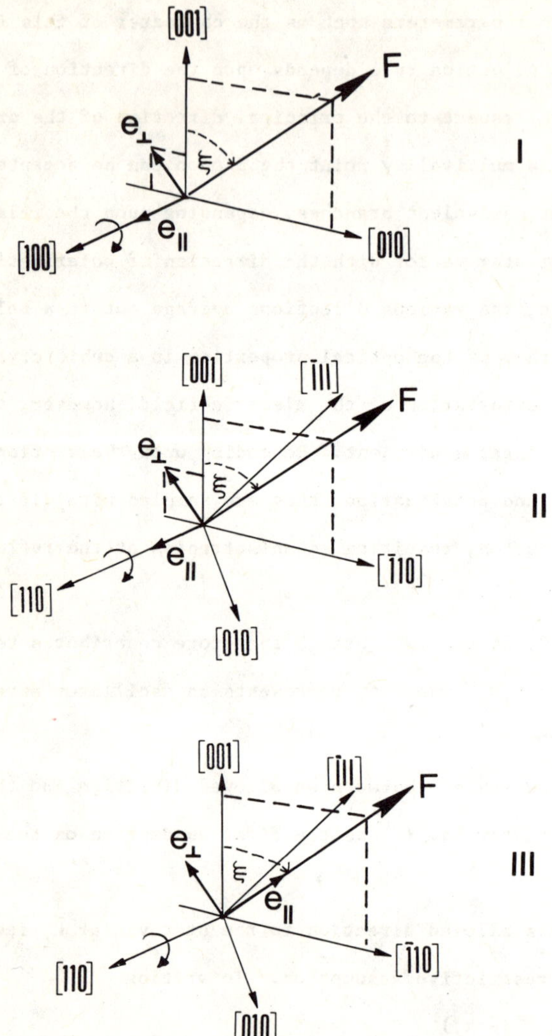

Fig. 1. Three possible trajectories of the field vector F and the polarization vectors $e_{\parallel}$, $e_{\perp}$ on which a symmetry analysis can be based. The vector triplet is fixed in its relative position as the trajectory coordinate ξ varies through its range.

As a third assumption - again most disputable - we request that the selection rule Eq. (8) be independent of the presence and strength of the electric field. This assumes the electric field to break the symmetry exclusively on the level of the electro-optical effect, leaving the optical sampling process unaffected. This assumption can be accepted in a first-order approximation only. It must be modified in a more refined treatment. However, in the final section of this lecture it will be shown that this assumption holds well if the symmetry information is retrieved from the proper experimental signature.

Adding the partial contributions of the various branches, we now obtain

$$\Delta\varepsilon_2(\bar{F},\bar{e}) \sim \sum_i f_i D_i \tag{9}$$

We will next investigate this expression for several trajectories of the electric field $\bar{F}$ and the polarization vector $\bar{e}$.

d. Trajectories of Field and Polarization

As an incomplete sample of numerous other possibilities, we follow $\Delta\varepsilon_2(\bar{F},\bar{e})$ along three trajectories of the field vector, defining components of the polarization parallel and perpendicular to a preferred direction. The three selected trajectories are of experimental significance and form a basic set for the investigation of others. They are plotted in Fig. 1 as functions of a trajectory coordinate ξ, together with the definition of parallel and perpendicular components of the polarization vector. In the first two cases these components are defined with respect to the axes of rotation generating the trajectory, and in the third case with respect to the field direction.

Note that in the first two trajectories the sampling light is incident along the field direction, whereas in the third case the field vector is perpendicular to the direction of light incidence. The three selected

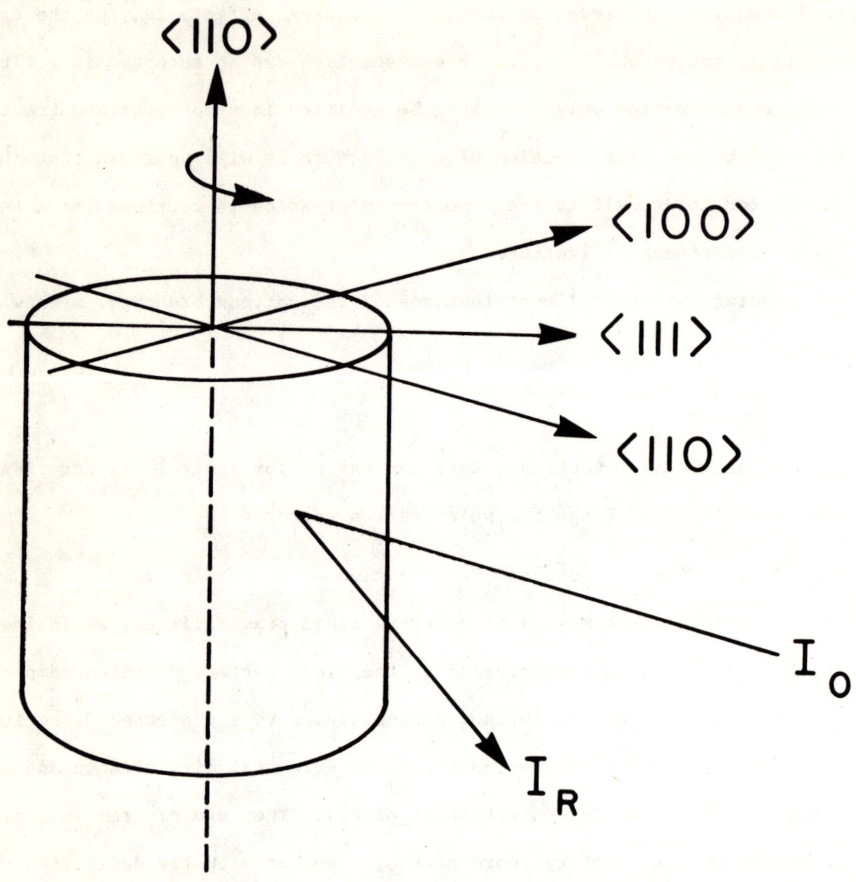

Fig. 2. Cylindrical Ge sample geometry generating trajectory II upon rotation, with the electric field oriented radially.

trajectories therefore correlate to an experimental configuration of particular simplicity. Trajectories I and II are realized in longitudinal electroreflectance from a sample of cylinder symmetry (Fig. 2) rotated around [001] and [110], respectively. Trajectory III represents transverse electroreflectance from a (110) surface.

For a location of the critical point along the three principal directions, Δ, Λ, and Σ, the angular functions $f_i^{\parallel} = p_i(\xi)$ and $f_i^{\perp} = n_i(\xi)$ of the oscillator strength for the parallel and perpendicular components of the incident light are listed for the three trajectories in Tables I, II, and III. Together with the symmetry-derived equivalence relations for the D_i and their derivatives, the line shape of $\Delta\varepsilon_2(\bar{F},\bar{e})$ can qualitatively be synthesized along a given trajectory and compared with experiment.

e. Symmetry Analysis of a Spectrum

In order to record an electroreflectance spectrum that can be symmetry-analyzed on the basis of the information contained in Tables I, II, and III, the experimentalist must select his mode of operation in two ways.

He must first decide on the experimental configuration as described by one of the three trajectories.

He must then - and this is the more important decision - select the experimental quantity to be recorded along the trajectory. A variety of parameters or their combinations can be chosen. The size of the most prominent peak(s), its spectral position, its shift with increasing field, and the ratio of its size in parallel versus its size in perpendicular polarization, are a few of the possible signatures of an electroreflectance spectrum. Different signatures bring out the angular anisotropies to a varying degree. Invariance to field variations along the trajectory is

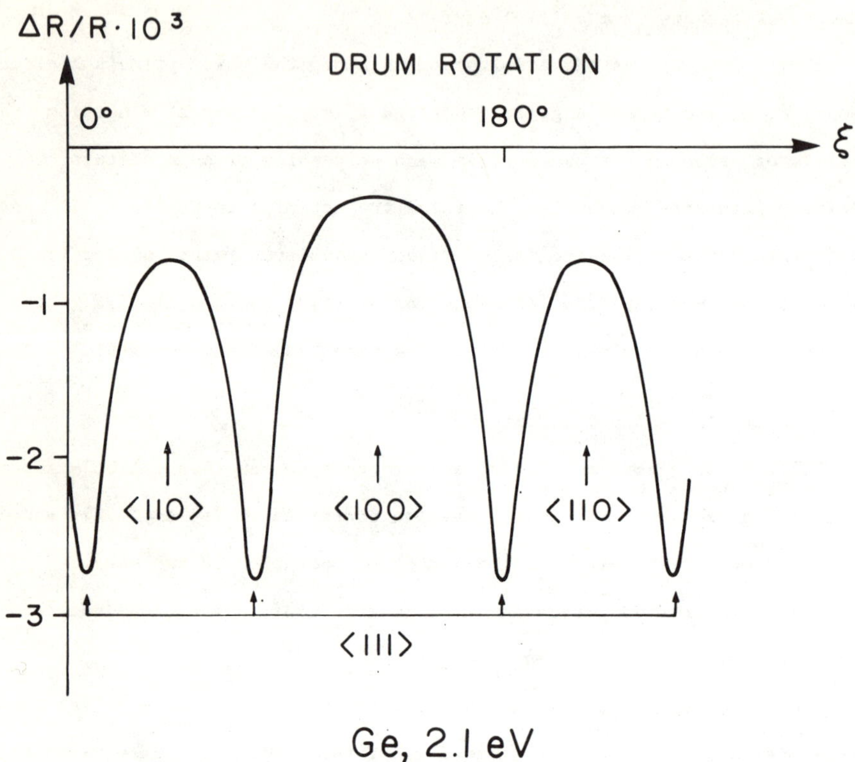

Fig. 3. Electroreflectance response at the 2.1-eV peak of Ge upon rotation of the cylindrical sample of Fig. 2 through a full turn (after Ref. 42).

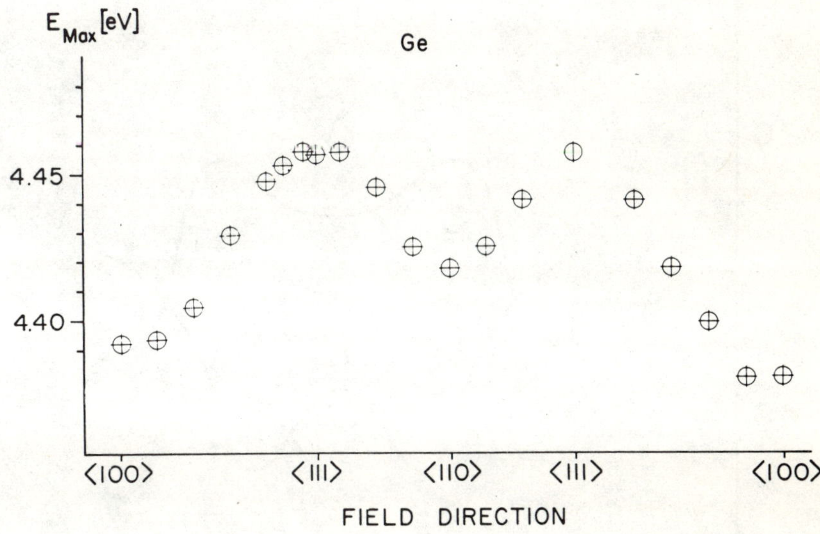

Fig. 4. Spectral position of a 4.4-eV peak in Ge upon rotation of the cylindrical sample of Fig. 2 through a full turn (after Ref. 42).

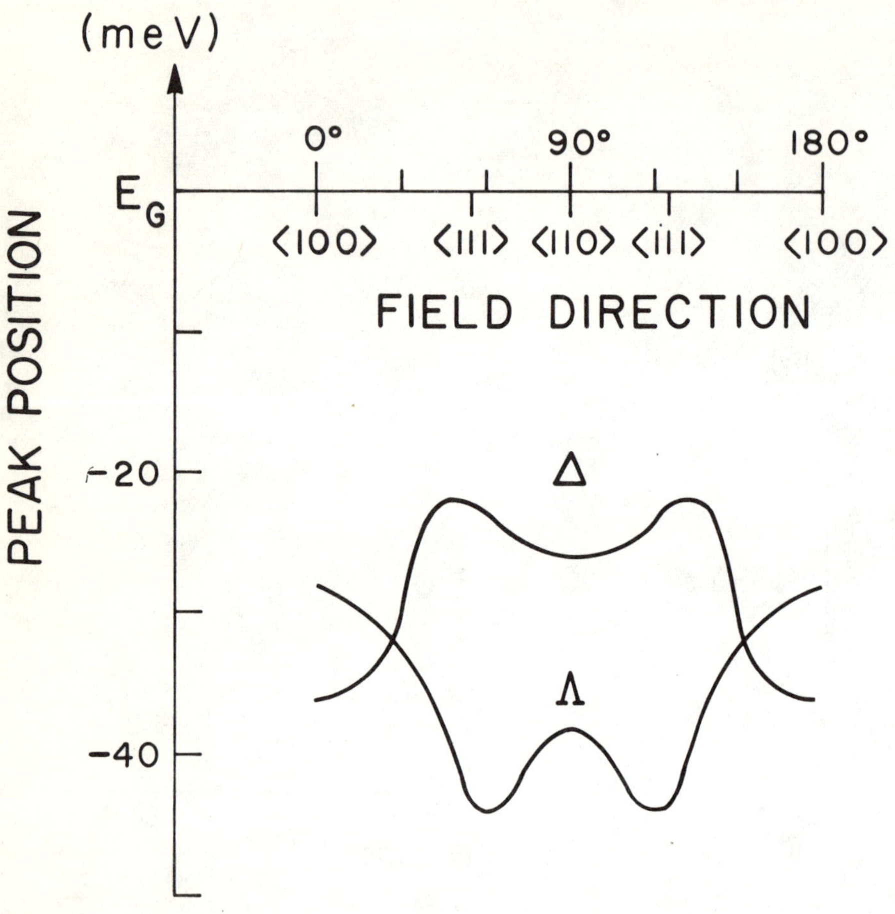

Fig. 5. Spectral position of the electroreflectance response of two hypothetical critical points located along directions Δ and Λ in the Brillouin zone, as calculated from Franz-Keldysh theory (after Ref. 38, 39).

required and no assumptions about the reduced effective mass ratio must be necessary for the symmetry analysis. As shown on a few examples in the next section, the proper choice of the signature determines the success of the analysis.

In following the selected signature $S(\xi)$ through the trajectory, traces similar to the ones plotted in Fig. 3, 4, and 5 are obtained. These traces necessarily reflect the symmetry operations of the lattice. Beyond this trivial requirement, however, the signature displays features representative of location and type of the correlated critical point. Lattice symmetry can require, for example, that the signature is polarization-independent in certain principal directions. The ratio of the peak size for the two directions of polarization must then go through one as the field lines up with these directions. How this value one is approached from either side of the trajectory, however, i.e., whether the derivative is zero, positive, or negative, reflects information beyond the requirements of lattice symmetry. Similarly, the sign of the signature, additional extrema, and zero-crossings between symmetry points along the trajectory are of diagnostic value.

To retrieve this information, the signature along the trajectory must be synthesized from Eq. (9) and Tables I, II, and III. We obtain a general expression that contains the D_i in an unspecified form

$$S(\xi) = S\left[D_i(\xi),\ f_i(\xi),\ \xi\right] \tag{10}$$

The experimental information is inserted into this expression at the p points for which the signature or its derivative assume simple values along the trajectory. All the trigonometrical functions assume numerical values at a given point of the trajectory, and we obtain a set of equations in the D_i and their derivative with respect to ξ.

$$S(\xi_q) = S\left[D_i(\xi_q),\ \left.\frac{\partial D_i}{\partial \xi}\right|_{\xi_q}\right], \quad q = 1,2...p \tag{11}$$

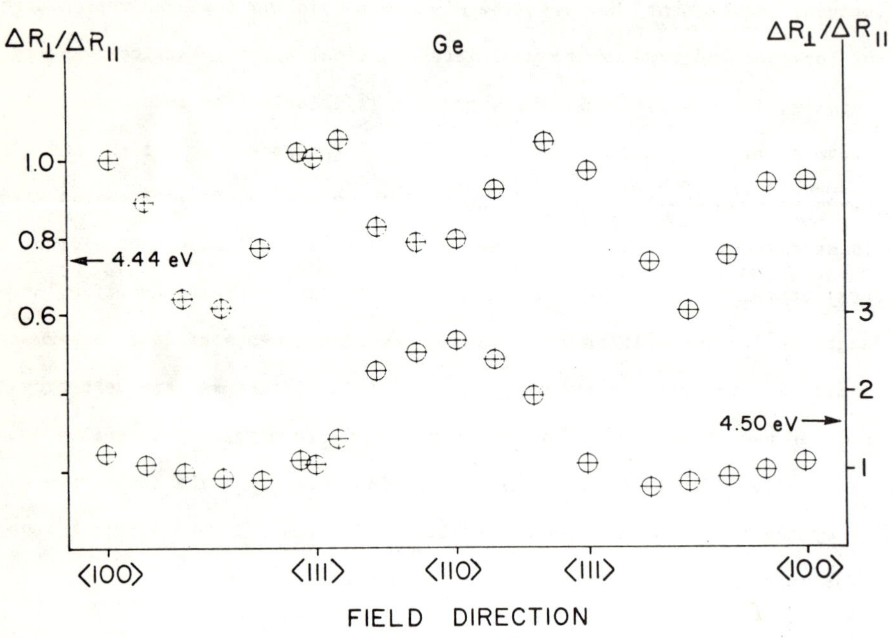

Fig. 6. The ratio $\Delta R_{\perp}/\Delta R_{\parallel}$ determined along trajectory II for the two components of the 4.4-eV structure in Ge. Note the different angular pattern for the two components, indicating different symmetry features of the correlated critical points (after Ref. 42).

This set of equations is underdetermined. We can derive from Eq. (11) a number m of relations between the D_i and their derivatives at directions of high symmetry.

$$\Pi(\xi_j) = \Pi\left[D_i(\xi_j), \left.\frac{\partial D_i}{\partial \xi}\right|_{\xi_j}\right], \quad j = 1,2\ldots m \tag{12}$$

These relations, Eq. (12), must be compatible with the equivalence relations listed in Table I, II, and III, for a location of the critical point along Δ, Λ, or Σ. If the relations, Eq. (12), derived from experiment are consistent with one, and only one, of the three principal directions, the analysis has resulted in an unambiguous identification of the structure generating the signature. If the experimental results, Eq. (12), are compatible with more than one location, we must select a different signature or vary the trajectory in order to remove the ambiguity.

We demonstrate this symmetry analysis by a simple example. We select the experimental configuration of trajectory III and record along it the ratio $\Delta R_\perp/\Delta R_\parallel$ of the electroreflectance response for two orthogonal directions of polarization. In this procedure the same electro-optical effect is sampled in two different ways. The ratio of the two readings is largely invariant to the factors that complicate ordinary line shape interpretation, such as field variations along the trajectory. Compared to the spectral line shape itself, the ratio signature is nearly invariant to variations of the surface conditions. As shown in Fig. 6, it assumes a wide range of values as the field swings through the trajectory, and its value one at points of high symmetry can be clearly recognized.

We can compare the experimental trace with a signature synthesized on the basis of Eq. (9) and Table III. Accepting only the symmetry assumptions described above, we obtain three general expressions $S_\Delta(\xi)$, $S_\Lambda(\xi)$, and $S_\Sigma(\xi)$, for the signature along the trajectory, if the correlated critical point is located along Δ, Λ, or Σ. For the electric field along directions

TABLE IV

	0	$\frac{\nu}{2}$	$\frac{\pi}{2}$
Δ	$\frac{1}{2}\left[1 + D(0)/D\left(\frac{\pi}{2}\right)\right]$	1	$\frac{2}{1 + D\left(\frac{\pi}{2}\right)/D\left(\frac{\pi}{4}\right)}$
Λ	1	$\frac{1}{8}\left[5 + 3 \cdot D(0)/D\left(\frac{2}{3}\nu\right)\right]$	$2 \cdot \frac{1 + D\left(\frac{\nu}{3}\right)/D\left(\frac{\pi}{2}\right)}{3 + D\left(\frac{\nu}{3}\right)/D\left(\frac{\pi}{2}\right)}$
Σ	$\frac{1}{2} \cdot \frac{3 + D\left(\frac{\pi}{2}\right)/D\left(\frac{\pi}{4}\right)}{1 + D\left(\frac{\pi}{2}\right)/D\left(\frac{\pi}{4}\right)}$	$\frac{1}{2} \cdot \frac{3 + 5 \cdot D\left(\frac{\nu}{3}\right)/D\left(\frac{\pi}{2}\right)}{3 + D\left(\frac{\nu}{3}\right)/D\left(\frac{\pi}{2}\right)}$	$\frac{2 + D\left(\frac{\pi}{2}\right)/D\left(\frac{\pi}{3}\right) + D(0)/D\left(\frac{\pi}{3}\right)}{3 + D\left(\frac{\pi}{2}\right)/D\left(\frac{\pi}{3}\right)}$

Table IV: Values of the signature $\Delta R_{\perp}/\Delta R_{\parallel}$ assumed along trajectory III at $\xi = 0$, $\frac{\nu}{2}$, and $\frac{\pi}{2}$, for three locations of the correlated critical point along Δ, Λ, or Σ. The angle between two directions $\langle 111 \rangle$ is ν, and cylindrical symmetry of the interband mass tensor is assumed for reasons of simplicity, even for the Σ-point.

⟨100⟩, ⟨111⟩, and ⟨110⟩, as described by the values 0, $\frac{\nu}{2}$, and $\frac{\pi}{2}$ of the trajectory coordinate ξ, these three angular functions reduce to the expressions listed in Table IV. In a mutually exclusive manner, the presence and location of value one of the signature classifies the critical point, unless degeneracies of the electro-optical function cause accidental values of one in addition to the symmetry points.

In this example the signature is discussed on a very simple basis. More information can be obtained by evaluating the derivative of the signature along the trajectory, using the relations between the derivatives of the electro-optical function D along high-symmetry directions.[42]

III. SIGNATURES OF ELECTROREFLECTANCE SPECTRA

In this final section experimental aspects of the analysis are discussed. The anisotropies on which the symmetry analysis is based stand out more or less clearly along the trajectory, depending upon the proper choice of an experimental signature. The factors that render spectral line shape interpretation useless also influence angular sweeps, but to a varying degree. Useful signatures display symmetry-induced anisotropies even in the presence of severe field variations along the trajectory, and no assumptions about the interband-mass ratio are required. We will inspect several signatures with respect to their merits for a symmetry analysis.

a. Signature of the Size

The size $\Delta R/R$ of an electroreflectance peak is the fundamental building block of any analysis. Unfortunately, this size is drastically affected by field variations along the trajectory, rendering it useless as a signature. The anisotropies induced by field variations can completely hide symmetry-induced features.

Results obtained by Fischer and coworkers[40] on a cylindrical Ge sample (Fig. 2) demonstrate this point. The anisotropies shown in Fig. 3 are observed as the 2.1-eV peak is recorded along trajectory II. The drastic angular variations of the size are not caused by interband effects exclusively, however. Boddy and Brattain[54] showed previously that the quiescent surface potential of etched Ge surfaces of various orientation grows in the sequence ⟨100⟩, ⟨110⟩, ⟨111⟩. The size variations plotted in Fig. 3 therefore reflect the different slope of the master curve of the surface potential in the different crystalline directions rather than symmetry-induced anisotropies of the electro-optical effect. This dominance of surface effects over interband effects suggests rejection of the size signature for use in the symmetry analysis.

b. Signature of the Spectral Position

As electric field and incident light beam vary their directions along the trajectory, the contributions from the different branches of the star are sampled with varying weights. Along the trajectory the total effect therefore varies in fractional composition. As a consequence, the maximum of the response occurs at different spectral positions as the trajectory is scanned. This spectral variation is considerably larger than the spectral shift induced at one particular photon energy by an increase in the field. As shown in Fig. 4, the spectral position of the 4.4-eV peak recorded around the Ge cylinder of Fig. 2, displays a characteristic angular pattern with an amplitude of over 50 meV.

The experiment can be simulated using the Franz-Keldysh theory for the electro-optical effect.[38,39] The theoretical pattern, shown in Fig. 5 for two hypothetical critical points along the directions Δ and Λ, resembles the experimental trace in appearance and size. If variations of the electric field are introduced into the calculation, the two patterns of Fig. 5 expand or contract in a vertical direction, but inside wide limits do not wipe out the characteristic shape. The discriminatory value of the spectral-position signature seems considerably more invariant to field variations than the size signature.

Assumptions about the ratio of longitudinal to transverse interband mass are necessary, however, to make Fig. 5 unambiguous. Inversion of this ratio c transforms the Δ trace into the Λ trace and vice versa.

The signature of the spectral position is therefore of value in cases where reliable information of this mass ratio is available. If an analysis on a strictly ab initio basis is required, we must turn to the ratio signature discussed in the next paragraphs.

c. Signature of the Polarization Ratio

A signature of sufficient invariance to changes in the modulating field and/or the mass ratio consists of the ratio $\Delta R_{\perp}/\Delta R_{\parallel}$ of the response to light of the two polarizations $e_{\perp}$ and $e_{\parallel}$ (Fig. 1). The invariance results from the fact that modulating field and mass ratio both enter on the level of the electro-optical effect. The situation that exists in the dark and is unperturbed by the subsequent sampling procedure is then probed in two different ways and the ratio of the two results is plotted along the trajectory. In our first-order approximation that samples the nonequivalence of the various branches with a field-independent selection rule, this procedure should result in the desired invariance.

Experimental results for the ratio signature are plotted in Fig. 6. On the Ge cylinder of Fig. 2, the spectral line shape of the response near 4.5 eV was recorded for the two directions of polarization of the incident light shown under II in Fig. 1. The ratio of the peak sizes obtained for two positions of the polarizer was then formed for the two parts of the up-down sequence of which the response consists in this spectral range. Note that the "down" part of the structure near 4.44 eV responds in an angular pattern that is characteristically different from the angular line shape of the ratio for the "up" part.[40]

The striking difference in the symmetry-generated angular line shape of the two parts of the structure suggests their correlation to two separate critical points with different symmetry character. This result, based on symmetry arguments of wide validity, provides an experimental confirmation of a basic theoretical concept of long standing. Band models for semiconductors of diamond and zinc blende structure assign the strong ultraviolet reflectance peak common to all these materials to a near-degeneracy of two critical points of different symmetry character and location. The strongly dissimilar angular pattern of the two components of the 4.5-eV structure in Ge proves that they indeed relate to two separate contributions of different symmetry character.

The ratio signature is superior to the other signatures because it delivers symmetry information invariant to the influence of the electric field on selection rules. This simply follows from the lower symmetry of the crystal in the presence of the field. Transitions allowed in the unperturbed system are still allowed under the field. However, previously forbidden transitions may be allowed in the field-perturbed system. Switching on the field therefore moves the sampling process into the direction of decreased selectivity. Extreme lowering of the symmetry - occurring for fields in low-symmetry directions - can result in a complete breakdown of the selection rules. Sampling photons of parallel and perpendicular polarization are then accepted with equal probability and $f_i^{\perp} = f_i^{\parallel}$.

Defining a selectivity parameter ε, we can write

$$f_i^{\parallel,\perp}(\varepsilon) = f_i^{\parallel,\perp} - \varepsilon(f_i^{\parallel,\perp} - f_0) \quad , \tag{13}$$

where $\varepsilon = 0$ corresponds to the unperturbed crystal of high selectivity, and $\varepsilon = 1$ represents the complete breakdown of selection rules.

Writing the basic element of the ratio signature as

$$\frac{\Delta\varepsilon_2^{\perp}}{\Delta\varepsilon_2^{\parallel}} = \frac{\sum_i f_i^{\perp} \cdot D_i}{\sum_i f_i^{\parallel} \cdot D_i} \tag{14}$$

and introducing $f_i(\varepsilon)$ from Eq. (13), it is apparent that the resulting equation for ε,

$$\sum_i (f_i^{\perp} - f_i^{\parallel}) \cdot D_i = \varepsilon \cdot \sum_i (f_i^{\perp} - f_i^{\parallel}) \cdot D_i \tag{15}$$

can be satisfied only for

$$\sum_i (f_i^{\perp} - f_i^{\parallel}) \cdot D_i = 0 \tag{16}$$

and

$$\varepsilon = 1 \quad . \tag{17}$$

This means that symmetry points at which S = 1 are not affected by an increase in ε (Eq. (16)). The signature between the symmetry points deflates without crossing the line S = 1. The angular pattern is simply squashed in the vertical direction, as the symmetry-breaking effect of the electric field increases. For $\varepsilon = 1$, the angular line shape shrinks into the line S = 1 for the whole length of the trajectory.

The deflation of the angular pattern towards S = 1 without crossing this line before $\varepsilon = 1$ is reached, has a significant consequence. It means that the symmetry information derived from the angular line shape is not influenced by the symmetry-breaking effect of the electric field on the selection rule. No new structural features can be added as the symmetry is lowered. Consequently, a symmetry analysis of the ratio signature can ignore the influence of the electric field on the selection rules - bypassing analytical difficulties of considerable complexity.

ACKNOWLEDGMENTS

In the course of preparing this lecture I have indispensably profited from stimulating and helpful discussions with my co-workers N. Bottka, J. E. Fischer, R. Glosser, B. J. Parsons, and H. Piller. The analysis section in particular reflects in numerous points the original contributions of N. Bottka and J. E. Fischer to the subject.

REFERENCES

1. J. C. Phillips, in "Solid State Physics" (F. Seitz and D. Turnbull, eds.), Vol. 18, p. 55. Academic Press, New York, 1966.
2. F. Herman, R. L. Kortum, C. D. Kuglin and R. A. Short, in "Quantum Theory of Atoms, Molecules and the Solid State: A Tribute to John C. Slater" (P. O. Loewdin, ed.), p. 381. Academic Press, New York, 1966.
3. F. Herman, R. L. Kortum, C. D. Kuglin and J. L. Shay, in "II-VI Semiconducting Compounds" (1967 Intern. Conf., Providence, 1967)(D. G. Thomas, ed.), p. 503. W. A. Benjamin, Inc., New York, 1967.
4. E. O. Kane, Phys. Rev. 146, 558 (1966).
5. W. Saslow, T. K. Bergstresser, C. Y. Fong and M. L. Cohen, Solid State Commun. 5, 667 (1967).
6. J. C. Phillips, Phys. Rev. 146, 584 (1966).
7. D. D. Sell and E. O. Kane, to be published.
8. N. Bottka, J. E. Fischer, and B. O. Seraphin, to be published.
9. B. O. Seraphin, in "Semiconductors and Semimetals" (R. K. Willardson and A. Beer, eds.), Vol. VI. Academic Press, New York, 1969.
10. W. Paul, in "Proc. Intern. School of Physics 'Enrico Fermi', Course XXXIV" (J. Tauc, ed.), p. 257. Academic Press, New York and London 1966.
11. B. O. Seraphin, in "Proc. 7th Intern. Conf. Physics of Semiconductors, Paris, 1964" (M. Hulin, ed.), p. 165. Dunod, Paris, 1964.
12. K. L. Shaklee, M. Cardona and F. H. Pollak, Phys. Rev. Lett. 16, 48 (1966).
13. R. Ludeke and W. Paul, in "II-VI Semiconducting Compounds" (1967 Intern. Conf., Providence, 1967) (D. G. Thomas, ed.), p. 123. W. A. Benjamin, Inc., New York, 1967.
14. Y. Hamakawa, F. A. Germano and P. Handler, Phys. Rev. 167, 703 (1968).

15. M. Cardona, K. L. Shaklee and F. H. Pollak, Phys. Rev. 154, 696 (1967).

16. W. E. Engeler, H. Fritzsche, M. Garfinkel and J. J. Tieman, Phys. Rev. Lett. 14, 1069 (1965); W. E. Engeler, M. Garfinkel, J. J. Tieman and H. Fritzsche, in "Proc. Intern. Colloquium Optical Properties and Electronic Structure of Metals and Alloys, Paris, 1965" (F. Abelès, ed.) p. 189. North Holland Publishing Co., Amsterdam, 1966.

17. G. W. Gobeli and E. O. Kane, Phys. Rev. Lett. 15, 142 (1965).

18. F. H. Pollak, M. Cardona and K. L. Shaklee, Phys. Rev. Lett. 16, 942 (1966).

19. U. D. Gerhardt, D. Beaglehole and R. Sandrock, Phys. Rev. Lett. 19, 309 (1967).

20. D. E. Aspnes, Bull. Amer. Phys. Soc. 13, 470 (1968).

21. B. Batz, Solid State Commun. 4, 241 (1966).

22. B. Batz, Solid State Commun. 5, 985 (1967).

23. E. Matatagui and M. Cardona, Bull. Amer. Phys. Soc. 12, 1033 (1967); 1968, to be published.

24. W. J. Scouler, Phys. Rev. Lett. 18, 445 (1967).

25. A. Balzarotti and M. Grandolfo, Phys. Rev. Lett. 20, 9 (1968).

26. C. N. Berglund, J. Appl. Phys. 37, 3019 (1966).

27. A. Baldereschi and F. Bassani, Phys. Rev. Lett. 19, 66 (1967).

28. S. H. Groves, C. R. Pidgeon and J. Feinleib, Phys. Rev. Lett. 17, 642 (1966).

29. C. R. Pidgeon, S. H. Groves and J. Feinleib, Solid State Commun. 5, 677 (1967).

30. C. R. Pidgeon and S. H. Groves, in "II-VI Semiconducting Compounds" (1967 Intern. Conf., Providence, 1967) (D. G. Thomas, ed.), p. 1080. W. A. Benjamin, Inc., New York, 1967.

31. R. L. Aggarwal, L. Rubin and B. Lax, Phys. Rev. Lett. 17, 8 (1966).

32. R. L. Aggarwal and B. Lax, Bull. Amer. Phys. Soc. 11, 828 (1966).

33. J. Feinleib, C. R. Pidgeon and S. H. Groves, Bull. Amer. Phys. Soc. 11, 828 (1966).

34. M. Cardona, F. H. Pollak and K. L. Shaklee, J. Phys. Soc. Japan Suppl. 21, 89 (1966).

35. F. Cerdeira, R. Lettenberger and M. Cardona, Bull. Amer. Phys. Soc. 12, 1049 (1967).

36. B. O. Seraphin and N. Bottka, Phys. Rev. Lett. 15, 104 (1965).

37. Victor Rehn and D. Kyser, Phys. Rev. Lett. 18, 848 (1967).

38. D. E. Aspnes, Phys. Rev. 147, 554 (1966); ibid. 153, 972 (1967).

39. D. E. Aspnes, P. Handler and D. Blossey, Phys. Rev. 166, 921 (1968).

40. J. E. Fischer, N. Bottka and B. O. Seraphin, Phys. Rev. (to be published).

41. J. E. Fischer, N. Bottka and B. O. Seraphin, Bull. Amer. Phys. Soc. 14, 415 (1969).

42 N. Bottka, J. E. Fischer and B. O. Seraphin, Phys. Rev. (to be published).

43. B. O. Seraphin and N. Bottka, Phys. Rev. 145, 628 (1966).

44. A. Frova, P. J. Boddy and Y. S. Chen, Phys. Rev. 157, 700 (1967).

45. Y. Hamakawa, F. A. Germano and P. Handler, J. Phys. Soc. Japan Suppl. 21, 111 (1966).

46. Y. Hamakawa, P. Handler and F. A. Germano, Phys. Lett. 25A, 617 (1967).

47. Y. Hamakawa, F. A. Germano and P. Handler, Phys. Rev. 167, 709 (1968).

48. B. O. Seraphin, R. B. Hess and N. Bottka, J. Appl. Phys. 36, 2242 (1965).

49. B. O. Seraphin, Surface Sci. 8, 399 (1967); ibid. 13, 136 (1969).

50. F. Evangelisti and A. Frova, Solid State Commun. 6, 621 (1968).

51. D. E. Aspnes and A. Frova, Solid State Commun. 7, 155 (1969).

52. B. O. Seraphin and N. Bottka, Solid State Commun. 7, 497 (1969).

53. N. Bottka and U. Rössler, Solid State Commun. 5, 939 (1967).

54. P. J. Boddy and W. H. Brattain, J. Electrochem. Soc. 110, 570 (1963).

X

SOLID STATE RESONANCE SPECTROSCOPY AT SUBMILLIMETER WAVELENGTHS

Kenneth J. Button

Francis Bitter National Magnet Laboratory*
Massachusetts Institute of Technology

INTRODUCTION

Quantum spectroscopy can be carried out in semiconductors if one makes use of an applied external magnetic field to create Landau quantum levels as described by other contributors to this volume, notably Kaplan and Wallis. This is illustrated schematically in Fig. 1 which implies, at the far left, that the density of states, D(ℰ) is not quantized in the absence of a magnetic field but, at the far right, the density of allowed states for electrons and holes is sharply peaked at quantum levels which are equally separated in energy. Then, various quantum transitions can take place and are given the different names illustrated in the center of Fig. 1. All of these types of quantum spectroscopy have been widely used for more than fifteen years in the fruitful study of the nature, shape and interrelationships of the energy bands of semiconductors. It is only necessary to provide for the population of a lower quantum state with electrons or holes and then to provide photons from a microwave or optical source of radiation which correspond in energy to the transition between the populated quantum state and a higher state which is not fully populated. The difference in energy between the quantum states can often be adjusted conveniently by changing the intensity of the applied magnetic field because the separation

---

*Supported by the U. S. Air Force Office of Scientific Research

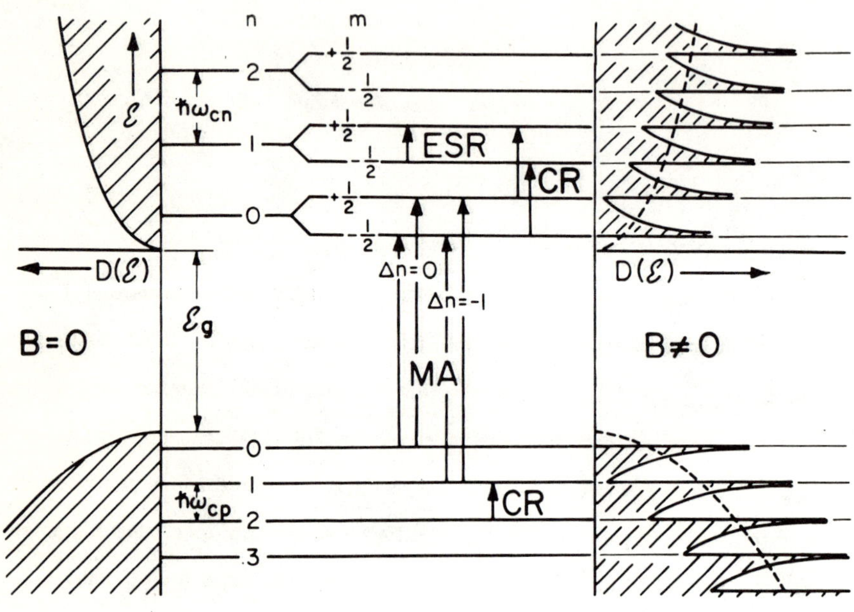

Fig. 1. In the absence of an applied magnetic field electrons and holes occupy a continuum of states within an energy band as shown at the left. When a magnetic field is applied to a semiconductor, the continuum changes to a set of sharply quantized levels as shown at the right. In the center, various quantum transitions are illustrated in such a way as to reveal these selection rules. (P. Grosse, "Die Festkörpereigenschaften von Tellur," Springer Verlag, Berlin, 1969.)

of the Landau quantum levels is proportional to the applied field intensity. The separation is also a measure of the curvature of the constant energy surface on which charge carriers are allowed to move in $\vec{k}$ space. This fact has been denoted in the past by the assignment of an effective mass m* to the charge carrier where the definition of the effective mass is given in terms of the curvature by $1/m^* = (1/\hbar^2)(\partial^2\mathcal{E}/\partial k^2)$. The cyclotron resonance (CR) quantum levels are separated by an energy

$$\hbar\omega_c = \frac{\hbar eH}{m^*c}$$

where H is the intensity of the applied magnetic field and ω_c is called the cyclotron frequency. In practice, this is the frequency of the radiation source which supplies the photons of energy $\hbar\omega_c$ for a cyclotron resonance transition.

As one can see from Fig. 1, electron spin resonance (ESR) can be used to measure the g-factor of a charge carrier in a semiconductor, cyclotron resonance will supply the effective mass (curvature of the constant energy surface) and magnetoabsorption (MA) will provide information about the energy gap, $\mathcal{E}_g$ and the reduced effective masses of electrons and holes. Combinations of these transitions can also be observed and they are given special names such as combination resonance which requires an energy which is the sum of CR and ESR because it involves a cyclotron resonance transition plus a "spin flip."

This simple explanation given above is adequate only for tutorial purposes because only the conduction band of semiconductors (Fig. 2, upper left) can be depended upon to remain reasonably uncomplicated in a practical case. The valence bands of even the simplest elemental semiconductors such as germanium (Fig. 2, lower left) interact with each other making the quantum spectroscopy a much more elaborate process to carry out. Indeed, at the right of Fig. 2 no attempt has been made to show the Landau quantum levels in the elemental semiconductor tellurium because these have yet to be worked out theoretically. The theory will make use of recent cyclotron resonance results which will be shown later in this chapter. Since cyclotron resonance provides the most direct and

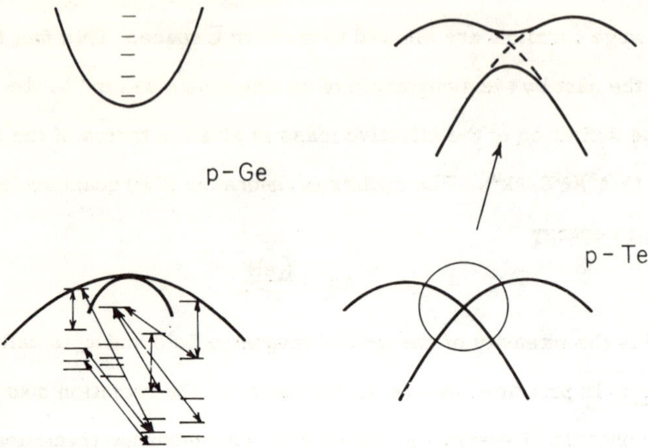

Fig. 2. At the left, the direct energy gap for germanium is shown. The conduction band (upper left) is almost parabolic and therefore the quantized energy levels in the presence of a magnetic field are almost equally separated harmonic oscillator levels. At lower left, two valence bands interfere at the band edge resulting in a complicated set of energy levels. At lower right the valence bands of tellurium also interfere with each other but can be thought of as two separate valence bands as indicated in the upper right corner.

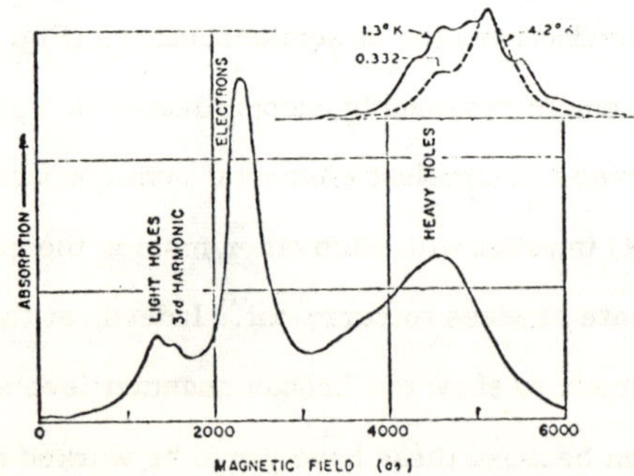

Fig. 3 Cyclotron resonance absorption in silicon using microwave radiation to provide the quantum transitions.

unambiguous experimental data for the description of semiconductor energy bands, the major portion of this paper will deal with that technique.

CYCLOTRON RESONANCE

As one can see from Fig. 1, cyclotron resonance is a low energy transition, and, in principle, can be carried out with the use of microwave photons[1] whereas magnetoabsorption transitions must take place across the semiconductor gap which require the use of near infrared or visible optical wavelength photons.[2] Figure 3 shows an example of a cyclotron resonance spectrum observed fifteen years ago in silicon[3] at a microwave frequency of 23 Gc which corresponds to a wavelength of about one centimeter. The absorption peaks simply represent the absorption of the microwave photons whenever the applied magnetic field has been increased to a point where Landau quantum levels are separated by an energy corresponding to the monochromatic photon energy of the microwave klystron. Provision was made for populating the initial states of the transitions in both the conduction band (electrons) and in the valence bands (holes) by illuminating the silicon specimen with white light which kicks electrons out of atomic bonding orbits, leaving holes, and giving the electrons sufficient kinetic energy to become conduction electrons. Both of these types of carriers are then eligible to circulate in an orbit about the direction of the applied magnetic field just as a proton or electron circulates in a classical particle accelerator or cyclotron. Thus, Fig. 3 shows absorption lines for both electrons and holes where the electron absorption lines correspond to the quantum transitions in the conduction band denoted by CR in the upper part of Fig. 1 and the absorption lines corresponding to the "light" and "heavy" holes presumably correspond to the two different valence bands of silicon which are very similar to those shown in the lower left of Fig. 2. Thus the situation would seem to have been quite satisfactory fifteen years ago leaving very little reason for spectroscopists to do more than to carry out routine measurements of the effective masses of electrons and holes in various semiconductors. One small problem remained, namely, the interaction between degenerate valence bands such as that shown in the lower left of Fig. 2 which causes the broad absorption line for holes at the right of Fig. 3 to break up into sharper structure. This was predicted in 1954 by Luttinger and Kohn[4] and observed by Fletcher, Yager and Merritt[5] in 1955, again using a microwave frequency of 23 Gc.

The Conditions for Observing Quantum Effects

Although Fletcher, Yager, and Merritt confirmed the prediction of Luttinger and Kohn that cyclotron resonance associated with interacting valence bands really consists of a large number of absorption lines (see inset Fig. 3) rather than one broad absorption (lower right, Fig. 3) the fine structure was not sufficiently well-resolved to provide accurate data for theoretical analysis of the detailed contours of constant energy surfaces near the valence band edge at the bottom of the energy gap. Three simple conditions need to be satisfied in order to resolve a multitude of quantum transitions which ordinarily are observable only as a broad envelope of unresolved transitions. The three conditions are $\hbar\omega \gg kT$ (the quantum condition), $\omega\tau \gg 1$ (the condition for a sharp absorption line), and $\hbar\omega > \mathcal{E}_F$ where $\mathcal{E}_F$ is the Fermi energy, $\hbar\omega$ is the photon energy of the radiation source, kT is the thermal energy, T is the absolute temperature of the specimen and τ is the time between collisions for a charged particle in its cyclotron orbit. The time τ must be sufficiently long that the charged particle may travel <u>at least</u> $\frac{1}{2}\pi$ of a revolution, that is, $\omega_c \tau > 1$, in order to see a resonant absorption line at all. In the case of more than one resonant absorption line in a spectrum we require $\omega\tau \gg 1$, that is, each line must be reduced in width toward its "natural" line width so that line overlap is minimized. The first condition $\hbar\omega \gg kT$ is called the quantum condition because the initial and the final quantum level must be sharply defined compared with the distance between them. The distance between quantum levels when the absorption line is observed is simply $\hbar\omega$, the photon energy that is absorbed. At the same time, the thermal broadening of the initial

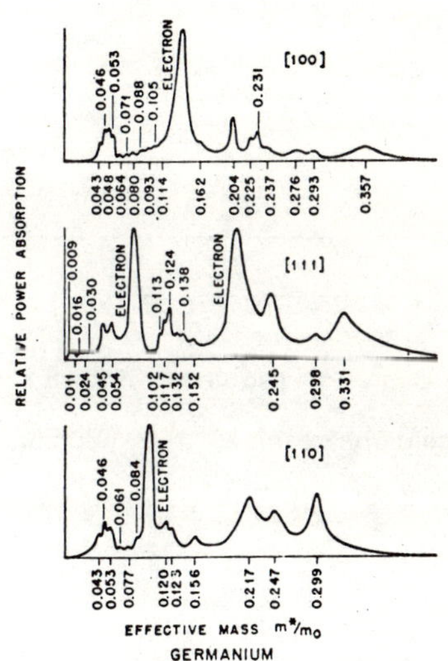

Fig. 4. Cyclotron resonance absorption lines corresponding to the transitions indicated at the lower left of Fig. 2. The absorption marked "electron" corresponds to the transitions at the upper left of Fig. 2. (After Stickler, Zeiger and Heller, Phys. Rev. 127, 1077 (1962).

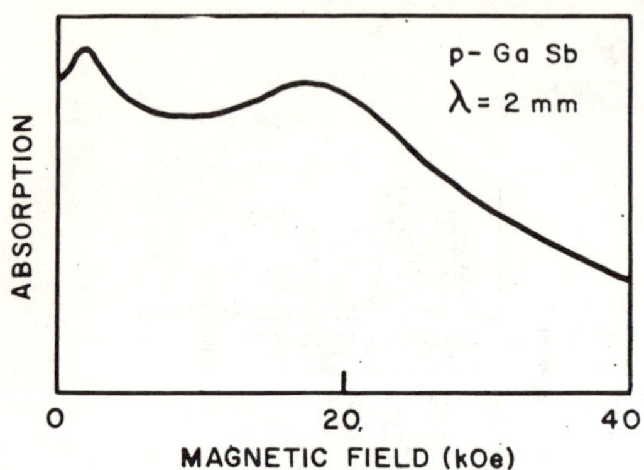

Fig. 5. Two envelopes of unresolved cyclotron resonance transitions. If the individual transitions could be resolved, the spectrum could be similar to that shown in Fig. 6.

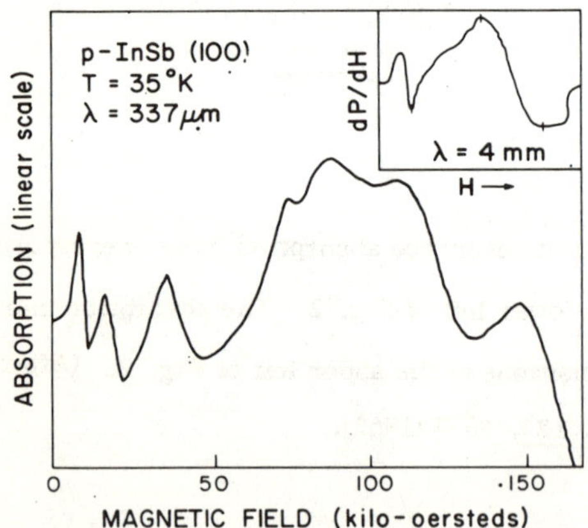

Fig. 6. The unresolved envelopes of quantum transitions were observed at millimeter wavelengths as shown in the inset. Seven separate quantum transitions were observed at a wavelength of 337 micrometers.

and final states is proportional to the thermal energy, kT. Thus, the quantum condition is $\hbar\omega \gg kT$. For the frequency of 23 Gc used to obtain the spectra in Fig. 3, the photon energy is approximately equal to kT at T = 1.3°K. In order to obtain discrete absorption lines suitable for theoretical analysis, it would be necessary either to lower the temperature, which is inconvenient, or to increase the frequency. Unfortunately, the frequency can be increased conveniently only by a factor of three or four using microwave techniques. Of course, this has been done by a number of workers. The results of Stickler, Zeiger and Heller[6] are shown in Fig. 4 where they have observed the fine structure of quantum transitions in the valence bands of germanium at 1.2°K and a photon energy corresponding to a wavelength of about 2 millimeters which is five times shorter wavelength than that used to take the data of Fig. 3. Further improvements in technique carried out principally by J. C. Hensel[7] have produced data for the valence bands of germanium and silicon which was of some value to the theorist. Nevertheless, the quantum condition is only marginally satisfied at millimeter wavelengths and it is really necessary to increase the frequency to the submillimeter wavelength range. Millimeter waves have proved to be of little value in the study of the valence bands of the compound semiconductors as illustrated by the spectrum for p-type gallium antimonide[8] in Fig. 5. This unresolved envelope of quantum transitions was also characteristic of observations in p-type indium antimonide[9] at a wavelength of 2 millimeters as shown by the inset of Fig. 6. Figure 6 also shows the structure that is revealed under the envelope when cyclotron resonance in the valence bands of indium antimonide is observed by using higher photon energy,[10] namely, that corresponding to a wavelength of 337 micrometers (~ ⅓ millimeter).

It should be recalled that each of the seven absorption lines observable in Fig. 6 correspond to transitions such as those shown in the lower left corner of Fig. 2 where two valence bands are degenerate and it is impossible to distinguish a "light" or "heavy" hole in the limit as one makes the observation near the point of degeneracy of the two bands.

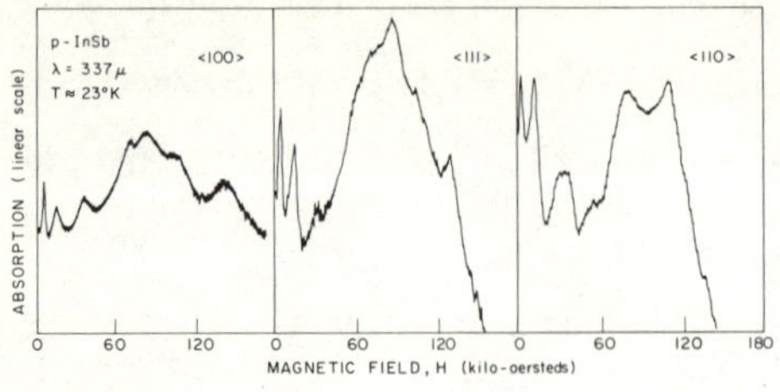

Fig. 7. Different transition probabilities and energies are observed for the magnetic field along different crystal directions.

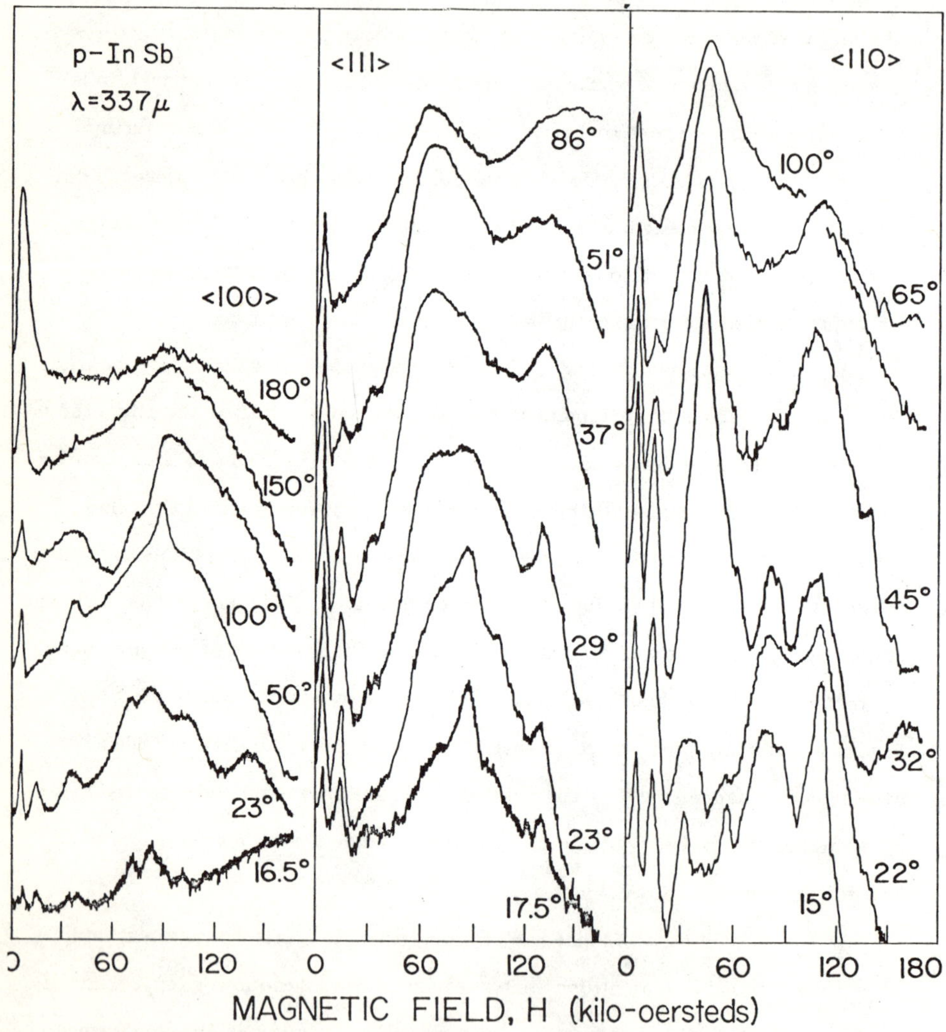

Fig. 8. Changes in the spectra with temperature.

Figure 7 shows the anisotropy in the valence bands of indium antimonide where, the transitions have a different intensity and, in one or two cases, a different position in the spectrum depending upon which principal direction of the specimen is parallel to the applied magnetic field. As one can see in Fig. 7 the peak of the envelope appears at a different field but this does not define the anisotropy of the valence bands because the peak of the envelope is determined by the intensity of individual transitions rather than by the position. An unfortunate controversy is still going on among microwave spectroscopists who are studying spectra such as that shown in the inset of Fig. 6. They are arguing with each other about the anisotropy of "the heavy hole" where they define the mass of the heavy hole as the peak of the envelope. It is very important to understand that the peak of the envelope is meaningless and that anisotropy can only be determined by observing the individual quantum transitions as shown in Fig. 7.

The Temperature Dependence of the Quantum Effects

There are always several striking features about the temperature dependence of spectra representing quantum transitions as shown in Fig. 8. At the extreme lower left corner of Fig. 8 we see that the cyclotron resonance absorption lines are very feeble at low temperature because the "holes" are largely "frozen out." We can see what that means by looking at the spectrum immediately above the feeble one, namely, the spectrum representing the temperature of 23°K. Each absorption line has become very much stronger because, at this slightly higher temperature, electrons have been thermally excited to the acceptor level of this p-type semiconductor leaving holes in the valence band levels. The cyclotron resonance absorption line represents the excitation of electrons from lower-lying levels to these sites where holes exist and this process takes place by absorption of photons from the source of radiation. This change of absorption intensity of the various absorption lines at low

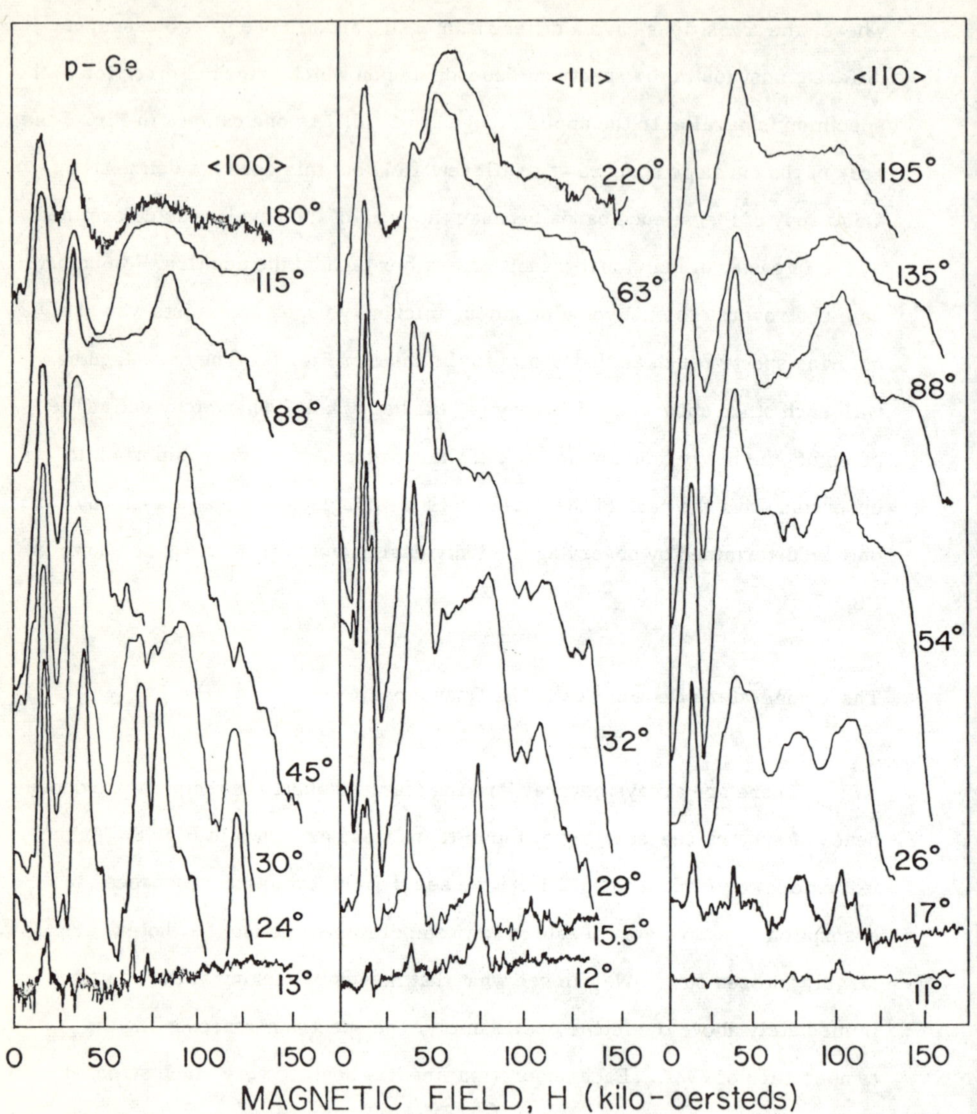

Fig. 9. Changes in spectra with temperature at wavelength of 337 micrometers.

temperature helps to identify the transition theoretically because the change of population of initial and final states with temperature causes transitions to appear and disappear in accordance with theory.

Another feature of the temperature dependence is noticeable when the temperature is increased to a value approximately equal to or larger than 43°K where $\hbar\omega = kT$ and the quantum condition is no longer satisfied. Then we see that the fine structure representing individual quantum transitions has largely merged into an envelope of unresolved transitions. Our spectra are then reduced to the previous case discussed above when the millimeter wave and microwave observations were presented. We see that the two uppermost spectra in the left column of Fig. 8 are very similar to the millimeter wave spectrum of Fig. 5.

It is also clear from Fig. 8 that the anisotropy of cyclotron resonance in p-type indium antimonide can be determined by comparing the low temperature curves in the three different columns which represent different orientations of the principal axes of the specimen with respect to the applied magnetic field direction. Whenever a spectral line appears at a different value of magnetic field for a different crystallographic direction that line is said to exhibit anisotropy. The anisotropy of a material can only be determined when individual quantum transitions have been resolved for different directions of orientation.

Figure 9 shows similar spectra for p-type germanium. In this case, the lowest temperature spectra clearly illustrate which quantum transitions appear first making it possible to remove some of the ambiguity in the identifications previously assigned before this temperature dependence had been observed. Since the theoretical analyses for both germanium and indium antimonide are still in progress, the spectra of both Fig. 8 and Fig. 9 have not yet been published.[10]

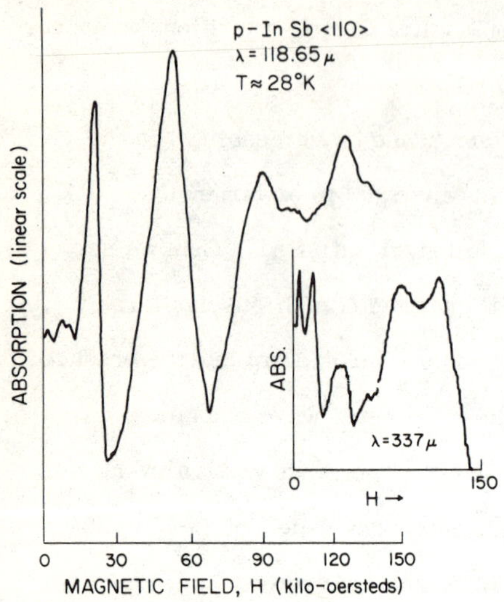

Fig. 10. Comparison of spectra for two different laser wavelengths.

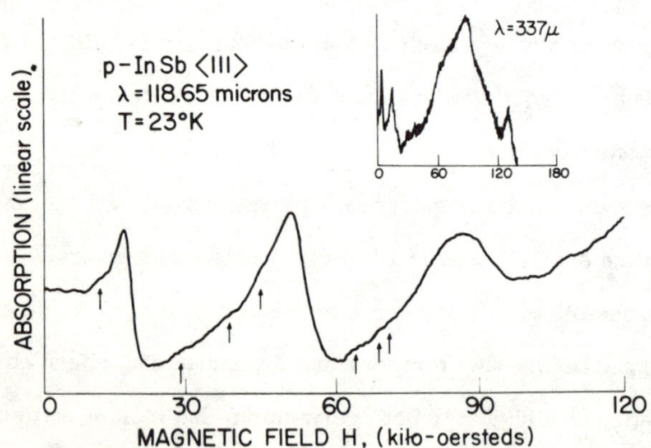

Fig. 11. Comparison of spectra for two different laser wavelengths with the magnetic field ⟨111⟩ direction.

Cyclotron Resonance at High Frequency and High Magnetic Field

It has long been recognized that cyclotron resonance spectra are much more clearly resolved and observed whenever one can increase the frequency and the magnetic field at which the experiment is performed. This helps to satisfy the second condition, $\omega\tau > 1$, more generously. This is one of the reasons for the original attempts[11] when two different groups tried to do cyclotron resonance at infrared frequencies by using the highest magnetic fields available and the smallest known effective electron masses, namely, the electrons in indium antimonide and in bismuth. Their most significant findings at these extreme values of high frequency and high intensity magnetic field were fortuitous. They both accidentally discovered the interband MA transitions and also they discovered that the cyclotron resonance technique is very valuable for measuring the change of curvature as one makes observations higher up into the energy band at higher frequencies.

It was the advent of the submillimeter molecular gas lasers which made the observation of cyclotron resonance at high frequencies practical in the sense that relatively large effective masses of both holes and electrons of almost all semiconductors can be observed without resorting to the use of pulsed magnets in order to obtain the ultra high magnetic fields needed for observation of cyclotron resonance at frequencies near the visible optical range. Various molecular gas lasers are now used in quantum spectroscopy[12] and an example of spectra for two different lasers is shown in Fig. 10. The inset shows the spectrum for p-type indium antimonide that was taken by using a continuous-wave HCN laser operating at a wavelength of 337 micrometers. The figure also shows a spectrum from the same specimen taken by using a continuous-wave H_2O laser operating at a wavelength of 118.6 micrometers. Much more detail is observable in the spectrum taken at 118.6 micrometers but every feature appears at a field intensity which is three times larger than that used for the spectrum in the inset. The effort to go to higher frequency is worthwhile, however, because one can

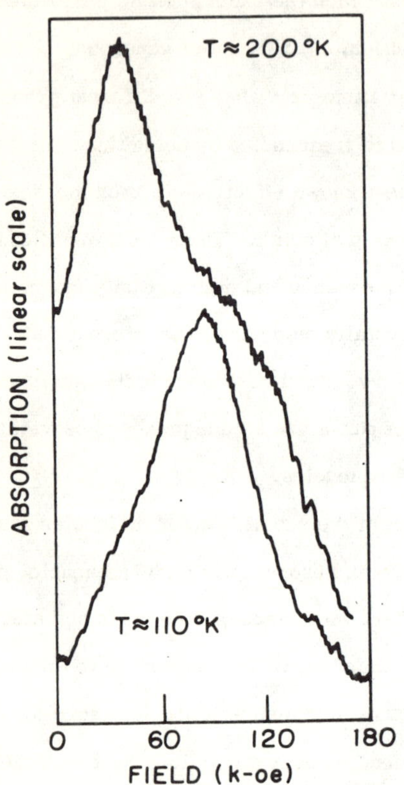

Fig. 12. The upper curve shows cyclotron resonance of the electron in tellurium. The lower curve is the envelope of unresolved transitions in the valence bands when $\hbar\omega < kT$.

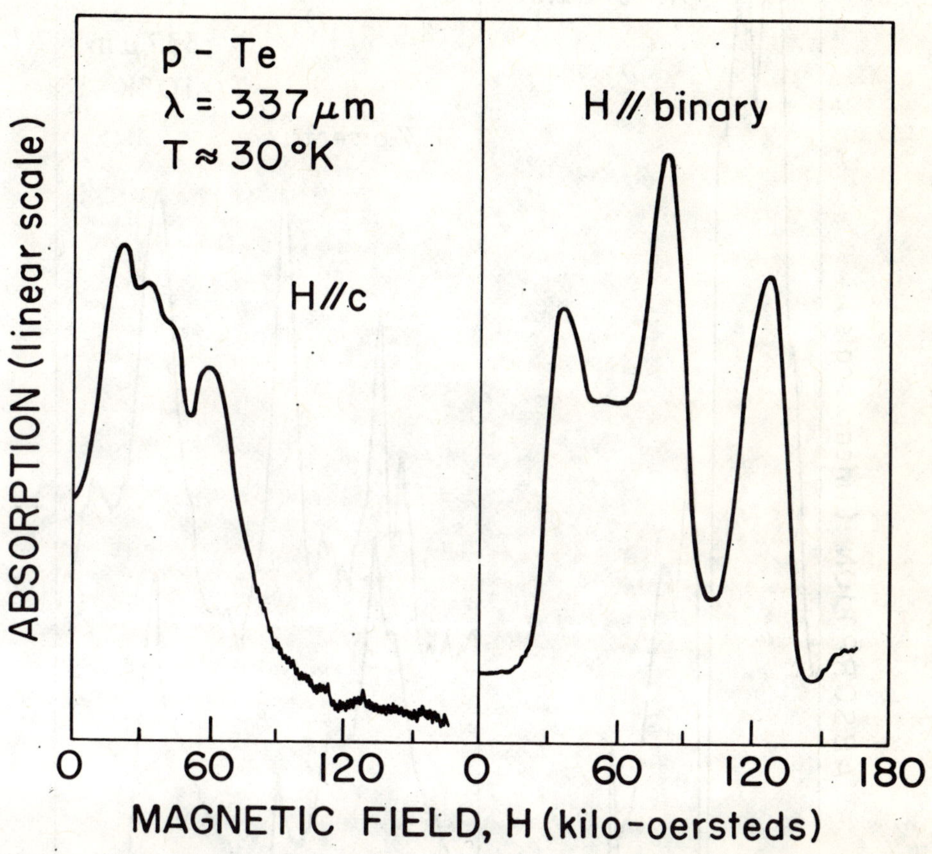

Fig. 13. Some details of the envelope of unresolved transitions can be seen when the temperature is decreased to make $\hbar\omega \approx kT$.

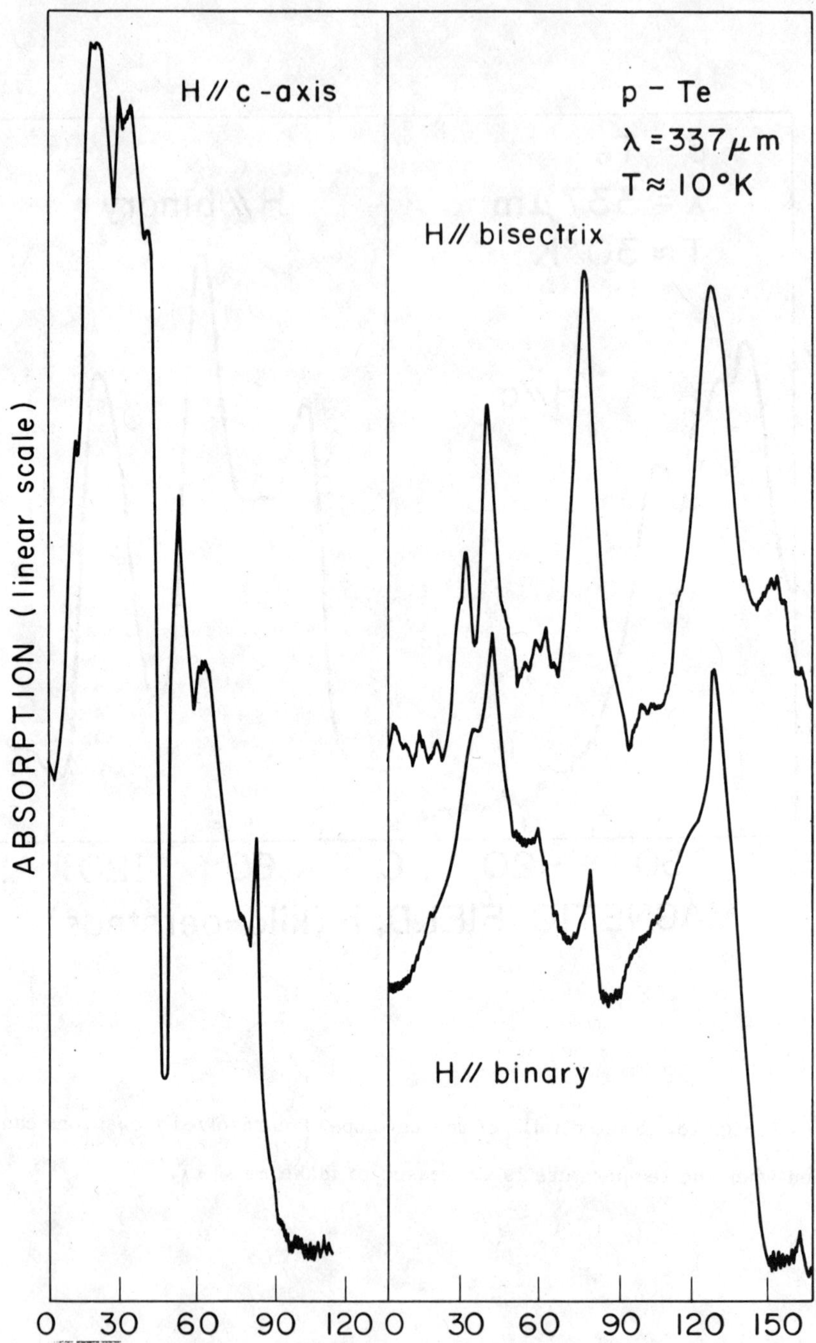

Fig. 14. The fine structure appears at low temperature where $\hbar\omega \gg kT$.

see that the third absorption line in the inset appears as two absorption lines, well resolved, in the spectrum at 118.6 micrometers because of the higher resolution at the higher frequency. Furthermore, Fig. 11 shows that one is able to study the line shape of the quantum transitions in much greater detail whenever the frequency is increased.

Cyclotron Resonance in Tellurium

Since it has been impossible to prepare n-type tellurium, cyclotron resonance of the electron in tellurium has not been observed until recently[13] when experiments could be performed in p-type tellurium at a sufficiently high frequency and temperature to permit thermal excitation of the electrons across the gap. The result is shown in Fig. 12 where both curves were obtained by using an HCN laser as a source of radiation.. The lower curve shows the typical high-temperature envelope of unresolved quantum transitions in the valence band of p-type tellurium but as the temperature is increased to 200°K electrons are excited thermally across the forbidden energy gap of tellurium ($\mathcal{E}_g$ = 0.33 electron volt) and these conduction electrons exhibit a cyclotron resonance absorption corresponding to an effective mass of about 0.135. Returning our attention to the lower curve of Fig. 12 for a moment, this unresolved envelope consists of three strong groups of absorptions which are easily resolved at only a slightly lower temperature as shown on the right hand side of Fig. 13. These are the principal groups of transitions which can be observed when $\hbar\omega \approx kT$. If however, we satisfy our three conditions for the observation of quantum effects generously, we see a great deal more detail appear as shown at the lower temperature in Fig. 14. The reason for all of this detail is similar, in principal, to the explanation given earlier for the detail in the valence bands of germanium and indium antimonide. Indeed, as we see on the right hand side of Fig. 2, the valence bands of tellurium are even slightly more complicated than those of germanium because the displacement of the extremum means that, in effect, we shall be observing a

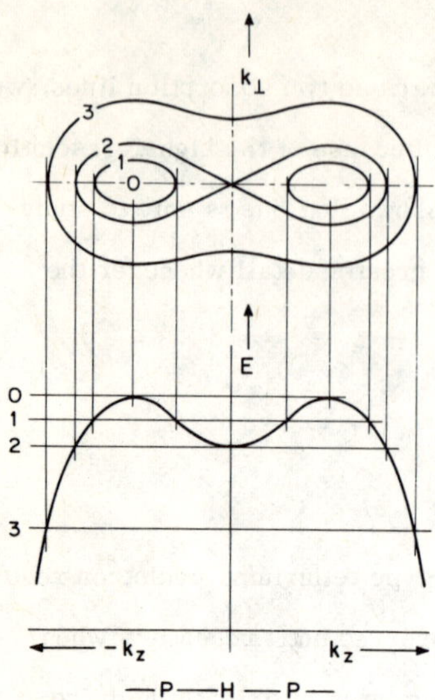

Fig. 15. The constant energy orbits (upper) have been constructed by projection from the upper valence band of tellurium. This valence band was constructed as shown at the right of Fig. 2.

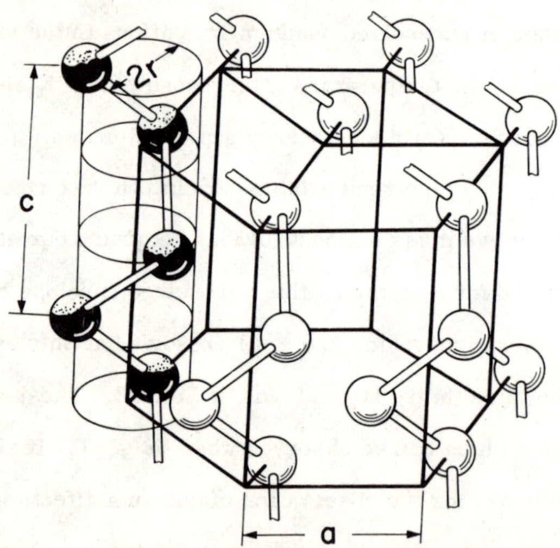

Fig. 16. The arrangement of tellurium atoms in the crystal is essentially hexagonal accounting for the clear difference in the spectra of Fig. 14 where, at the left the magnetic field is applied along the c direction but at the right the magnetic field is applied perpendicular to the c direction. (P. Grosse, "Die Festkörpereigenschaften von Tellur," Springer Verlag, Berlin, 1969.)

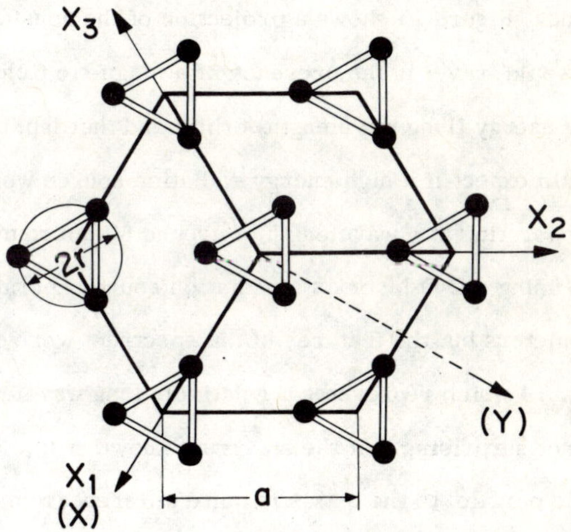

Fig. 17. The view of tellurium atoms along the c direction showing the trigonal symmetry which is repeated every three layers.

different sort of valence band at low frequencies compared to that at high frequencies. The reason for this is suggested at the upper right of Fig. 2 where we have resolved the two interacting bands into one parabolic band and another having two extrema.[14] Thus at low frequencies one would observe phenomena associated with the upper band at the edge of the energy gap whereas at frequencies corresponding to those provided by submillimeter molecular gas lasers, one would see the effects of the parabolic band as well and, in fact, at wavelengths shorter than about 100 micrometers one would expect to see the spectra characteristic of the parabolic band. Figure 15 shows a projection of the constant energy surfaces on which holes would travel in the presence of a magnetic field, the orbit labelled 1 being the low energy (long wavelength) orbits and that labelled 3 would be the orbit one would expect if a high energy radiation source were used such as a water vapor laser operating at a wavelength of around 80 micrometers. Observations were made[13] using a D_2O laser as a radiation source operating at a wave length of 172 micrometers but the features of the spectrum were quite similar to those shown in Fig. 14 which represents a relatively long wavelength.

It is not surprising that the spectrum shown at the left of Fig. 14 for the magnetic field parallel to the c-axis is quite different from the two spectra shown at the right of Fig. 14 for the magnetic field perpendicular to the c-axis because tellurium is not basically cubic as in the case of germanium and indium antimonide but is hexagonal and therefore exhibits a very high degree of anisotropy. The crystal structure of tellurium is shown in Fig. 16, which shows that the crystal not only has a fundamental hexagonal symmetry but that the layers of atoms also rotate in their positions along the c-axis, repeating the cycle of rotation every three layers. This results in a twisting which provides a view such as that shown in Fig. 17 if one looks along the c-axis. The consequences of this twisting has not yet shown up in the cyclotron resonance data although an effort was made to reveal it by a careful comparison of spectra for the magnetic field parallel to the binary direction and the magnetic field parallel to the bisectrix direction as shown at the right of Fig. 14. No essential differences in the positions of the important

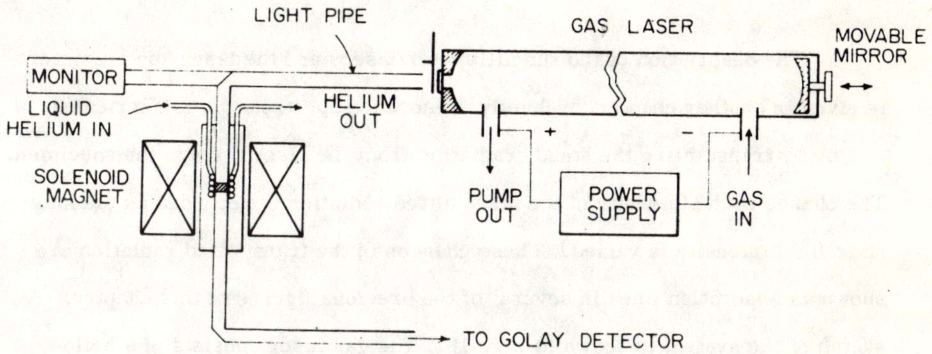

Fig. 18. The monochromatic submillimeter radiation emerges from the coupling hole in the left-hand mirror of the gas laser and is guided by a hollow metal pipe through the specimen to the detector. The specimen is located at the center of a magnet which provides a continuous magnetic field. The specimen is cooled by the liquid helium which passes around the metal light pipe.

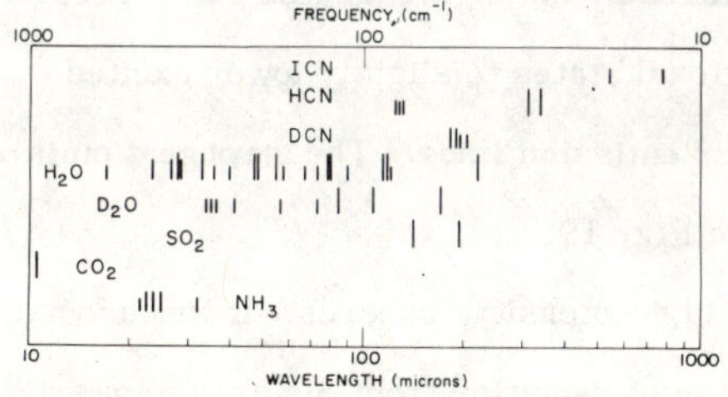

Fig. 19. A few of the most commonly used laser emission lines.

features are noted. Apparently the forthcoming theoretical effort to calculate the positions of the Landau levels and the selection rules for quantum transitions will be rather complicated.

THE LASER SPECTROMETER

The description of the submillimeter lasers and the laser spectrometers is given in another chapter.[15] Briefly, resonance spectroscopy is carried out simply by transmitting the steady radiation from the laser through the specimen. The change in the intensity of the transmitted radiation is recorded as the magnetic field intensity is varied. These changes in the transmitted radiation are shown as absorption lines in several of the previous figures of this chapter. A sketch of the system is shown in Fig. 18. The gas laser consists of a hollow glass tube about three meters long and ten centimeters in diameter. An appropriate gas such as H_2O, D_2O, CD_3CN (for DCN) NH_3 and CH_3 (for HCN), CO_2 or SO_2 is pumped through the tube maintaining a vapor pressure of about $\frac{1}{2}$ millimeter of mercury by adjusting a needle valve at the inlet. A continuous power supply is used to create a discharge in the gas. About five kilovolts is required to initiate the discharge but it is usually maintained at about one kilovolt with a flow of one or two amperes. The current flow is much less than one ampere for the CO_2 laser, however. The photons emitted by the highly excited gas represent relaxations from excited vibrational-rotational states to slightly lower excited states. There are hundreds of useful laser emission lines. The strongest emission lines of interest to us are tabulated in Fig. 19.

The output of the laser consists of high intensity, coherent, monochromatic radiation because of the stimulated emission of radiation from the excited gas molecules. This stimulated emission is caused by photons reflected repeatedly through the gas by the mirrors at each end of the laser shown in Fig. 18. Some of the radiation escapes through a hole in the center of the left-hand mirror and this radiation is used for the spectroscopy described above.

CONCLUSION

Although it has been possible to carry out submillimeter spectroscopy with conventional dispersion grating spectrometers,[16] very little work has been done on the optical properties of materials in this wavelength range during the past hundred years. The advent of microwave spectroscopy thirty years ago has not measurably improved our ability, but has increased our desire, to study the optical properties of materials at submillimeter wavelengths. Although Fourier transform spectroscopy[17] was developed many years ago, its practical use awaited the availability of inexpensive digital computation techniques which became available during the past ten years. Nevertheless, solid state spectroscopy in the submillimeter range has been revitalized by the advent of the molecular gas laser spectrometer. Thus it appears that a combination of techniques including molecular gas lasers, Fourier transform spectroscopy and tunable Raman lasers[18] will permit us finally to make progress in an area that has long been neglected.

BIBLIOGRAPHY

[1]G. Dresselhaus, A. F. Kip, and C. Kittel, Phys. Rev. 92, 827 (1953); B. Lax, H. J. Zeiger, R. N. Dexter, and E. S. Rosenblum, Phys. Rev. 93, 1418 (1954); R. N. Dexter, H. J. Zeiger, and B. Lax, Phys. Rev. 95, 557 (1954); R. N. Dexter, B. Lax, A. F. Kip, and G. Dresselhaus, Phys. Rev. 96, 222 (1954); R. N. Dexter and B. Lax, Phys. Rev. 96, 223 (1954).

[2]S. Zwerdling, B. Lax, L. M. Roth, and K. J. Button, Phys. Rev. 114, 80 (1959); S. Zwerdling, B. Lax, K. J. Button and L. M. Roth, J. Phys. Chem. Solids 9, 320 (1959); S. Zwerdling, K. J. Button, B. Lax, and L. M. Roth, Phys. Rev. Letters 4, 173 (1960).

[3]R. N. Dexter, H. J. Zeiger, and B. Lax, Phys. Rev. 104, 637 (1956).

[4]J. M. Luttinger and W. Kohn, Phys. Rev. 97, 869 (1955); J. M. Luttinger, Phys. Rev. 102, 1030 (1956).

[5]R. C. Fletcher, W. A. Yager, and F. R. Merrit, Phys. Rev. 100, 747 (1955).

[6]J. J. Stickler, H. J. Zeiger, and G. S. Heller, Phys. Rev. 127, 1077 (1962).

[7]J. C. Hensel, Bull. Am. Phys. Soc. 6, 115 (1961); J. C. Hensel, Proc. Int. Conf. Semiconductor Physics, Exeter, 1962, p. 281; J. C. Hensel and G. Feher, Phys. Rev. 129, 1041 (1963); J. C. Hensel, H. Hasegawa, and M. Nakayama, Phys. Rev. 138, A225 (1965).

[8]R. A. Stradling, Phys. Letters 20, 217 (1966).

[9]D. M. S. Bagguley, M. L. A. Robinson, and R. A. Stradling, Phys. Rev. Letters 6, 143 (1963); M. L. Robinson, Phys. Rev. Letters 17, 963 (1966); H. T. Tohver and G. Ascarelli, Bull. Am. Phys. Soc. 13, 94 (1968); H. T. Tover and G. Ascarelli, in Proceedings of the Ninth International Conference on Physics of Semiconductors, Moscow, 1968; A. L. Mears and R. A. Stradling (private communication).

[10]K. J. Button, Benjamin Lax, and C. C. Bradley, Phys. Rev. Letters 21, 350 (1968). K. J. Button, A. Brecher, B. Lax, and C. C. Bradley, Phys. Rev. (to be published).

[11] E. Burstein, G. S. Picus, and H. A. Gebbie, Phys. Rev. 103, 825 (1956); R. J. Keyes, S. Zwerdling, S. Foner, H. H. Kolm, and Benjamin Lax, Phys. Rev. 103, 1804 (1956).

[12] K. J. Button, H. A. Gebbie, and B. Lax, IEEE J. Quantum Electronics, QE-2, 202 (1966); C. C. Bradley, K. J. Button, B. Lax, and L. G. Rubin, IEEE J. Quantum Electronics, QE-4, 733 (1968); K. M. Evenson, H. P. Broida, J. S. Wells, R. J. Mahler, and M. Mizushima, Phys. Rev. Letters 21, 1038 (1968).

[13] Kenneth J. Button, G. Landwehr, C. C. Bradley, P. Grosse, and Benjamin Lax, Phys. Rev. Letters 23, (7 July 1969).

[14] P. Grosse, "Die Festkörpereigenschaften von Tellur," Springer Verlag, Berlin, 1969.

[15] Kenneth J. Button, "Molecular Gas Lasers and Their Applications," Proceedings of the Chania School on Short Laser Pulses and Coherent Interactions, Gordon and Breach, N. Y. 1969.

[16] G. R. Wilkinson and D. H. Martin, Chapter 3, "Grating Spectroscopy" in "Spectroscopic Techniques" edited by D. H. Martin, North Holland Publishing Co., Amsterdam, 1967.

[17] P. L. Richards, Chapter 2, "Fourier Transform Spectroscopy" in "Spectroscopic Techniques" edited by D. H. Martin, North Holland Publishing Co., Amsterdam, 1967.

[18] J. M. Yarborough, S. S. Sussman, H. E. Puthoff, R. H. Pantell and B. C. Johnson, Appl. Phys. Letters (in press).

[illegible]. Parker, C. S. Hsieh, and [illegible]. Cooley, Phys. Rev. 109, 538 (1958); [illegible]

Kaye, S. Zwerdling, S. Fisher, [illegible], and Benjamin Lax, [illegible] 1964 (195[illegible]).

[illegible]. Batterson, [illegible]. A. [illegible], and B. Lax, IEEE J. Quantum Electronics [illegible] 202 (1966); C. C. Bradley, [illegible], and [illegible], IEEE J. Quantum Electronics, QE-[illegible], 28 (1968); [illegible]

R. J. Moller and M. Mizushima, Phys. Rev. Letters 21, 1048 (1968).

Kennedy, [illegible] Patton, O. Gandhi, C. C. Bradley, [illegible] Gross, and Benjamin Lax, Phys. Rev. Letters 23, (July 1969).

[illegible], "Die [illegible]", Springer-Verlag, Berlin, 1963.

Kenneth J. Button, "Molecular Constants [illegible]", Proceedings of the [illegible] Symposium on [illegible] and [illegible], N. Y., 1970.

G. R. Wilkinson and D. H. Martin, Chapter 3, [illegible] in "Spectroscopic Techniques" edited by D. H. Martin (North-Holland Publishing, Amsterdam, 1967).

[illegible], Chapter 7, [illegible] in "Spectroscopic Techniques" edited by D. H. Martin (North-Holland Publishing Co., Amsterdam, 1967).

[illegible], [illegible], and [illegible], Appl. Phys. Letters (in press).

XI

THEORY OF ELECTRON-PHOTON-PHONON INTERACTION IN A MAGNETIC FIELD

R. F. Wallis

Naval Research Laboratory
Washington, D. C. 20390 U.S.A.

1. Free Carriers

Free current carriers in semiconductors can interact with lattice vibrations through a variety of mechanisms. In polar semiconductors the macroscopic electric field provides a strong coupling to longitudinal optical phonons of long wavelength. This coupling leads to a number of striking phenomena, particularly when a magnetic field is present. The magnetic field provides a resonance, cyclotron resonance, which can be tuned by varying the magnetic field. When the cyclotron frequency is passed through the frequency of long wavelength longitudinal optical phonons, the cyclotron resonance line splits into two lines. Furthermore, the line width changes drastically as one passes through the interaction region.

We present a classical treatment of the interaction of free carriers with longitudinal optical phonons in the presence of a constant external magnetic field. Let us consider an n-type polar semiconductor with two atoms per primitive unit cell in a magnetic field in the z-direction. The sample is taken to be a thin plate in the x-z plane. For this configuration, only the long-wavelength optical phonons polarized in the y-direction (perpendicular to the plane of the plate) have a macroscopic electric field. We seek the normal modes of the coupled electron-phonon system.

Consider the equations of motion

$$m^* \frac{dv_x}{dt} + (e/c)\, v_y H = 0 \qquad (1)$$

$$m^* \frac{dv_y}{dt} - (e/c)\, v_x H = -eE_y \tag{2}$$

$$M_1 \frac{d^2 u_1}{dt^2} + k(u_1 - u_2) = e^* E_y \tag{3}$$

$$M_2 \frac{d^2 u_2}{dt^2} - k(u_1 - u_2) = -e^* E_y \tag{4}$$

where v_i is the i-th component of velocity of the electron, u_i is the y-component of displacement of the i-th ion in the unit cell, H is the magnitude of the external magnetic field, E_y is the y-component of the macroscopic electric field, m* is the electron effective mass, M_i is the mass of the i-th ion, e* is the ionic effective charge, and k is the short range coupling constant of the ions. If one introduces the difference variable $w = u_1 - u_2$, the lattice vibrational equations of motion can be transformed to

$$\overline{M} \frac{d^2 w}{dt^2} + kw = e^* E_y \tag{5}$$

where $\overline{M} = M_1 M_2/(M_1+M_2)$ is the reduced mass of the ions.

We now assume that each variable varies as $\exp(i\omega t)$. Equations (1), (2), and (5) now become

$$i\omega V_x + \omega_c V_y = 0 \tag{6}$$

$$i\omega V_y - \omega_c V_x = -(e/m^*) E_y^{(o)} \tag{7}$$

$$-\omega^2 W + \omega_o^2 W = (e^*/\overline{M}) E_y^{(o)} \tag{8}$$

where V_x, V_y, W and $E_y^{(o)}$ are the amplitudes of the appropriate variables and ω_c is the cyclotron frequency eH/m^*c. From the fact that that there is no externally applied electric field, we can conclude that

$$E_y + 4\pi P = 0 \tag{9}$$

where P is the polarization in the y-direction. The polarization can be written as the sum of three quantities

$$P = P_o - ney + Ne^*w \tag{10}$$

where P_o is the background lattice polarization, y is the y-component of displacement of the electrons, n is the concentration of electrons, and N is the number of unit cells in the crystal. Utilizing the result

$$E_y + 4\pi P_o = \varepsilon_o E_y, \tag{11}$$

where ε_o is the dielectric constant of the background lattice, we can rewrite Eq. (10) as

$$\varepsilon_o E_y - 4\pi ney + 4\pi Ne^*w = 0 \tag{12}$$

or, after differentiating with respect to t, as

$$i\omega\varepsilon_o E_y^{(o)} - 4\pi neV_y + i\omega 4\pi Ne^*W = 0 \tag{13}$$

Equation (13) can be rewritten as

$$E_y^{(o)} - (m^*/e)(\omega_p^2/i\omega)V_y + (\overline{M}/e^*)\Omega_p^2 W = 0 \tag{14}$$

where

$$\omega_p^2 = 4\pi ne^2/m^*\varepsilon_o \tag{15}$$

$$\Omega_p^2 = 4\pi Ne^{*2}/\overline{M}\varepsilon_o \tag{16}$$

Equations (6), (7), (8), and (14) constitute a set of linear homogeneous equations in the four variables V_x, V_y, W, and $E_y^{(o)}$. For a non-trivial solution, it is necessary that the determinant of coefficients must be zero:

$$\begin{vmatrix} i\omega & \omega_c & 0 & 0 \\ -\omega_c & i\omega & 0 & e/m^* \\ 0 & 0 & \omega^2-\omega_o^2 & e^*/\overline{M} \\ 0 & -(m^*\omega_p^2/ie\omega) & \overline{M}\Omega_p^2/e^* & 1 \end{vmatrix} = 0 \tag{17}$$

Expanding the determinant, one gets

$$(\omega^2 - \omega_c^2)(\omega^2 - \omega_\ell^2) - \omega_p^2(\omega^2 - \omega_o^2) = 0 \tag{18}$$

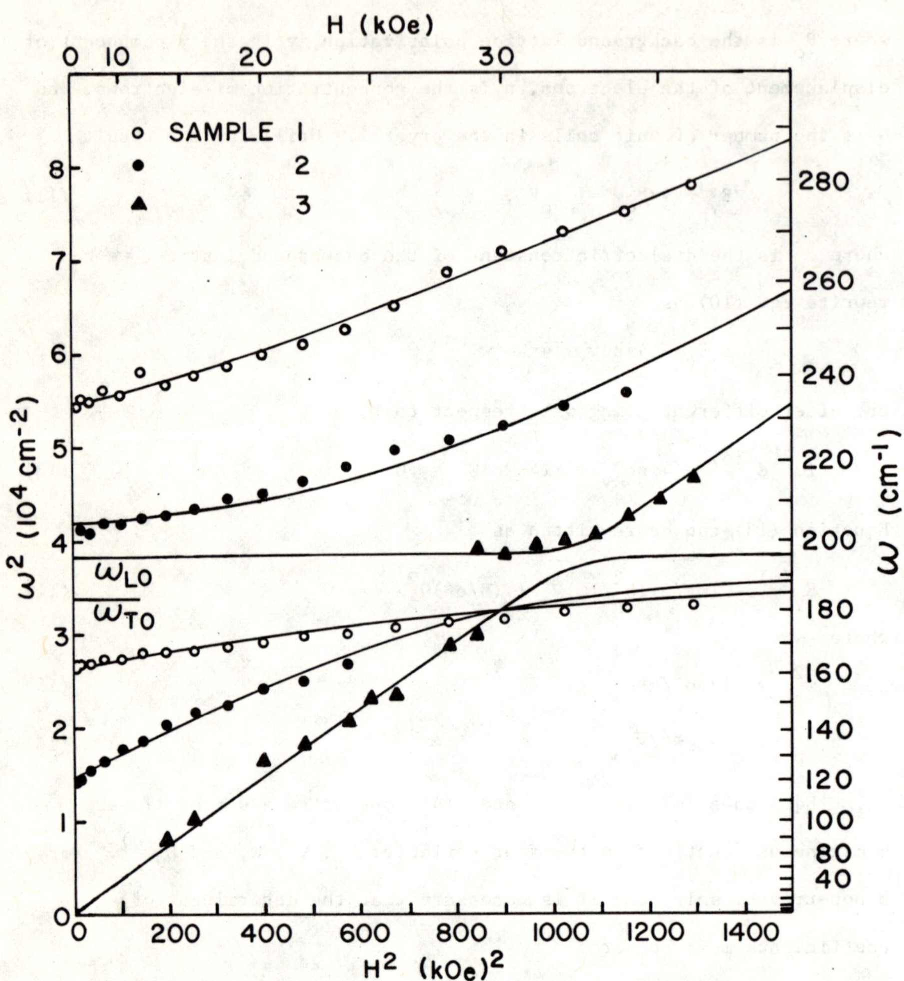

Fig. 1. Frequency positions of coupled modes as a function of magnetic field. The solid lines are calculated from Eq. (22). The points are experimental data for n-type InSb with carrier concentrations 1.4×10^{17} cm^{-3} (open circles), 5.5×10^{16} cm^{-3} (closed circles), and 1×10^{15} cm^{-3} (triangles).

where

$$\omega_\ell^2 = \omega_o^2 + \Omega_p^2 \tag{19}$$

The quantities ω_o and ω_ℓ are, respectively, the frequencies of long wavelength transverse and longitudinal optical vibrations in the absence of free charge carriers. Thus, from Eq. (14) with n = 0, we find

$$E_y^{(o)} = -\frac{\overline{M}}{e^*}\Omega_p^2 W. \tag{20}$$

Substituting into Eq. (8), we get

$$(\omega^2 - \omega_o^2 - \Omega_p^2)W = 0 \tag{21}$$

The longitudinal optical vibrations, which have a non-vanishing macroscopic electric field thus have the frequency $(\omega_o^2 + \Omega_p^2)^{1/2} = \omega_\ell \equiv \omega_{LO}$. The transverse optical vibrations, which have no macroscopic electric field, can be seen from Eq. (8) to have the frequency $\omega_o \equiv \omega_{TO}$.

Returning now to the case where free carriers are present, we can write the solutions of Eq. (18) as

$$\omega_\pm^2 = \frac{1}{2}\left\{\omega_v^2 + \omega_\ell^2 \pm \left[(\omega_v^2 - \omega_\ell^2)^2 + 4\,\omega_p^2\,\Omega_p^2\right]^{1/2}\right\} \tag{22}$$

where

$\omega_v^2 = \omega_c^2 + \omega_p^2$. The quantities ω_+ and ω_- are the frequencies of the normal modes of the coupled electron-phonon system in a magnetic field. Their squares are plotted as functions of the square of the magnetic field in Fig. 1 for the case of n-type InSb. The interaction of the electrons and phonons causes a splitting of the modes when the unperturbed frequencies ω_ℓ and ω_v are equal. From Eq. (22) one can see that the splitting increases with carrier concentration through ω_p. The experimental data[1] in Fig. 1 for samples of different carrier concentration clearly reveal this effect.

The treatment given above can be readily extended to the calculation of the dielectric constant and optical constants if the electron and phonon relaxation times are constants. The optical absorption is strong for the (-) mode and weak for the (+) mode when $\omega_v < \omega_\ell$. The opposite is true when $\omega_v > \omega_\ell$. At $\omega_v = \omega_\ell$, both modes have strong absorption, and one can observe a double peak[1].

The assumption of constant relaxation times, however, is not consistent with the experimental observations of Summers et al[2] who find a sharp variation in line width. This behavior has been explained theoretically by Harper[3] using a quantum mechanical treatment. The line width is relatively small for the (-) mode when $\omega_v < \omega_\ell$. As ω_v is increased through ω_ℓ, the line width suffers a discontinuous jump to a very large value as the (+) mode takes over the absorption. Further increase of ω_v results in a decrease in the line width of the (+) mode. The discontinuity in width occurs when the optically excited electrons can be scattered by the electron-phonon interaction into a high density of final states with the simultaneous emission of longitudinal optical phonons.

2. Bound Carriers

In order to handle the bound carrier case, it is necessary to use quantum mechanics. We first consider the quantum states of a free carrier in a constant external magnetic field. Using a vector potential of the form $A = (-\frac{1}{2}yH, \frac{1}{2}xH, 0)$, the Schrödinger equation can be written in cylindrical coordinates ρ, ϕ, z as

$$\left[-\nabla^2 - i\gamma \frac{\partial}{\partial\phi} + \gamma^2\rho^2/4 \right] F = EF \tag{23}$$

where F is the modulating function of the effective mass theory, energy and length are measured in units of the effective Rydberg $R_y^* = m^*e^4/2\hbar^2\varepsilon_o^2$ and effective Bohr radius $a_o^* = \varepsilon_o\hbar^2/m^*e^2$, and $\gamma = \hbar\omega_c/2R_y^*$. The solutions to Eq. (23) have been discussed by Dingle[4] and may be written as

$$F = \Phi_{\ell m}(\sigma, \phi)\, L_z^{-\frac{1}{2}} \exp(ik_z z) \tag{24}$$

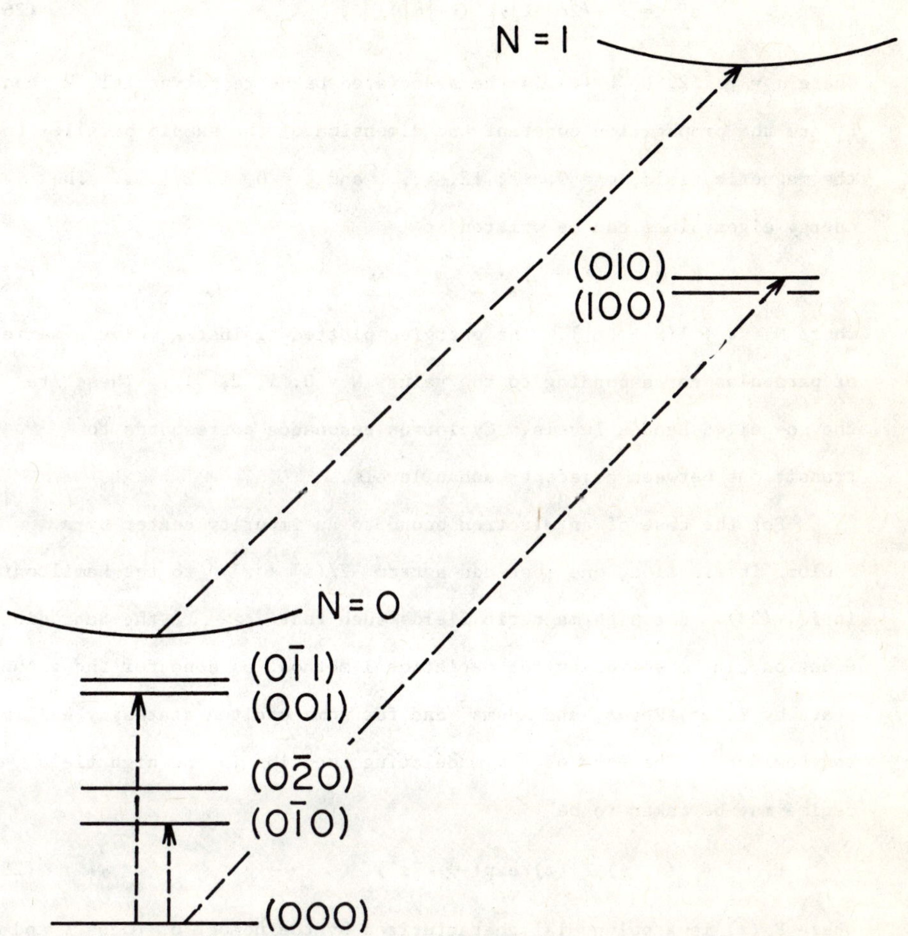

Fig. 2. Energy levels of a hydrogenic impurity electron in a high magnetic field. The parabolas are the first two Landau levels of the free electron. The bound states are labeled with quantum numbers discussed in the text.

with

$$\Phi_{\ell m}(\sigma, \phi) = c_1 e^{im\phi} \sigma^{\frac{1}{2}|m|} e^{-\frac{1}{2}\sigma} L_{\ell+|m|}^{|m|}(\sigma) \tag{25}$$

$$c_1^2 = (\gamma/2\pi)\, \ell!/\left[(\ell+|m|)!\right]^3, \tag{26}$$

where $\sigma = \gamma\rho^2/2$, $L_{\ell+|m|}^{|m|}(\sigma)$ is the associated Laguerre polynomial, k_z and L_z are the propagation constant and dimension of the sample parallel to the magnetic field, $m = 0, \pm 1, \pm 2, \ldots$, and $\ell = 0, 1, 2, \ldots$. The energy eigenvalues can be written as

$$E = 2\gamma(N + \tfrac{1}{2}) + k_z^2 \tag{27}$$

where $N = \ell + \frac{1}{2}(m + |m|)$. The energies plotted against k_z give a series of parabolas corresponding to the values $N = 0, 1, 2, \ldots$. These are the so-called Landau levels. Cyclotron resonance corresponds to transitions between adjacent Landau levels.

For the case of an electron bound to an impurity center by a coulomb interaction, one must add a term $-2/(\rho^2 + z^2)^{\frac{1}{2}}$ to the Hamiltonian in Eq. (23). For high magnetic fields such that $\gamma >> 1$, the Schrödinger equation can be solved by the variational method, as done for the ground state by Yafet, Keyes, and Adams[5] and for some excited states by Wallis and Bowlden[6]. The form of the modulating function in the high field regime may be taken to be

$$F = \Phi_{\ell m}(\sigma, \phi)\, P_\lambda(z) \exp(-\tfrac{1}{4}\gamma\varepsilon^2 z^2) \tag{28}$$

where $P_\lambda(z)$ is a polynomial characterized by the number of nodes λ and ε is a variational parameter. The character of the energy levels is illustrated in Fig. 2. From each Landau level are split off a number of discrete levels corresponding to bound states and labeled according to the values of ℓ, m, λ. These states all have binding energies relative to the parent Landau level edge which are small compared to the separation of adjacent Landau levels. The binding energies increase essentially logarithmically with magnetic field in the high field regime.

Theory of Electron-Photon-Phonon Interaction

Optical transitions between the various discrete levels have been discussed by Wallis and Bowlden[6]. Of particular interest is the transition between the ground state (000) and the excited state (010) shown in Fig. 2. This transition is referred to as impurity-shifted cyclotron resonance because it lies just to the high energy side of cyclotron resonance. In the remainder of this paper we shall be concerned with the phenomena which occur when the frequency of impurity-shifted cyclotron resonance is passed through the frequency of long-wavelength longitudinal optical phonons.

The Hamiltonian for longitudinal optical phonons can be written as

$$H = \tfrac{1}{2} P^2 + \tfrac{1}{2}\omega^2 Q^2 \tag{28}$$

where ω is the frequency, Q is the normal coordinate, and P is the conjugate momentum. The eigenfunctions ψ_n and energy eigenvalues E_n are those of a harmonic oscillator. The creation and annihilation operators b^+ and b are defined by

$$(\hbar\omega)^{\frac{1}{2}} b^+ = 2^{-\frac{1}{2}}(-iP + \omega Q) \tag{29}$$

$$(\hbar\omega)^{\frac{1}{2}} b = 2^{-\frac{1}{2}}(iP + \omega Q) \tag{30}$$

and have the properties

$$b^+ \psi_n = \sqrt{n+1}\, \psi_{n+1} \tag{31}$$

$$b\, \psi_n = \sqrt{n}\, \psi_{n-1} \tag{32}$$

The Hamiltonian can be written in terms of b^+ and b as

$$H_p = \hbar\omega\, (b^+ b + \tfrac{1}{2}) \tag{33}$$

The Hamiltonian expressing the interaction between the electrons and longitudinal optical phonons can be obtained in the following way. Utilizing Eqs. (19) and (20) and the Lyddanc-Sachs-Teller relation

$$\frac{\omega_\ell^2}{\omega_o^2} = \frac{\varepsilon_s}{\varepsilon_o} \tag{34}$$

we can write for the macroscopic electric field

$$E = -\omega_\ell \left[4\pi\left(\frac{1}{\varepsilon_o} - \frac{1}{\varepsilon_s}\right) \overline{M}/v_a\right]^{\frac{1}{2}} w \tag{35}$$

where $v_a = V/N$ is the volume of a unit cell and E and w in Eq. (35) are now vector quantities. The electric field at r due to an electron at r_e is given by

$$E_e = \frac{1}{\varepsilon_o} \nabla \frac{e}{|r-r_e|} \tag{36}$$

The energy associated with the electric field of the electron and phonons is

$$H = (\varepsilon_o/8\pi) \int (E + E_e)^2 \, d\tau \tag{37}$$

The interaction energy is

$$H_I = (\varepsilon_o/4\pi) \int E \cdot E_e \, d\tau$$

$$= -\varepsilon_o \, \omega_\ell \left[\frac{1}{4\pi} \left(\frac{1}{\varepsilon_o} - \frac{1}{\varepsilon_s}\right) \overline{M}/v_a\right]^{\frac{1}{2}} \int w \cdot E_e \, d\tau \tag{38}$$

The displacement w can be Fourier analyzed as

$$w = \frac{1}{\sqrt{N\overline{M}}} \sum_q e(q) \, e^{iq\cdot r} Q_q \tag{39}$$

where e(q) is the unit polarization vector for longitudinal waves $q/|q|$. Utilizing Eqs. (36) and (39) and carrying out an integration by parts, we obtain for the integral in Eq. (38)

$$\int e^{iq\cdot r} \nabla \frac{e}{|r-r_e|} \, d\tau$$

$$= - \frac{4\pi ieq}{|q|^2} e^{iq\cdot r_e} \tag{40}$$

The normal coordinate Q_q can be expressed in terms of creation and annihilation operators:

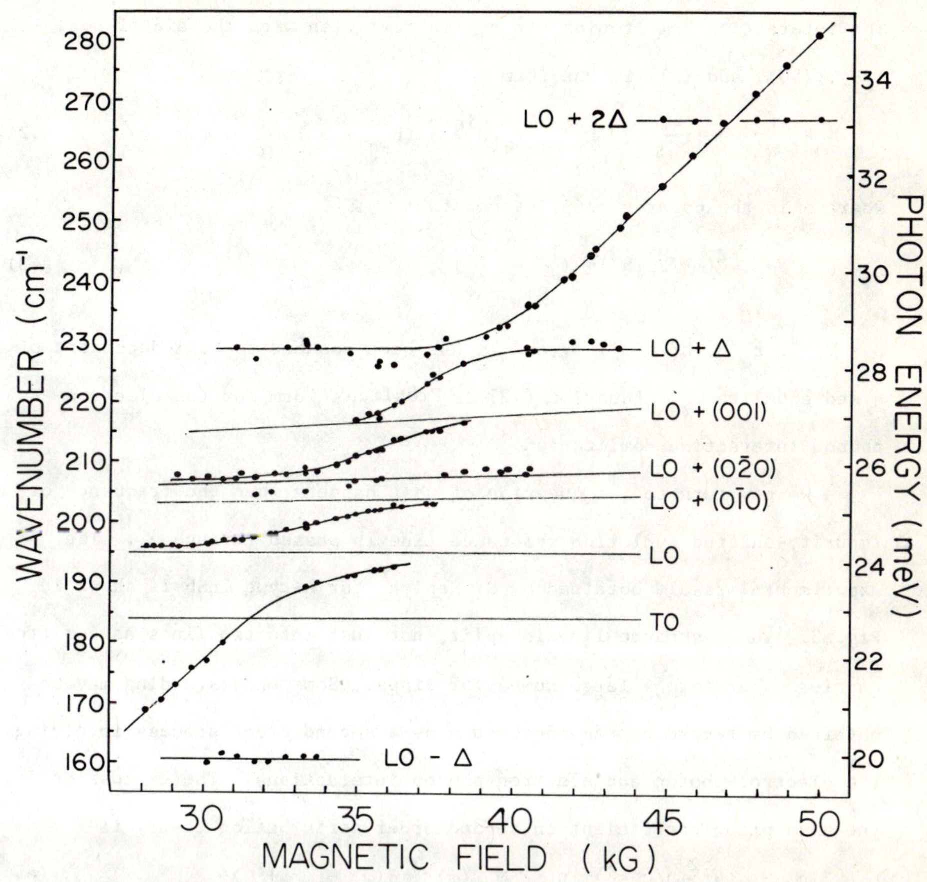

Fig. 3. Magnetic field dependence of the frequencies of coupled modes of impurity electrons and longitudinal optical phonons. The points are experimental data for n-type InSb. The solid curves are smooth fits to the data points.

$$Q_q = (\hbar/2\omega_q)^{\frac{1}{2}} (b^+_{-q} + b_q) \quad (41)$$

The interaction Hamiltonian can now be rewritten with the aid of Eqs. (39), (40), and (41) in the form

$$H_I = \hbar\omega_\ell \left(\frac{4\pi\alpha}{S}\right)^{\frac{1}{2}} \sum_q \frac{1}{r_o|q|} e^{iq\cdot r} (b^+_{-q} + b_q) \quad (42)$$

where α is the polaron coupling constant

$$\alpha = e^2 (m^*/2\omega_\ell \hbar^3)^{\frac{1}{2}} \left(\frac{1}{\varepsilon_o} - \frac{1}{\varepsilon_s}\right) , \quad (43)$$

$S = V/r_o^3$, $r_o = (\hbar/2m^*\omega_\ell)^{\frac{1}{2}}$, and ω_q has been assumed to be independent of q and equal to ω_ℓ. Equation (42) is Fröhlich's form for the electron-phonon interaction Hamiltonian.

We now turn to the question of what happens when the frequency of impurity-shifted cyclotron resonance line is passed through ω_ℓ. The experimental result obtained by R. Kaplan[7] for n-type InSb is shown in Fig. 3. The resonance line is split, not just into two lines as for free carriers, but into a large number of lines. Some understanding may be obtained by regarding the phenomenon as a second order process involving the electron-photon and electron-phonon interactions. The expression for the absorption coefficient in second order perturbation theory is

$$\eta = (4\pi^2 \omega n/c\hbar\varepsilon_o^{\frac{1}{2}}) \, |n_o \cdot M_{ki}(\omega)|^2 \, \delta(\omega_i - \omega_k + \omega) \quad (44)$$

for initial and final states i and k, respectively, and photons of frequency ω and unit polarization vector n_o. The second order matrix element is given by

$$M_{ki}(\omega) = \sum_j \left\{ \frac{\langle k|H_I|j\rangle C_{ji}}{E_i - E_j + \hbar\omega} + \frac{C_{kj} \langle j|H_I|i\rangle}{E_i - E_j} \right\} \quad (45)$$

where

$$C_{ji} = - \frac{e(E_j - E_i)}{\hbar\omega} \int F_j^* \, r \, F_i d\tau \quad (46)$$

and $E_i = \hbar\omega_i$ is the energy (electron plus phonon) of state i.

Under the conditions of interest, the first term in Eq. (45) is near resonance if the state (000) is the electronic state in i and (010) is the electronic state in j. Energy conservation requires that

$$\hbar\omega = E_k^e - E_i^e + \hbar\omega_\ell \tag{47}$$

where the superscript e denotes the electronic part of the energy. Equation (47) specifies the positions of the absorption peaks. If i is the ground state, the subsidiary peaks lie at higher frequencies than ω_ℓ. These are clearly seen experimentally. If k is the ground state, the subsidiary peaks lie at lower frequencies than ω_ℓ. There is some indication of such lines in the experimental data, but the lines are very weak.

The absorption coefficient relative to that of impurity-shifted cyclotron resonance can be calculated using the impurity-state wave functions in a high magnetic field given above. We need electron-phonon interaction matrix elements such as $\langle 010|e^{iq\cdot r}|0\bar{1}0\rangle$, $\langle 010|e^{iq\cdot r}|0\bar{2}0\rangle$, and $\langle 010|e^{iq\cdot r}|001\rangle$. We shall present the evaluation of these matrix elements in detail. In general, we can write

$$\langle \ell m\lambda|e^{iq\cdot r}|\ell' m'\lambda'\rangle = I_1 I_2 \tag{48}$$

where

$$I_1 = \int_0^\infty \int_0^{2\pi} \rho\, d\rho\, d\phi\, \Phi_{\ell m}(\sigma,\phi) e^{i|q_\perp|\rho\cos\phi}\, \Phi^*_{\ell' m'}(\sigma,\phi) \tag{49}$$

$$I_2 = \int_{-\infty}^{\infty} dz\, P_{\ell m\lambda}(z) e^{-\frac{1}{4}\gamma\varepsilon^2 z^2}\, e^{iq_z z}\, P_{\ell' m'\lambda'}(z)\, e^{-\frac{1}{4}\gamma\varepsilon^2 z^2} \tag{50}$$

and we have written

$$\begin{aligned} q\cdot r &= q_\perp \cdot r_\perp + q_z r_z \\ &= |q_\perp|\rho\cos\phi + q_z r_z \end{aligned} \tag{51}$$

$$|q_\perp|^2 = q_x^2 + q_y^2 \tag{52}$$

It should be noted that $q_\perp$ and q_z in Eqs. (49) and (50) are expressed in units of $1/a_o^*$.

The evaluation of I_1 is facilitated by setting $u = |q_\perp|\rho$ and using the expansion[8]

$$e^{iu\cos\phi} = J_o(u) + 2\sum_{n=1}^{\infty} (-1)^n J_{2n}(u)\cos 2n\phi$$

$$+ 2i \sum_{n=0}^{\infty} (-1)^n J_{2n+1}(u)\cos(2n+1)\phi \qquad (53)$$

Let

$$\Phi_{\ell m}(\sigma,\phi) = \chi_{\ell m}(\sigma) \frac{1}{\sqrt{2\pi}} e^{im\phi} \qquad (54)$$

Then, using Eq. (53), we can write the integral over ϕ in I_1 as

$$\frac{1}{2\pi} \int_o^{2\pi} d\phi\, e^{i(m-m')\phi} e^{iu\cos\phi}$$

$$= \sum_{n=0}^{\infty} (-1)^n J_{2n}(u)\, \delta_{|m'-m|,2n}$$

$$+ i \sum_{n=0}^{\infty} (-1)^n J_{2n+1}(u)\, \delta_{|m'-m|,2n+1} . \qquad (55)$$

For given m and m', there is only one term in Eq. (55) which contributes. The integral I_1 can now be expressed as

$$I_1 = \int_o^{\infty} \rho d\rho \chi_{\ell m}(\sigma)\, (-1)^{|m-m'|/2} J_{|m'-m|}(u)\chi_{\ell' m'}(\sigma) \qquad (56)$$

where $J_{|m-m'|}(u)$ is the Bessel function of the first kind.

From Eqs. (25) and (54) one sees that $\chi_{\ell m}(\sigma)$ involves the Laguerre polynomial, an explicit expression for which is

$$L_{\ell+k}^{k}(\sigma) = (-1)^k\, \Gamma(k+\ell+1) \sum_{p=0}^{\ell} (-1)^{\ell-p} \binom{k+\ell}{p} \sigma^{\ell-p}/(\ell-p)! . \qquad (57)$$

With the aid of Eq. (57), the integral I_1 can be reduced to the form

$$I_1 = (-1)^{\nu/2} 2\, c_1 c_1' (\gamma/2)^{\frac{1}{2}(k+k')} (-1)^{k+k'} \Gamma(k+\ell+1) \times$$

$$\times \Gamma(k'+\ell'+1) \sum_{s=0}^{\ell} \sum_{s'=0}^{\ell'} \frac{(-1)^{\ell-s} (-1)^{\ell'-s'}}{(\ell-s)!\,(\ell-s')!} \binom{k+\ell}{s} \binom{k'+\ell'}{s'} \times$$

$$\times (\gamma/2)^{\ell-s} (\gamma/2)^{\ell'-s'} G(\mu,\nu) \tag{58}$$

where

$$G(\mu,\nu) = \int_0^\infty d\rho\, \rho^{\mu-1} e^{-\sigma} J_\nu(u) \tag{59}$$

$$\mu = 2+k+k'+2(\ell-s)+2(\ell'-s') \tag{60}$$

$$\nu = |m-m'| \tag{61}$$

$$k = |m|, \; k' = |m'| \tag{62}$$

With the aid of Watson[8], p.394, one can evaluate $G(\mu,\nu)$ with the result

$$G(\mu,\nu) = \frac{\Gamma(\frac{1}{2}\nu + \frac{1}{2}\mu)(\frac{1}{2}q_\perp)^\nu (\gamma/2)^{\nu/2}}{2\Gamma(\nu+1)(\gamma/2)^{\mu/2}} \exp(-\frac{q_\perp^2}{2\gamma}) \times$$

$$\times {}_1F_1(\tfrac{1}{2}\nu - \tfrac{1}{2}\mu + 1;\; \nu + 1;\; q_\perp^2/2\gamma) \; . \tag{63}$$

where ${}_1F_1(\alpha;\beta;x)$ is the confluent hypergeometric function. This completes the evaluation of I_1.

The evaluation of I_2 given by Eq. (50) requires integrals of the form

$$K_n(q_z, \alpha) = \int_{-\infty}^{\infty} dz\, z^n e^{-\alpha z^2} e^{iq_z z} \; . \tag{64}$$

We note that

$$K_n(q_z,\alpha) = (-i)^n \frac{d^n}{dq_z^n} K_o(q_z,\alpha) \; . \tag{65}$$

Thus, we need only evaluate $K_o(q_z, \alpha)$ which is given by

$$K_o(q_z, \alpha) = \int_{-\infty}^{\infty} dz\, e^{-\alpha z^2} e^{iq_z z}$$

$$= (\pi/\alpha)^{\frac{1}{2}} \exp(-q_z^2/4\alpha) \quad (66)$$

We now tabulate a few of the integrals I_1 and I_2 of interest:

$$I_1(010, 000) = -i(q_\perp^2/2\gamma)^{\frac{1}{2}} \exp(-q_\perp^2/2\gamma) \quad (67)$$

$$I_2(010, 000) = \sqrt{2}(\varepsilon\varepsilon'/\varepsilon^2 + \varepsilon'^2)^{\frac{1}{2}} \exp -q_z^2/\gamma(\varepsilon^2 + \varepsilon'^2) \quad (68)$$

$$I_1(010, 0\bar{1}0) = -(q_\perp^2/2\gamma) \exp(-q_\perp^2/2\gamma) \quad (69)$$

$$I_2(010, 0\bar{1}0) = \exp(-q_z^2/2\gamma\varepsilon) \quad (70)$$

$$I_1(010, 001) = -i(q_\perp^2/2\gamma)^{\frac{1}{2}} \exp(-q_\perp^2/2\gamma) \quad (71)$$

$$I_2(010, 001) = i\, 2\sqrt{2}\, \frac{\varepsilon^{\frac{1}{2}} \varepsilon'^{\frac{3}{2}}}{(\varepsilon^2 + \varepsilon'^2)} \left[q_z^2/\gamma(\varepsilon^2+\varepsilon'^2)\right]^{\frac{1}{2}} \times$$

$$\times \exp\left[-q_z^2/\gamma(\varepsilon^2+\varepsilon'^2)\right] \quad (72)$$

$$I_1(010, 0\bar{2}0) = (-i/\sqrt{2})(q_\perp^2/2\gamma)^{\frac{3}{2}} \exp(-q_\perp^2/2\gamma) \quad (73)$$

$$I_2(010, 0\bar{2}0) = \sqrt{2}\, (\varepsilon\varepsilon'/\varepsilon^2 + \varepsilon'^2)^{\frac{1}{2}} \exp\left[-q_z^2/\gamma(\varepsilon^2 + \varepsilon'^2)\right] \quad (74)$$

We now return to the calculation of the absorption coefficient and restrict ourselves to very low temperatures so that the initial state i is the ground state of the system. Then, of the two terms in Eq. (45) for $M_{ki}(\omega)$, only the first shows a resonant behavior, since the energy denominator of the second term is at least as large as $\hbar\omega_\ell$. We therefore neglect the second term and write for the absorption coefficient

$$\eta = (4\pi^2\omega n/c\hbar\varepsilon_o^{\frac{1}{2}}) \sum_j \left| \frac{n_o \cdot C_{ji}}{E_i - E_j + \hbar\omega} \right|^2 \times$$

$$\times \sum_q |\langle k|H_I|j\rangle|^2\, \delta(\omega_i - \omega_k + \omega) \quad (75)$$

Our result for the absorption coefficient corresponds to the excitation of an impurity electron by the electromagnetic field to an intermediate state followed by transition of the electron to its final state through the electron-phonon interaction and accompanied by the emission of a longitudinal optical phonon. We consider in detail the case where the initial electronic state is (000), the intermediate state is (010), and the final state is $(0\bar{1}0)$. The sum over q in Eq. (75) can be performed by passing to an integral over q and using the preceding results for the matrix elements of H_I. For the case of interest,

$$\sum_q |\langle k|H_I|j\rangle|^2 = (h\omega_\ell)^2 \left(\frac{4\pi\alpha}{S}\right) \left(\frac{r_o}{a_o^*}\right) \frac{V}{8\pi^3} \times$$

$$\times \; 2\pi \int_0^\infty q_\perp dq_\perp \int_{-\infty}^{\infty} dq_z \frac{q_\perp^4}{q_\perp^2+q_z^2} e^{-q_\perp^2/\gamma} e^{-q_z^2/\gamma\varepsilon^2} \tag{76}$$

The integrals in Eq. (76) can be evaluated to give the result

$$\sum_q |\langle k|H_I|j\rangle|^2 = \frac{(h\omega_\ell)^2}{2\sqrt{\pi}} \left(\frac{r_o}{a_o^*}\right) \alpha\gamma^{\frac{1}{2}} \times$$

$$\times \; \frac{\varepsilon(2-5\varepsilon^2)}{8(1-\varepsilon^2)^2} + \frac{3\varepsilon^5}{16(1-\varepsilon^2)^{5/2}} \log \frac{1+\sqrt{1-\varepsilon^2}}{1-\sqrt{1-\varepsilon^2}} \tag{77}$$

$$\approx \frac{(h\omega_\ell)^2}{8\sqrt{\pi}} \left(\frac{r_o}{a_o^*}\right) \alpha\gamma^{\frac{1}{2}}\varepsilon \quad \text{for } \varepsilon^2 \ll 1 . \tag{78}$$

The approximation $\varepsilon^2 \ll 1$ in Eq. (78) is reasonably well satisfied in n-type InSb at high magnetic fields[6].

It should be noted that the matrix element C_{ji} in Eq. (75) has the same value as that for impurity-shifted cyclotron resonance. Hence, we may relatively easily calculate the ratio of the integrated intensity of the subsidiary peak to the integrated intensity of the impurity-shifted cyclotron resonance peak. This ratio is given approximately by

$$R = \frac{\alpha\gamma^{\frac{1}{2}}\varepsilon}{8\sqrt{\pi}} \left(\frac{\hbar\omega_\ell}{E_i - E_j + \hbar\omega}\right)^2 \left(\frac{r_o}{a_o^*}\right) . \tag{79}$$

For n-type InSb and typical experimental conditions, $\alpha = 0.02$, $\gamma = 14$, and $\varepsilon = 0.4$. Taking

$$E_i - E_j + \hbar\omega = E_{000}^{(e)} - E_{010}^{(e)} + \hbar\omega$$

$$= 0.1\, \hbar\omega_\ell ,$$

we find from Eq. (79) that $R \approx 0.04$. Similar results are obtained for final states $(0\bar{2}0)$ and (001). It is difficult to make a precise comparison with the photoconductivity data of Kaplan[7], but the order of magnitude seems reasonable.

The foregoing second-order perturbation theory is valid only if R is small - i.e., away from the region of resonance. The absorption coefficient in the resonance region can be fully treated only with more sophisticated techniques. One such approach is the use of Green's functions, which is currently under investigation. The calculation of the peak positions in the resonance region, however, may be carried out using Wigner-Brillouin perturbation theory as suggested to the writer by D. M. Larsen. The procedure is to first write down the second-order perturbative expression for the energy of the intermediate state j arising from the electron-phonon interaction:

$$E = E_j + \sum_k \frac{|\langle k|H_I|j\rangle|^2}{E_j - E_k} \tag{80}$$

The first order term is omitted because it plays no essential role. The energy E_j in the denominator on the right side of Eq. (80) is then replaced by the unknown energy E:

$$E = E_j + \sum_k \frac{|\langle k|H_I|j\rangle|^2}{E - E_k} \tag{81}$$

By restricting oneself to only a few final electronic states, Eq. (80)

becomes an algebraic equation of reasonable order in E and may be solved for the eigenvalues of the coupled electron-phonon system.

For the situation of interest, $E_j = E_j^{(e)}$ and $E_k = E_k^{(e)} + \hbar\omega_\ell$. If we restrict our attention to a single state k, then Eq. (81) takes the form

$$E = E_j^{(e)} + \frac{A}{E - E_k^{(e)} - \hbar\omega_\ell} \tag{82}$$

where A is the interaction constant. The solutions of Eq. (82) are

$$E = \tfrac{1}{2}\Big\{E_j^{(e)} + E_k^{(e)} + \hbar\omega_\ell \pm \Big[(E_j^{(e)} - E_k^{(e)} - \hbar\omega_\ell)^2 + 4A\Big]^{\frac{1}{2}}\Big\} \tag{83}$$

In the limit $(E_j^{(e)} - E_k^{(e)} - \hbar\omega_\ell)^2 >> 4A$, the solutions can be written as

$$E - E_k^{(e)} = \begin{cases} E_j^{(e)} - E_k^{(e)} \\ \hbar\omega_\ell \end{cases} \tag{84}$$

Now $E_j^{(e)} - E_k^{(e)}$ is essentially linear in the magnetic field, while $\hbar\omega_\ell$ is independent of the field. In the region where these zero order curves would cross, the interaction causes the typical repulsion and non-crossing of the curves. By adding additional states k, one can get the more complicated pattern of non-crossing curves as shown by the solid lines in Fig. 3.

References

1. R. Kaplan, E. D. Palik, R. F. Wallis, S. Iwasa, E. Burstein, and Y. Sawada, Phys. Rev. Letters 18, 159 (1967).

2. C. Summers, P. G. Harper, and S. D. Smith, Solid State Commun. 5, 615 (1967).

3. P. G. Harper, Proc. Phys. Soc. 92, 793 (1967).

4. R. B. Dingle, Proc. Roy. Soc. A211, 500 (1952).

5. Y. Yafet, R. W. Keyes, and E. N. Adams, J. Phys. Chem. Solids 1, 69 (1956).

6. R. F. Wallis and H. J. Bowlden, J. Phys. Chem. Solids 7, 78 (1958).

7. R. Kaplan and R. F. Wallis, Phys. Rev. Letters, 20, 1499 (1968).

8. G. N. Watson, Theory of Bessel Functions, Cambridge University Press (1948).

XII

FAR-INFRARED MAGNETOOPTICAL STUDIES OF SEMICONDUCTORS USING FOURIER SPECTROSCOPY AND PHOTOCONDUCTIVITY TECHNIQUES

R. Kaplan

Naval Research Laboratory
Washington, D. C. 20390 U.S.A.

I. INTRODUCTION

Magnetooptical investigations have added greatly to our knowledge of many of the properties of semiconductor materials. In recent years, increasingly more attention has been devoted to magnetooptical studies in the far-infrared region of the electromagnetic spectrum. This has occurred largely because of the improvement of spectroscopic techniques, and the greater availability of high quality material for samples. The desire for a far-infrared spectroscopic capability in semiconductor investigations stems in general from two sources. First there are phenomena, such as lattice vibrations and impurity excitations, whose natural frequencies correspond to far-infrared wavelengths. Second, there are phenomena which may be observed over a wide spectral region, but which can be studied most advantageously in the far-infrared. Cyclotron resonance falls in the second category, since carrier mobility and effective mass may dictate the use of a particular spectral region.

The present paper is concerned with the application of Fourier transform spectroscopy and the technique of resonant photoconductivity to several areas of current interest in the magnetooptics of semiconductors. The use of Fourier spectroscopy in far-infrared investigations has become widespread during the last few years, largely due to the increasing availability of high-speed electronic computer facilities. For certain applications, Fourier spectroscopy has distinct advantages over competing methods such as grating, laser, and microwave spectroscopy. Since the theoretical basis of Fourier spectroscopy is treated extensively in the literature, this subject will be omitted here*. The use of photoconductivity as a technique for studying the resonant absorption of radiation in the presence of a magnetic field will be described. In this technique the sample under investigation serves in effect as its own detector, a procedure which has a number of advantages.

As examples of the application of Fourier spectroscopy and resonant photoconductivity to magnetooptical investigations in semiconductors, recent work at the Naval Research Laboratory related to impurity states and electron-phonon interactions will be described. Observations of the optical spectra of weakly-bound donor impurities in InSb and GaAs, and the interactions of free and localized electrons with optical phonons in InSb, will be discussed in detail.

II. APPLICATION OF FOURIER SPECTROSCOPY TO FAR-INFRARED MAGNETOOPTICAL STUDIES OF SEMICONDUCTORS

A number of experimental techniques are available for the study of magnetooptical effects in the far-infrared. These may conveniently be classified according to whether the magnetic field strength or the photon energy is varied to obtain a spectrum. In the first class, radiation of fixed energy is most commonly obtained from gas lasers or various microwave devices. In the second class a grating monochrometer or Fourier spectrometer is generally used in conjunction with a broadband radiation source. Each of these techniques has advantages for particular applications. Fourier spectroscopy is probably the most advantageous for a wide variety of far-infrared magnetooptical studies in semiconductors. The major reasons for this may be summarized as follows.

1. In many experiments the effects of interest cannot be studied by field-sweep techniques. For example, in the case of impurity

*See, for example, Applied Optics 8, No. 3 (1969) for a recent survey of Fourier and other spectroscopic techniques, and references to published work.

transitions to be considered later, certain transition energies depend very weakly on magnetic field strength. The corresponding absorption lines thus appear extremely broad in a spectrum obtained by sweeping the field. Furthermore, essential changes in sample conditions may occur during field sweep, examples being the field-induced localization of charge carriers, and alteration in the population distribution over a set of energy levels. These effects are much less important for narrow spectral lines such as may be observed in magnetic resonance experiments. For the latter, the high resolution obtainable with a laser or microwave source may be essential. Such high resolution is unnecessary for most far-infrared magnetooptical studies in semiconductors. Thus for such studies, in view of the foregoing, the use of grating or Fourier spectroscopy appears to be indicated.

2. The relative merits of grating and Fourier spectroscopy have been debated at great length. To some extent the choice depends on such factors as the availability of high-speed digital computers, and the experience of the individual spectroscopist. For the magnetooptical experiments undertaken recently at the Naval Research Laboratory, a relatively portable instrument capable of a resolution of about 0.5 cm^{-1} in the spectral range 5-300 cm^{-1} has been required. The portability was necessary because the instrument has been used with a variety of magnets, including the rather bulky Bitter-type high-field solenoids. A grating instrument capable of the required performance would tend to be much larger and heavier than the corresponding Fourier spectrometer. Furthermore, the latter can cover large spectral regions without change of components, requires little filtering or order-sorting, and speeds up the process of data-taking due to its multiplex advantage over a grating spectrometer. These attributes are especially helpful when broad spectral regions must be scanned, a common requirement in the present

application. The single major disadvantage with the Fourier spectrometer has been the time delay between data-taking and Fourier analysis of the interferograms obtained.

The adaptation of a Fourier spectrometer for use in magnetooptical investigations is a fairly straightforward procedure, and has by now been accomplished in a number of laboratories. Since the use of cryogenic detectors such as InSb and Ge bolometers greatly increases the spectrometer sensitivity, it is natural to employ superconducting magnets, which can be mounted in the same liquid helium Dewars as the detectors. The light is coupled from the spectrometer to the detector by a metal light pipe, which passes through the magnet. The sample is positioned at the center of the magnet so that the field dependence of its absorptivity or photoconductivity may be observed. Arrangements for the study of sample reflectivity are also possible. By means of suitable tapers and mirrors, the direction of propagation of the light incident on the sample may be varied. Thus propagation parallel or normal to the field direction may be achieved; these are the Faraday and Voigt orientations, respectively. Use of wire grid polarizers in the Voigt orientation makes possible the selection of incident radiation with electric vector $\mathbf{E}$ either parallel or normal to the direction of the magnetic field $\mathbf{H}$. Sample temperature may be varied by mounting the sample on a block containing a resistive heater, in a closed tube immersed in liquid helium. Thermal contact between the sample block and the walls of the tube is provided by helium gas at low pressure. Sample temperatures as high as about 80°K are easily maintained by proper selection of helium gas pressure and current in the resistive heater. Further details of this system have been described elsewhere[1].

The usual spectroscopic practice is to obtain spectra for the sample-in and sample-out conditions, and compute the ratio. In observing magnetic field-induced effects, however, it is often more meaningful to compare field-on and field-off spectra with the sample in the beam. This is also a much simpler procedure for samples mounted in a cryogenic apparatus, and reduces the advantage of a two-beam (i.e. sample and reference beam) instrument. Most of the spectra to be presented in the following discussions are either uncorrected field-on runs, or ratios of successive field-on and field-off spectra.

III. USE OF RESONANCE PHOTOCONDUCTIVITY IN MAGNETOOPTICAL STUDIES OF SEMICONDUCTORS

For the purposes of this work, photoconductivity may be taken to include any process in which the conductivity of a sample is affected by the absorption of light. The light alone may not be sufficient to produce the desired effect; for example the absorption of phonons may also be required. In general, a photoconductive signal may result from either a change in the number or the mobility of the charge carriers in a sample. If the photoconductivity results from a resonant absorption of radiation, then the wavelength dependence of the photoconductivity and the absorption are likely to be similar. Thus, for example, cyclotron resonance and impurity excitations can be studied in a number of semiconductors by observations of resonant photoconductivity. In certain circumstances, this approach is preferable to the use of the normal transmission techniques. A few advantages of the resonant photoconductivity technique are the following.

First, the photoconductivity approach makes a detector unnecessary. This is especially useful when the magnetic field dependence of the transition energy of a resonance must be studied over a wide spectral region, whose coverage in a transmission experiment would require the use of more than one detector. For example, electron cyclotron resonance in InSb has been studied[2] by means of its resonant photoconductivity in the range 20-480 cm^{-1} (3-100 kG). Second, extremely weak absorptions are sometimes more easily observed in photoconductivity, since the latter in many circumstances is essentially a null-background technique. That is, zero signal is observed except at the wavelengths corresponding to the absorption process. Third, the photoconductivity technique often makes it possible to separate competing absorption processes, not all of which lead to a change in sample resistance. For example, fairly weak electronic absorption processes have been studied[3] up to within 3 or 4 cm^{-1} of the extremely strong transverse optical lattice absorption in InSb. Fourth, photoconductivity spectra can yield information not available from transmission experiments. A typical case is the observation[4] of sharp oscillations in the photoconductivity of a number semiconductors, not seen in absorption, corresponding to the creation of an integral number of optical phonons when certain energy conditions are satisfied. These are a few typical applications of resonant photoconductivity as a spectroscopic technique. Of course the technique is not generally applicable to all absorption studies, but when it is, it often proves very useful.

When resonant photoconductivity spectra are obtained by means of a Fourier spectrometer, quite interesting recorded signals are observed. Aside from a dc component, the signal amplitude $F(\Delta)$ as a function of path difference Δ is given by[5]

$$F(\Delta) = \int_{-\infty}^{\infty} E(\nu) \cos(2\pi\nu\Delta) d\nu \qquad (1)$$

where ν and $E(\nu)$ are the wavenumber and spectral distribution of the detected radiation, respectively. Since the use of a narrow-band detector and "white" source leads to the same form for $E(\nu)$ as a monochromatic source and wide-band detector, for either case $F(\Delta)$ is a simple cosine wave. If the sample under investigation has a single photoconductivity peak of Lorentzian shape, whose resonance wavenumber is ν_o and half-width at half-maximum is $\delta\nu$, then $E(\nu)$ has the form

$$E(\nu) \sim \frac{\nu_o \delta\nu}{(\nu - \nu_o)^2 + (\delta\nu)^2}. \qquad (2)$$

The relatively broad-band effects due to source, filters, beam splitter, etc. have been omitted from Eq. 2. It is readily shown that the interferogram obtained in this case is a damped cosine wave:

$$F(\Delta) \sim \cos(2\pi\nu_o\Delta)\exp(-2\pi\Delta\delta\nu). \qquad (3)$$

Thus the parameters ν_o and $\delta\nu$, which characterize the optical transition responsible for the photoconductivity, can be obtained directly from the interferogram.

Studies of cyclotron resonance and other transitions of electrons in InSb often reveal a doublet absorption. The two components arise from the simultaneous presence of free and localized electrons in the samples, as will be discussed later in greater detail. Both transitions lead to photoconductivity peaks whose widths and shapes are comparable, and whose separation $\nu_1 - \nu_2$ is small compared to ν_1 and ν_2, the resonant wavenumbers of the two transitions. Furthermore, by adjusting the sample temperature, the intensities of the two photoconductivity peaks can be made equal. In this situation the interferogram $F(\Delta)$ is a simple sum of two terms like those in Eq. (3), which combine to give the result:

$$F(\Delta) \sim \cos\left[2\pi\Delta\,\frac{(\nu_1+\nu_2)}{2}\right]\cos\left[\pi\Delta(\nu_1-\nu_2)\right]\exp(-2\pi\Delta\delta\nu). \qquad (4)$$

As would be expected, a beating appears in the envelope of the damped cosine wave. The period of the beating gives the separation in wavenumber of the doublet components, while the period of the cosine wave gives the average resonant wavenumber. If the conditions given above on

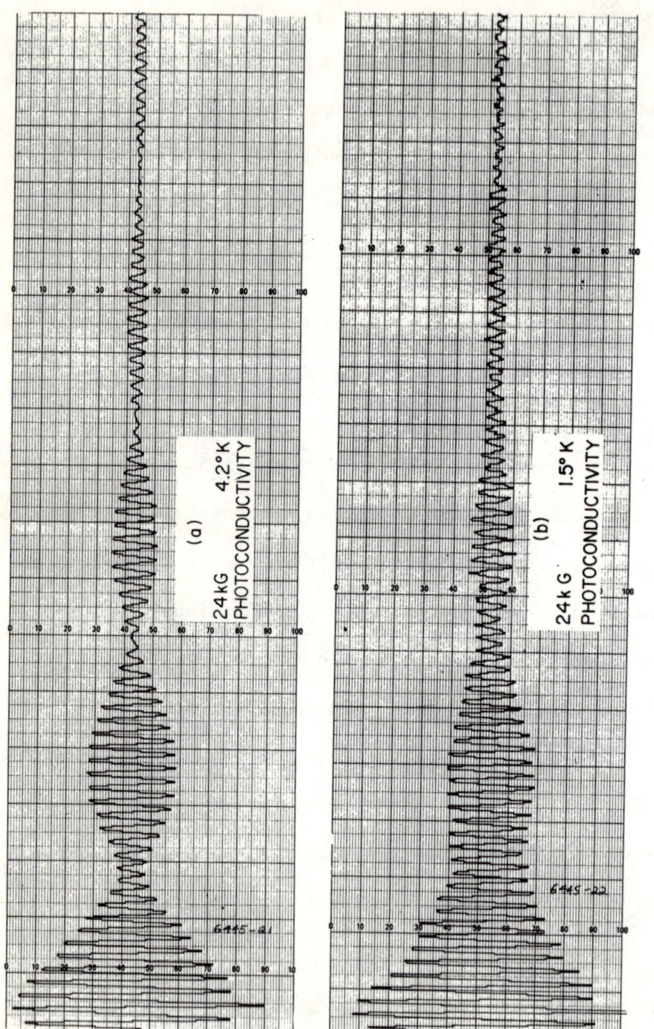

Fig. 1. Interferograms obtained in studies of resonant photoconductivity in InSb by means of Fourier transform spectroscopy. At 4.2°K the sample contains both free and localized electrons, and a doublet transition occurs near the cyclotron resonance energy. At 1.5°K most of the free carriers have been removed, greatly reducing the strength of one of the doublet components.

the widths and relative intensities of the peaks are relaxed, additional terms appear in Eq. (4), one result being that the amplitude of the beating effect is reduced. In Fig. 1, interferograms are shown for the cyclotron resonance of electrons in InSb, as studied by resonant photoconductivity. The doublet components were at about equal strength for a sample temperature of 4.2°K. As the temperature was lowered, more of the electrons became localized, and one of the doublet components became dominant.

For photoconductivity studies using Fourier spectroscopy it is necessary only to attach electrical leads to the samples via low-noise contacts. It is often useful to leave the detector in place, so that successive transmission and photoconductive spectra can be compared.

IV. MAGNETOOPTICAL SPECTRA OF HYDROGENIC IMPURITIES: InSb and GaAs

One general area to which the techniques described above have been applied, is the study of the magnetooptical spectra of impurity atoms in semiconductors. This study is of interest in its own right, but also because of its involvement with other areas of interest, such as carrier transport and freezeout, and the development of infrared detectors. The case of donor impurities in host crystals such as InSb and GaAs is relatively simple. In these crystals, the conduction band energy has its minimum and is isotropic at zero wavevector. The theory appropriate to this situation will now be outlined, and then the experimental results for donor impurities in InSb and GaAs will be discussed.

The theory of impurity states and spectra in a magnetic field is concerned basically with two interactions, the Coulomb interaction between the donor electron and the positively charged impurity ion, and the interaction between the donor electron and the applied magnetic field. In the absence of a magnetic field, for the case of an isotropic conduction band with its minimum at zero wave vector, the hydrogenic model can be applied directly. According to the effective mass theory, the electron moves in the Coulomb field of the charged impurity ion with an effective mass m^*. Since the charged impurity is imbedded in a dielectric medium whose dielectric constant is κ, the Coulomb potential at a distance r from the impurity is $e^2/\kappa r$. (A singly charged impurity ion is assumed.) The localized states of the electron then have the energies

$$E_n = -\frac{m^* e^4}{2\hbar^2 \kappa^2} \frac{1}{n^2} = -\frac{R_y^*}{n^2}\ ; \quad n = 1, 2, \cdots . \tag{5}$$

The ionization energy, or effective Rydberg energy R_y^*, is typically orders of magnitude smaller than that of atomic hydrogen. The usual s, p, d, $\cdots$ hydrogenic states are obtained for the values of the quantum number $\ell = 0, 1, 2, \cdots$. Allowed transitions from the ground state are $1S \rightarrow np$, $n = 2, 3, \cdots$. Transitions of this type have been observed[6,7] in GaAs, and Eq. (5) has been verified to within experimental error of a few percent for this case.

The effects of an applied magnetic field on the impurity states and spectra have been studied theoretically by a number of workers[8-12]. These studies have usually been restricted to the high or low field regimes, the distinction being based on the relative strengths of the Coulomb and magnetic interactions. It is convenient to define a parameter γ, as the ratio of the zero-point magnetic interaction energy to the Coulomb binding energy:

$$\gamma = \frac{\hbar\omega_c}{2R_y^*} . \tag{6}$$

In Eq. (6), ω_c is the angular cyclotron frequency. For the low field case, $\gamma \ll 1$, the magnetic interaction is a small perturbation, and the usual hydrogenic Zeeman effect is observed. The effects of spin splitting may also be present if the spin g-value is sufficiently large. In the high-field regime, $\gamma >> 1$, the Coulomb interaction is a small perturbation on the Landau states of a free carrier in a magnetic field. The discrete impurity states are found to cluster at energies slightly below that of each Landau level; i.e. slightly below the energies $\hbar\omega_c(N+\frac{1}{2})$, $N = 0, 1, \cdots$. The high field impurity states and spectra thus bear little resemblance to their counterparts at low fields.

Calculations of the high field states adhere to the cylindrical symmetry imposed by the magnetic field. With the field in the z-direction, the appropriate effective mass equation in cylindrical coordinates is

$$\left\{-\nabla^2 - i\gamma \frac{\partial}{\partial\phi} + \frac{\gamma^2\rho^2}{4} - \frac{2}{(\rho^2+z^2)^{\frac{1}{2}}}\right\} F(\rho, \phi, z) = EF(\rho, \phi, z). \qquad (7)$$

Equation (7) is expressed in dimensionless units, in the sense that it does not contain the material parameters m* and κ. Energy is in units of R^*_y, magnetic field in terms of γ. Solutions may be applied to donor states in any host crystal whose conduction band has the characteristics listed previously.

It has been shown[8] that in the limit $\gamma \to \infty$, exact solutions of Eq. (7) can be written in the form

$$F_{NM\lambda} = \Phi_{NM}(\rho,\phi)\, f_{NM\lambda}(z) . \qquad (8)$$

The functions $\Phi_{NM}(\rho,\phi)$ are the transverse part of the usual free carrier solutions in a magnetic field in the absence of the Coulomb interaction, while the $f_{NM\lambda}(z)$ satisfy a one-dimensional Schrödinger equation with an averaged Coulomb potential. The energies corresponding to the $\gamma \to \infty$ wave functions, Eq. (8), are

$$E_{NM\lambda} = \hbar\omega_c(N+\tfrac{1}{2}) + \varepsilon_{NM\lambda} . \qquad (9)$$

For the bound impurity states, the $\varepsilon_{NM\lambda}$ assume discrete negative values whose magnitudes are small compared to $\hbar\omega_c$. Thus as previously described, the energy level system in the high field limit resembles a free carrier Landau ladder with a set of discrete levels just beneath each rung of the ladder.

For comparison with experimental data, theoretical results are required for the entire range of magnetic field from $\gamma = 0$ to large but finite γ. As mentioned earlier, for $\gamma << 1$ the normal hydrogenic atom results can be used. For large but finite γ, the effective mass equation is not separable, so an approximate method of solution must be found. A number of workers have obtained estimates of the discrete ground and excited state energies by means of variational calculations, using as trial functions, product functions having the general form of Eq. (8). For example, the ground state variational function is typically taken as

$$F_{ooo} \sim \exp(-\rho^2/8a^2)\ \exp(-z^2/8b^2) \tag{10}$$

with a^2 and b^2 the parameters which minimize the energy. This wave function becomes unrealistic as γ decreases toward unity.

Having calculated the impurity states and energy levels for $\gamma << 1$ and $\gamma >> 1$, it would be interesting to follow the transformation that occurs between the low and high field regimes in the neighborhood of $\gamma = 1$. Only one calculation[12] for intermediate fields has been made, for the energies of the ground and two excited states. In this work, trial functions which more nearly approached hydrogen atom wave functions at low fields were used. For the ground state, for example,

$$F_{ooo} \sim \exp(-\rho^2/8a^2)\ \exp(-s\left[\rho^2 + \alpha z^2\right]^{1/2}) \ . \tag{11}$$

In the limit $\gamma \to 0$, $\alpha \to 1$, and $a^2 \to \infty$, the ground state function approaches the form $\exp(-sr)$, as required. However, the greater complexity of these

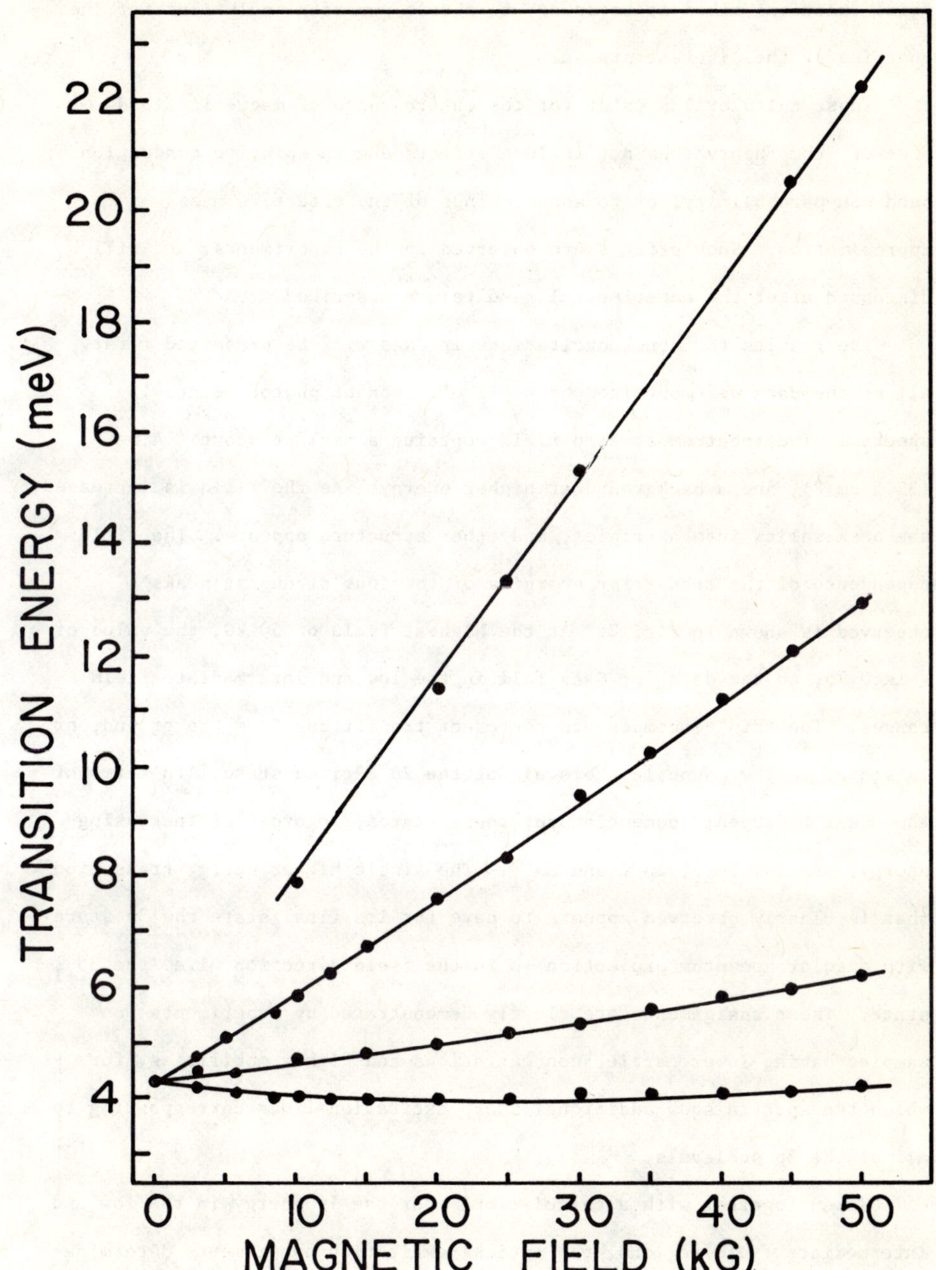

Fig. 2. Field dependence of the transition energies of the four main peaks observed in the photoconductivity spectra of epitaxially grown n-type GaAs.

trial functions makes it necessary to obtain numerical solutions for the energies of the discrete states.

Thus, calculations exist for the entire range of magnetic fields. However, the theory does not include effects due to spin, to conduction band non-parabolicity, or to shortcomings of the effective mass approximation. Such effects are observed in the experiments, and will be discussed after the experimental results are described.

The results for donor excitations in GaAs will be presented first. All of the data were obtained by means of resonant photoconductivity spectra. The spectrum at zero field contains a peak at about 4.4 meV (35.5 cm^{-1}), and a background at higher energy. As the field is increased, the peak splits into a triplet, and other structure appears. The field dependence of the transition energies of the four strongest peaks observed is shown in Fig. 2. At the highest field of 50 kG, the value of γ is 0.75, so the data for GaAs fall in the low and intermediate field ranges. The triplet components represent transitions from the ground, or 1s state, to the magnetic sublevels of the 2p excited state. In terms of the usual hydrogenic nomenclature, these states, in order of increasing energy, are 1s, $2p_{-1}$, $2p_0$, and $2p_{+1}$. The single higher energy transition that is clearly observed appears to have for its final state the 3p state with angular momentum projection $+\hbar$ in the field direction, i.e. the $3p_{+1}$ state. These assignments are clearly demonstrated by experiments[7] on samples having lower carrier concentrations and higher mobilities, for which the spectra show additional sharp excitation lines corresponding to all of the 3p sublevels.

Taken together with a calculation[12] for the 1s energy in the low and intermediate field regions, the optical excitation experiments determine the excited state energies as a function of magnetic field. This procedure is necessary since the experiments do not determine the ground

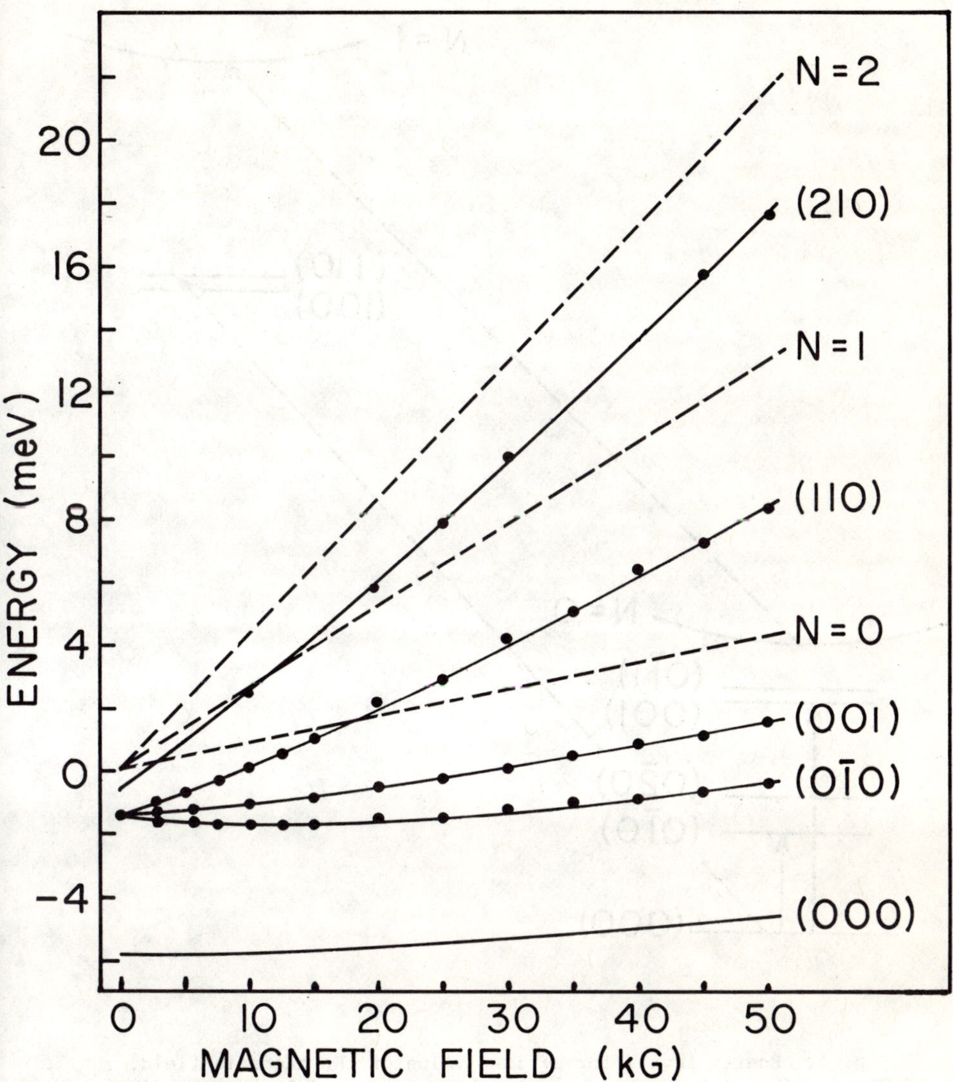

Fig. 3. Field dependence of the first three Landau levels (dashed lines), impurity ground state energy (000), and the excited state energies as determined from the photoconductivity spectra of GaAs.

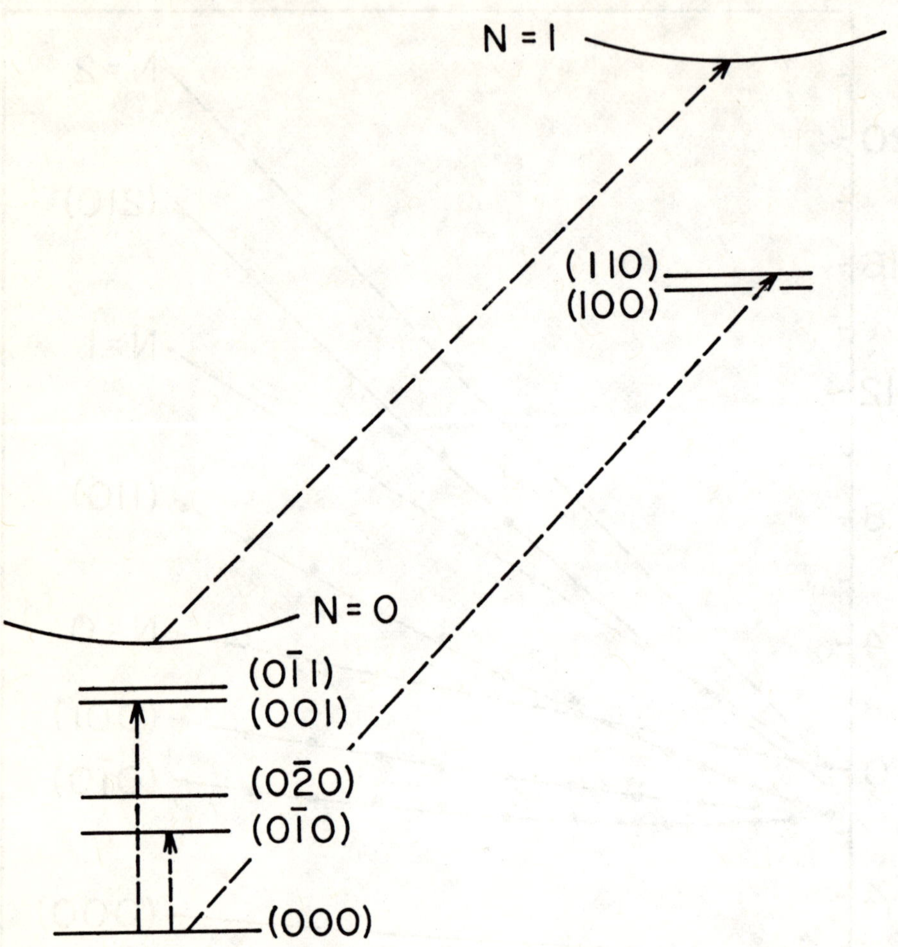

Fig. 4. Energy levels for a fixed value of the magnetic field, $\gamma > 1$. The two lowest Landau levels are shown as curved lines, the discrete impurity states as flat lines. Free carrier cyclotron resonance and three transitions from the impurity ground state are indicated by the arrows.

state binding energy directly. In Fig. 3 the data points represent the result of adding the observed transition energies to the calculated energy of the ground state. The Landau level energies $(N + \frac{1}{2})\hbar\omega_c$ have been plotted for N = 0, 1, and 2 using the value $m^* = 0.0675\ m_o$, and are indicated by dashed lines. For fields greater than 20 kG for GaAs, corresponding to values of γ greater than about 0.3, only the lower two of of the 1s → 2p triplet components remain below the N = 0 Landau level. The other triplet component accompanies the N = 1 Landau level to higher energy. This situation is characteristic of the high field region, as will next be seen. The quantum numbers which label the impurity states in Fig. 3 are in fact those used to describe states in the high field region.

The experiments on InSb were carried out in the field range 10-100 kG, which corresponds to values of γ between 7 and 70. At fields much lower than 10 kG, the donor electrons are no longer localized at 4.2°K in the InSb currently available, thus restricting the experiments to higher fields. The transitions studied are indicated by the arrows in Fig. 4, which includes some of the impurity levels calculated[9] for the high field case. In this case the states are identified by the quantum numbers $(NM\lambda)$, where N is the principal quantum number, M indicates the orbital angular momentum about the field in units of $\hbar$, and λ is related to the z-dependence of the solutions $f_{NM\lambda}(z)$ in the averaged Coulomb potential of the impurity atom. Since occupation both of free and localized states at a given field and temperature is often observed, the free carrier cyclotron resonance transition, (N=0) → (N=1), may appear together with the impurity transition (000) → (110). This produces the doublet spectrum discussed earlier. The other two impurity transitions, (000) → $(0\bar{1}0)$ and (000) → (001), are observed in the very far infrared at all magnetic fields used. A typical resonant photoconductivity spectrum for these two

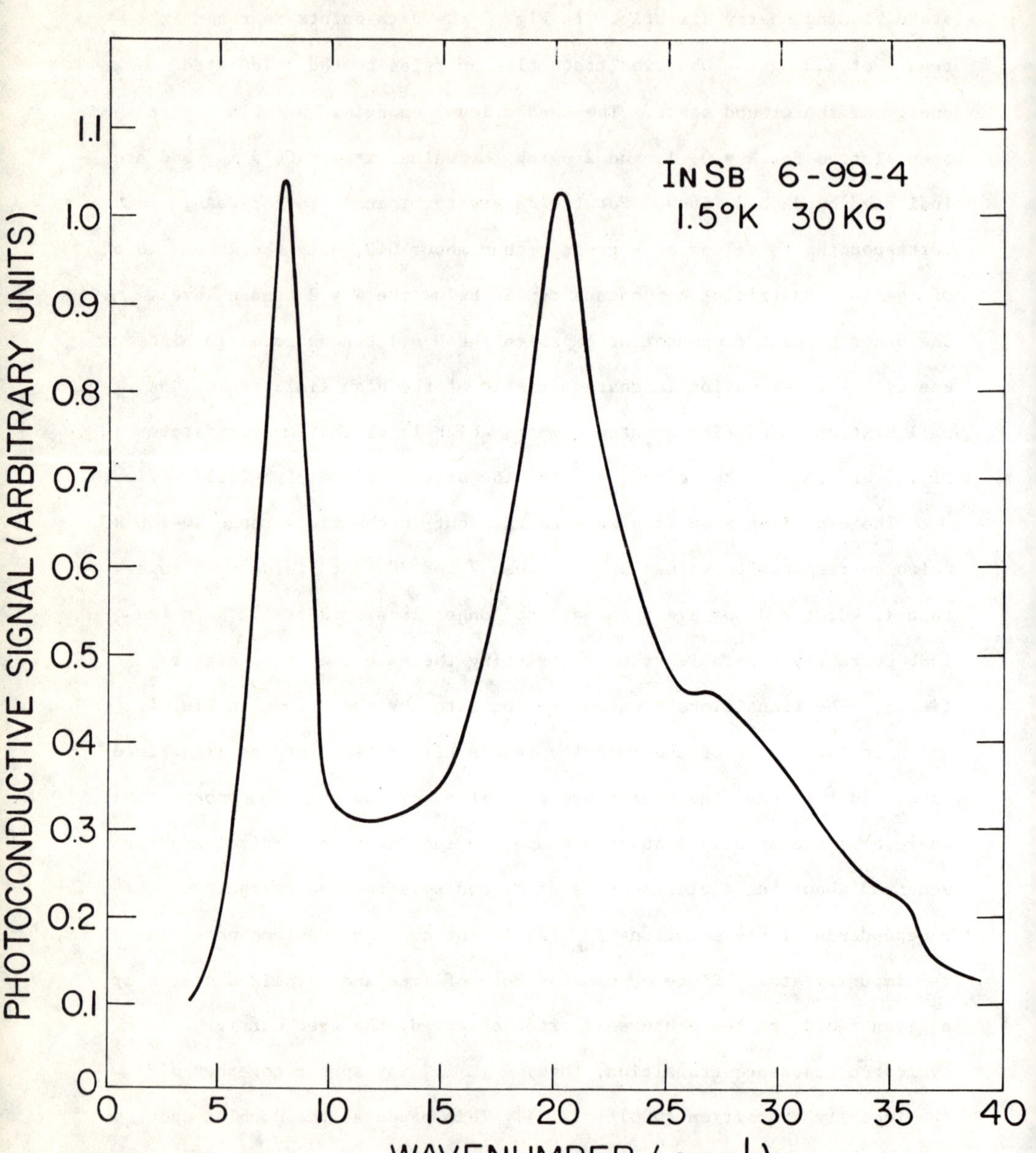

Fig. 5. Photoconductive spectrum for InSb showing peaks due to the donor impurity transitions (000) → (0$\bar{1}$0) and (000) → (001).

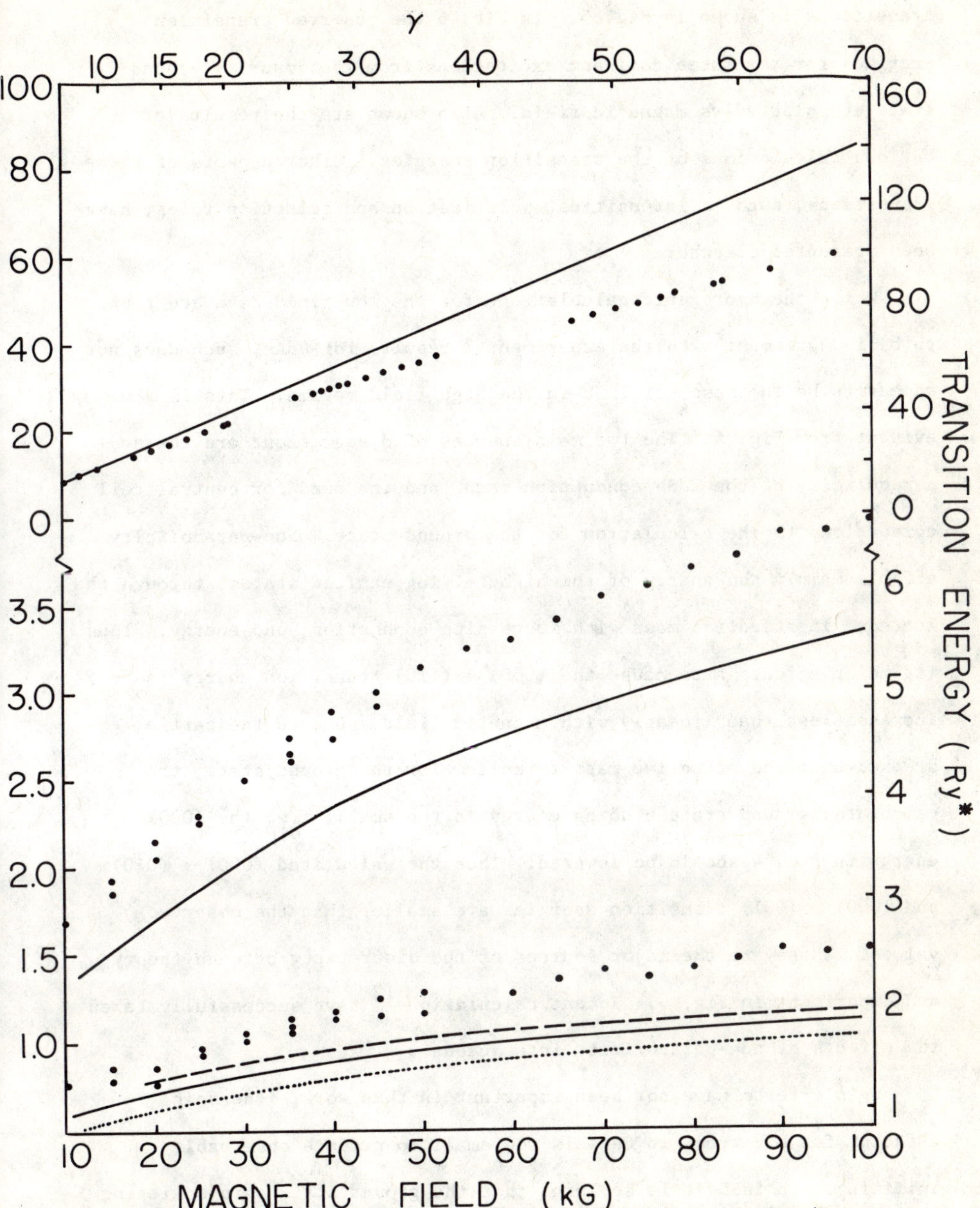

Fig. 6. Field dependence of the energies of the three predominant donor impurity transitions in InSb. The observed energies are indicated by the black circles. The curves show the following theoretical results: solid lines, ref. 9; dotted line, ref. 11; dashed line, ref. 12 (non-parabolic calculation). Note change in energy scale for the (000) → (110) transition.

transitions is shown in Fig. 5. In Fig. 6 the observed transition energies for the three dominant excitations from the impurity ground state are plotted vs magnetic field. Also shown are the results of various calculations of the transition energies. Other aspects of these transitions, such as intensities, polarization and selection rules, have been presented elsewhere[2].

While the hydrogenic calculations for the low-field case are found to be in agreement with the experimental results for GaAs, such does not appear to be the case for InSb in the high field regime. This is evident from Fig. 6. The two main sources of disagreement are the non-parabolicity of the InSb conduction band, and the need for central cell corrections to the calculation for the ground state. Non-parabolicity affects mainly the energy of the higher-lying excited states, through the increase in effective mass with increasing conduction band energy. Thus its major effect is to cause the (000) $\rightarrow$ (110) transition energy to increase less than linearly with magnetic field. Due to the partial breakdown of the effective mass formalism for the ground state, the calculated ground state binding energy is too small; i.e. the (000) energy in Fig. 4 should be lowered. Thus the calculated (000) $\rightarrow$ $(0\bar{1}0)$ and (000) $\rightarrow$ (001) transition energies are smaller than the observed values. These are the major sources of the discrepancy between theory and experiment in Fig. 6. Recent calculations[12] have successfully taken the effects of non-parabolicity into account.

Spin effects have not been important in this work. The spin g-factor for electrons in GaAs is too small to produce observable splitting. In InSb it is so large that the ground state spin splitting is much greater than kT over most of the temperature range used, so that the upper spin state was not populated. Impurity spin effects are

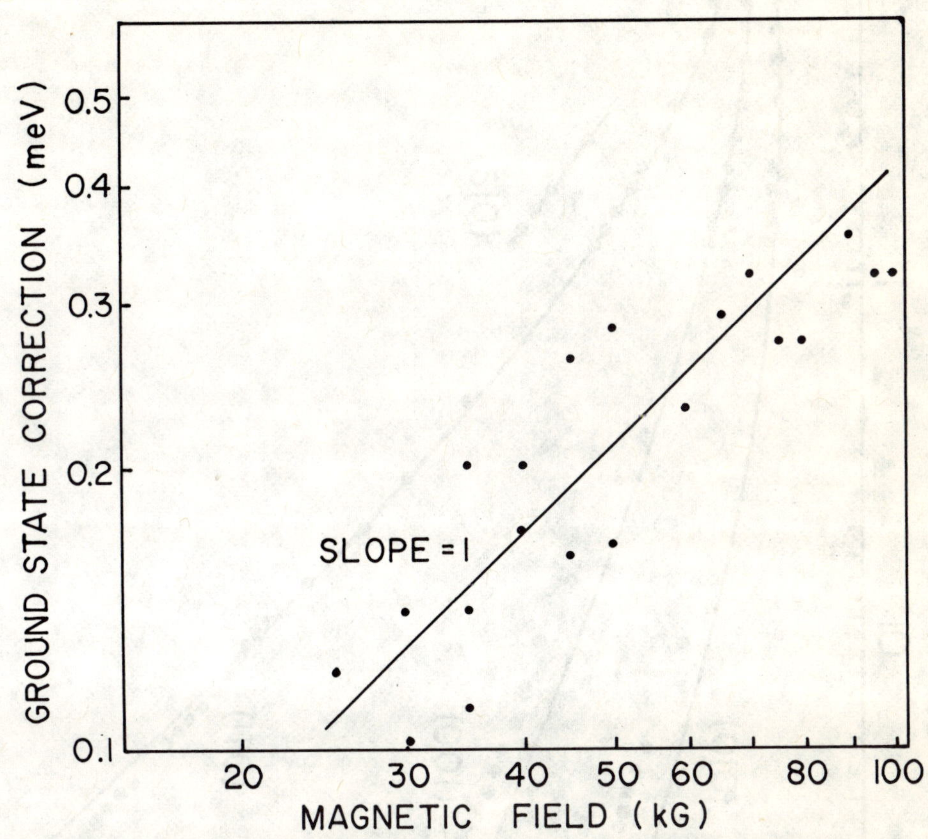

Fig. 7. Field dependence of the ground state energy correction deduced from a comparison between theoretical and experimental transition energies for (000) → (0$\bar{1}$0).

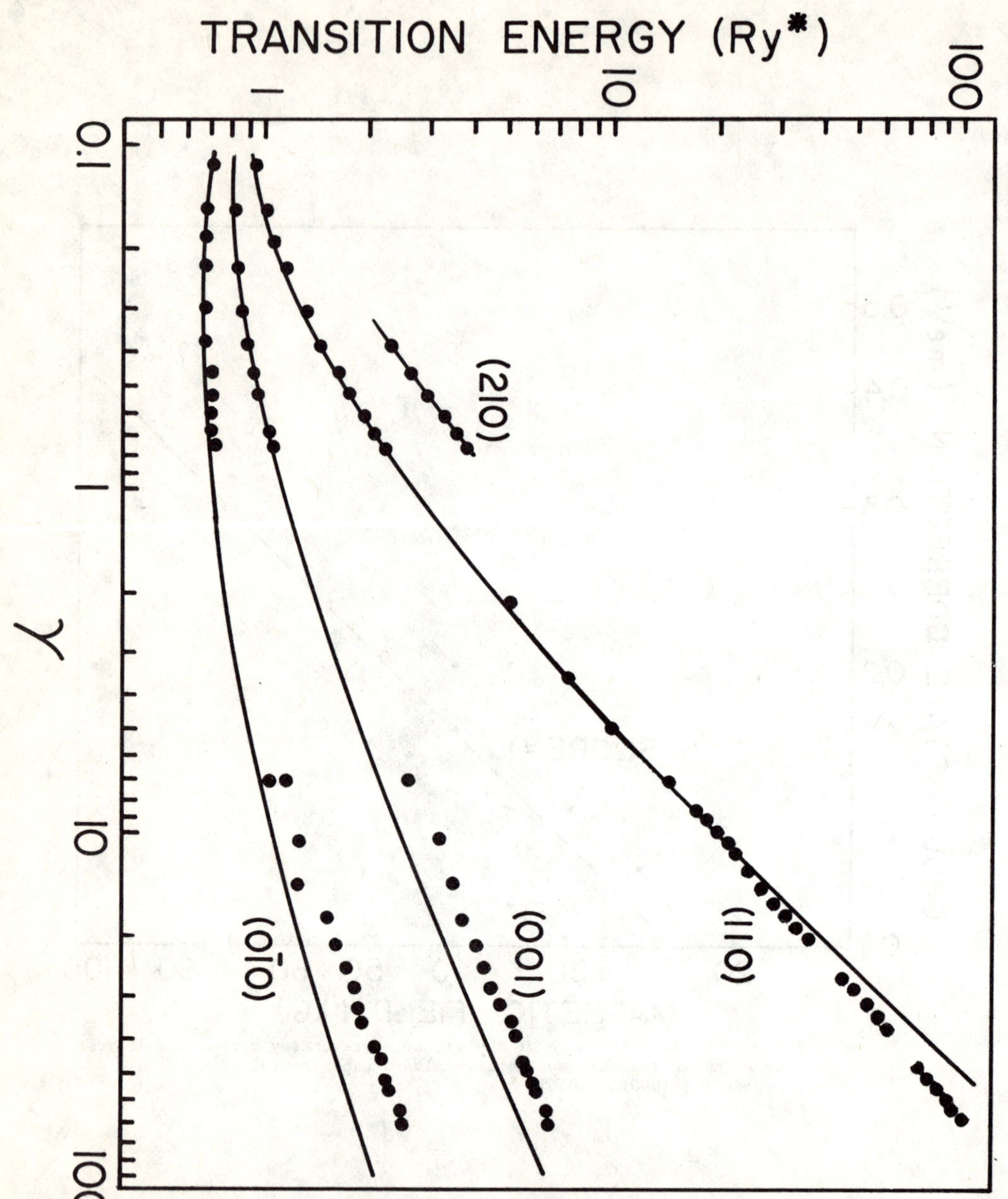

Fig. 8. Field dependence of the transition energies of hydrogenic donor impurities over a wide range of magnetic field. Energy is in units of the effective Rydberg R_y^*, field in terms of $\gamma = \hbar\omega_c/2R_y^*$. The data for $\gamma < 1$ were obtained from GaAs, and for $\gamma > 1$, from InSb.

observed in combined resonance experiments[13], but the transitions described in the present work conserve spin.

The magnitude of the central cell correction to the ground state energy has not been calculated. An estimate of the magnitude of the correction required may be obtained from knowledge of the electronic charge density in the region of the charged impurity ion core. Since this charge density increases with increasing field in the high field region[8], the central cell correction should do likewise. As a rough estimate, the charge density at the impurity site may be assumed to vary with the cyclotron radius λ_o as $1/\lambda_o^2$, since most of the compression of the localized electron charge density occurs in the plane transverse to the field. Since λ_o varies inversely as the square root of the field, $\lambda_o = (c\hbar/eH)$, the central cell correction should depend linearly on H. The difference between the observed $(000) \rightarrow (0\bar{1}0)$ transition energy, and the best calculation of this quantity including non-parabolicity, may be taken as a measure of the required central cell correction. It is plotted _vs_ H in Fig. 7. The points fall reasonably along a line of slope unity, indicating that the simple model described above for the central cell problem is qualitatively correct.

As described earlier, the results of the hydrogenic atom calculations can be expressed in dimensionless form suitable for comparison with experimental results on different impurity-host crystal systems. In Fig. 8 the curves represent transition energies calculated from effective mass theory for parabolic conduction band (except for the $(000) \rightarrow (001)$ energy which has not been calculated for $\gamma \lesssim 1$). The data points below $\gamma = 1$ are for GaAs, and above $\gamma = 1$ for InSb. It is hoped that experiments on donors in InAs will provide data for the intermediate field region. Deviations from the simple theory seem to appear only at the higher values of γ, for the reasons discussed above. It is of interest to follow the evolution of the low field states into their high field counterparts; this has been discussed elsewhere[6,14].

Thus a fairly complete picture of the hydrogenic atom spectrum from zero field to the limit of very large fields has now been achieved. In the experimental side of this work, the techniques of Fourier spectroscopy and resonant photoconductivity have been extremely helpful.

V. PHONON-ASSISTED IMPURITY EXCITATIONS IN InSb

A subject of considerable interest in studies of the optical properties of semiconductors is the electron-phonon interaction. This interaction is responsible for many optical effects, one of which has been investigated recently at the Naval Research Laboratory. It provides a good example of an application of the resonant photoconductivity technique to a problem for which normal transmission techniques are much less adequate.

The process of interest is the phonon-assisted excitation of charge carriers bound to impurity atoms. In this process, photon absorption leads to the creation of one or more optical phonons, with the simultaneous excitation of the charge carrier to a discrete impurity excited state. Transitions of this type have previously been studied[15,16] in semiconductors in which the impurity binding energy is considerably larger than the optical phonon energy $\hbar\omega_o$.

In InSb the electron-optical phonon interaction is responsible for the free carrier polaron effect that splits the cyclotron resonance transition[17,18] at fields such that $\hbar\omega_c \approx \hbar\omega_o$. When the free carriers are localized by reducing the sample temperature and the experiment is repeated, quite complicated spectra are observed. Obtaining useful transmission spectra is difficult because the spectral region of interest

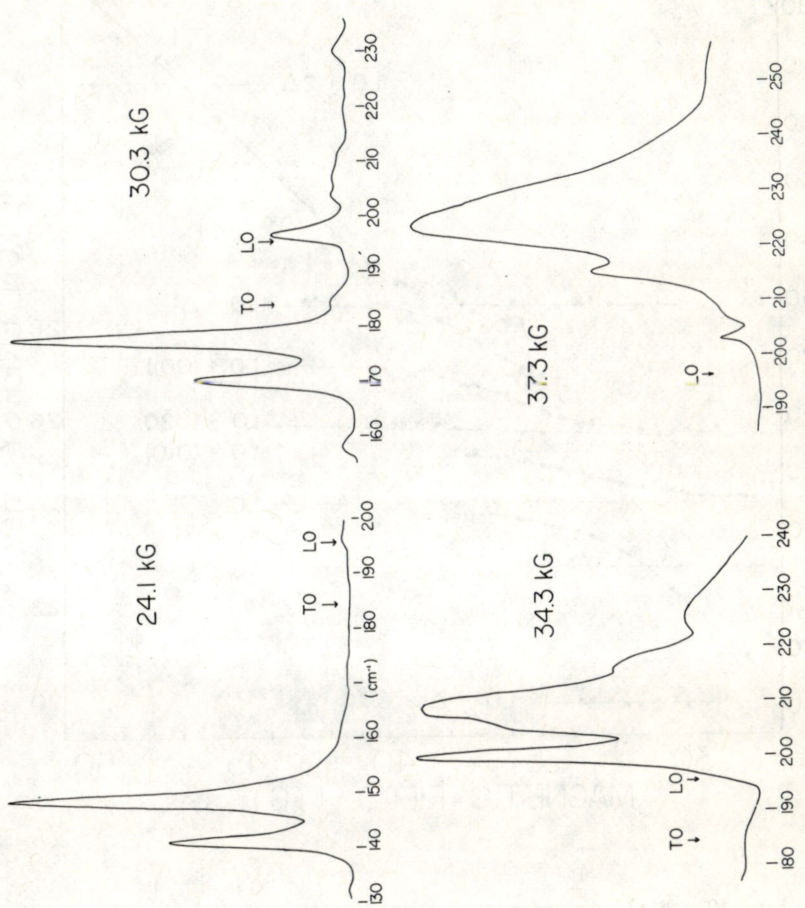

Fig. 9. Photoconductivity spectra from n-type InSb in magnetic fields between 20 and 40 kG. At 30 kG, $\hbar\omega_c \approx \hbar\omega_o$. The transverse optical (TO) and longitudinal optical (LO) phonon energies are indicated. The two spectra at lower fields were obtained at 4.2°K, the others at 1.5°K.

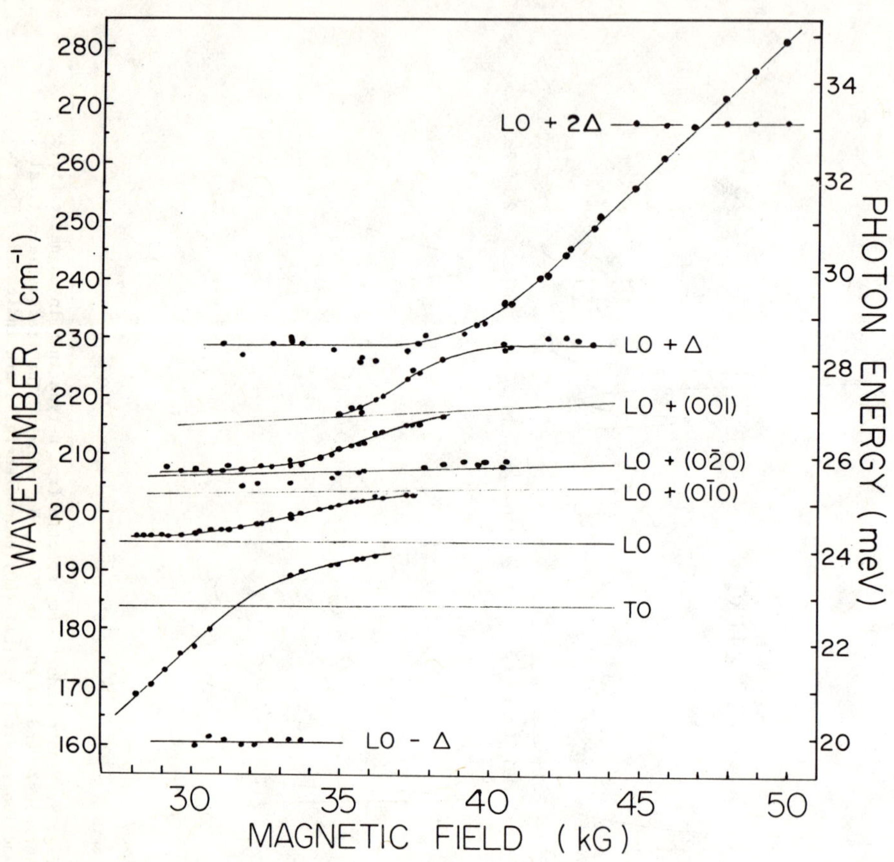

Fig. 10. Field dependence of the transition energies of donor impurities in InSb in the vicinity of $h\omega_o$, in the range 30-50 kG.

includes the strong lattice absorption at the transverse optical phonon frequency ω_{TO}, and the Reststrahlen band between ω_{TO} and ω_o. Thin samples must be used if they are not to be opaque over a wide spectral region, but this precludes observation of the weaker impurity excitation lines. However, with the resonant photoconductivity technique very thin samples may be used. Experimental details will be omitted as they have been given elsewhere[3], but some typical photoconductivity spectra for a 30 micron thick InSb sample containing about 7×10^{13} cm^{-3} excess donor impurities are shown in Fig. 9. The sample temperature for the first two spectra was 4.2°K, and the doublet described earlier is evident below ω_{TO}. The other spectra were obtained at 1.5°K, where essentially all of the electrons are localized, and all the peaks represent impurity transitions. At fields for which $\omega_c \approx \omega_o$ a large number of transitions is observed. The field dependence of the transition energies is shown in Fig. 10. Discontinuities are found at energies corresponding to the sum of $\hbar\omega_o$ and the excitation energies of the donors from their ground state to various excited states. Transitions whose energies involve the quantity Δ in Fig. 10 are believed to have a different origin from the others.

A theoretical treatment of the phonon-assisted impurity excitation spectra has been given by Dr. R. F. Wallis, and is to be found elsewhere in this volume. This treatment uses second order perturbation theory and is based on the Fröhlich-type electron-phonon interaction. A different approach[19], based on the use of Wigner-Brillouin perturbation theory, could also be applied to this problem.

REFERENCES

1. R. Kaplan, Appl. Opt. 6, 685 (1967).
2. R. Kaplan, to be published in Phys. Rev., May 15, 1969.
3. R. Kaplan and R. F. Wallis, Phys. Rev. Letters 20, 1499 (1968).
4. H. J. Stocker, H. Levinstein, and C. R. Stannard, Jr., Phys. Rev. 150, 613 (1966).
5. J. Strong and G. A. Vanasse, J. Opt. Soc. Amer. 49, 844 (1959).
6. R. Kaplan, M. A. Kinch, and W. C. Scott, to be published in Solid State Commun.
7. G. E. Stillman, C. M. Wolfe, and J. O. Dimmock, to be published in Solid State Commun.
8. Y. Yafet, R. W. Keyes, and E. N. Adams, J. Phys. Chem. Solids 1, 137 (1956).
9. R. F. Wallis and H. J. Bowlden, J. Phys. Chem. Solids 7, 78 (1958).
10. R. J. Elliot and R. J. Loudon, J. Phys. Chem. Solids 15, 196 (1960).
11. H. Hasegawa and R. E. Howard, J. Phys. Chem. Solids 21, 179 (1961).
12. D. M. Larsen, J. Phys. Chem. Solids 29, 271 (1968).
13. B. D. McCombe and R. Kaplan, Phys. Rev. Letters 21, 756 (1968).
14. H. Hasegawa, "Effects of High Magnetic Fields on Electronic States in Semiconductors - The Rydberg Series and the Landau Levels," in Physics of Solids in High Magnetic Fields, E. D. Haidemenakis, editor, Plenum Press, New York, 1969.
15. R. E. Nahory and H. Y. Fan, Phys. Rev. 156, 825 (1967).
16. J. R. Hardy, S. D. Smith, and W. Taylor, in Proceedings of the International Conference on the Physics of Semiconductors, Exeter, July 1962, edited by A. C. Stickland (The Inst. of Phys. and The Phys. Society, London, England, 1962), p.521.
17. D. H. Dickey, E. J. Johnson, and D. M. Larsen, Phys. Rev. Letters 18, 599 (1967).

18. C. J. Summers, P. G. Harper, and S. D. Smith, Solid State Commun. 5, 615 (1967).

19. D. M. Larsen and E. J. Johnson, J. Phys. Soc. Japan 21 Suppl., 443 (1966).

XIII

THE UANTUM THEORY OF ATOMIC VIBRATION

F. A. Johnson

INTRODUCTION

In this lecture we shall first deal with the adiabatic principle and the concept of an effective nuclear potential energy function. We will next consider the expansion of this effective nuclear potential energy function in terms of small nuclear displacements and derive explicite expressions for these in terms of the nuclear-nuclear and nuclear-electron interactions.

THE ADIABATIC APPROXIMATION

We can write the Hamiltonian for an assembly of atoms (a molecule or a crystal) as

$$H = H_N + H_e \tag{1}$$

where the nuclear part H_N is given by

$$H_N = \sum_a \frac{1}{2M_a} \underset{\sim}{P}_a^2 + \tfrac{1}{2} \sum_{ab}{}' V_{nn} (\underset{\sim}{x}_a - \underset{\sim}{x}_b) \tag{2}$$

ie. the kinetic energy of the nuclei plus their coulomb energy and where the electronic part H_e is given by

$$H_e = \sum_j \frac{1}{2m} \underset{\sim}{P}_j^2 + \sum_{ja} V_{en} (\underset{\sim}{r}_j - \underset{\sim}{x}_a) + \tfrac{1}{2} \sum_{jj'}{}' V_{ee} (\underset{\sim}{r}_j - \underset{\sim}{r}_{j'}) \tag{3}$$

ie. the kinetic energy plus the electron-nuclear and the electron-electron coulomb energies: the primes on the summations are to exclude self-energy terms. We now wish to find the solutions to the Schrodinger equation

$$H\chi_q = E_q \chi_q \tag{4}$$

We first assume that we know the solutions of

$$H_e \psi (\underset{\sim}{k},\underset{\sim}{r},\underset{\sim}{x}) = \mathcal{E} (\underset{\sim}{k},\underset{\sim}{x}) \psi (\underset{\sim}{k},\underset{\sim}{r},\underset{\sim}{x}) \tag{5}$$

ie. the wave functions for the electrons for the particular nuclear configuration (here we use $\underset{\sim}{r}$ to represent collectively all the electronic co-ordinates $\underset{\sim}{r}_j$ and $\underset{\sim}{x}$ to represent collectively all the nuclear co-ordinates $\underset{\sim}{x}_a$). We now expand χ_q in terms of the ψ's as follows:

$$\chi_q = \sum_k \Phi_q (\underset{\sim}{k},\underset{\sim}{x}) \psi (\underset{\sim}{k},\underset{\sim}{r},\underset{\sim}{x}) \tag{6}$$

and note that the Φ's depend only on the nuclear co-ordinates for a given electronic state $\underset{\sim}{k}$. If we substitute equations (1) and (6) into (4) we obtain:

$$(H_N + H_e) \sum_{k'} \Phi_q(\underset{\sim}{k}', \underset{\sim}{x}) \psi (\underset{\sim}{k}',\underset{\sim}{r},\underset{\sim}{x}) = E_q \sum_k \Phi_q(\underset{\sim}{k},\underset{\sim}{x}) \psi (\underset{\sim}{k},\underset{\sim}{r},\underset{\sim}{x}) \tag{7}$$

We can now pick out terms by multiplying through by $\psi^* (\underset{\sim}{k},\underset{\sim}{r},\underset{\sim}{x})$ and integrating over all $\underset{\sim}{r}$'s. Doing this step by step the kinetic energy part of H_N gives:

$$- \sum_a \frac{\hbar^2}{2M_a} \nabla_a^2 \Phi_q (\underset{\sim}{k},\underset{\sim}{x})$$

$$- 2 \sum_{k'a} \frac{\hbar^2}{2M_a} \left\{ \int \psi^* (\underset{\sim}{k},\underset{\sim}{r},\underset{\sim}{x}) \nabla_a \psi(\underset{\sim}{k}',\underset{\sim}{r},\underset{\sim}{x}) d^3\underset{\sim}{r} \right\} \nabla_a \Phi_q(\underset{\sim}{k}',\underset{\sim}{x})$$

$$- \sum_{k'a} \frac{\hbar^2}{2M_a} \left\{ \int \psi^* (\underset{\sim}{k},\underset{\sim}{r},\underset{\sim}{x}) \nabla_a^2 \psi(\underset{\sim}{k}',\underset{\sim}{r},\underset{\sim}{x}) d^3\underset{\sim}{r} \right\} \Phi_q(k',x)$$

where ∇_a acts on $\underset{\sim}{x}_a$ only. The potential part of H_N gives:

$$\frac{1}{2} \sum_{ab}{}' V_{nn} (\underset{\sim}{x}_a - \underset{\sim}{x}_b) \Phi_q (\underset{\sim}{k}, \underset{\sim}{x})$$

and from equation (5) the H_e part gives:

$$\mathcal{E} (\underset{\sim}{k}, \underset{\sim}{x}) \Phi_q (\underset{\sim}{k}, \underset{\sim}{x})$$

and the right of equation (7) gives:

$$E_q \Phi_q (\underset{\sim}{k}, \underset{\sim}{x})$$

thus we can write equation (7) as:

$$[H_N + \mathcal{E}(\underset{\sim}{k}, \underset{\sim}{x})] \Phi_q (\underset{\sim}{k}, \underset{\sim}{x}) + \sum_{k'} [A(\underset{\sim}{k}, \underset{\sim}{k}') + B(\underset{\sim}{k}, \underset{\sim}{k}')] \Phi_q (\underset{\sim}{k}', \underset{\sim}{x})$$

$$= E_q \Phi_q (\underset{\sim}{k}, \underset{\sim}{x}) \tag{8}$$

where

$$A(\underset{\sim}{k}, \underset{\sim}{k}') = -2 \sum_a \frac{\hbar^2}{2M_a} \left\{ \int \psi^* (\underset{\sim}{k}, \underset{\sim}{r}, \underset{\sim}{x}) \nabla_a \psi(\underset{\sim}{k}', \underset{\sim}{r}, \underset{\sim}{x}) d^3\underset{\sim}{r} \right\} \nabla_a \Phi_q (\underset{\sim}{k}', \underset{\sim}{x})$$

$$B(\underset{\sim}{k}, \underset{\sim}{k}') = - \sum_a \frac{\hbar^2}{2M_a} \left\{ \int \psi^* (\underset{\sim}{k}, \underset{\sim}{r}, \underset{\sim}{x}) \nabla_a^2 \psi(\underset{\sim}{k}', \underset{\sim}{r}, \underset{\sim}{x}) d^3\underset{\sim}{r} \right\} \Phi_q (\underset{\sim}{k}', \underset{\sim}{x})$$

The adiabatic approximation is to assume that we can neglect $A(\underset{\sim}{k}, \underset{\sim}{k}')$ and $B(\underset{\sim}{k}, \underset{\sim}{k}')$ whenever $\underset{\sim}{k} \neq \underset{\sim}{k}'$. The $A(\underset{\sim}{k}, \underset{\sim}{k}')$ terms involves coupling between the ψ's and the Φ's i.e. between the electronic wave functions and the nuclear wave functions while the $B(\underset{\sim}{k}, \underset{\sim}{k}')$ terms are similar to center of mass type corrections since they involve the amount of nuclear kinetic energy associated with electronic wave functions $\psi(\underset{\sim}{k}', \underset{\sim}{r}, \underset{\sim}{x})$ through their dependence on $\underset{\sim}{x}$. The $B(\underset{\sim}{k}, \underset{\sim}{k}')$ terms will thus be very small and can readily be dropped. We note too that the diagonal terms $A(\underset{\sim}{k}, \underset{\sim}{k})$ are zero if $\psi(\underset{\sim}{k}, \underset{\sim}{r}, \underset{\sim}{x})$ is real since by differentiating the normalising condition we have

$$\int \psi^*(\underset{\sim}{k},\underset{\sim}{r},\underset{\sim}{x})\, \nabla_a \psi(\underset{\sim}{k},\underset{\sim}{r},\underset{\sim}{x})\, d^3\underset{\sim}{r} \; + \; + \int \psi(\underset{\sim}{k},\underset{\sim}{r},\underset{\sim}{x})\, \nabla_a \psi^*(\underset{\sim}{k},\underset{\sim}{r},\underset{\sim}{x})\, d^3\underset{\sim}{r} = 0$$

thus each term is zero if $\psi(\underset{\sim}{k},\underset{\sim}{r},\underset{\sim}{x})$ is real and $\psi(\underset{\sim}{k},\underset{\sim}{r},\underset{\sim}{x})$ will be real for the ground state, thus, provided the nuclear displacements are small enough so that electronic excitations from the ground state $\underset{\sim}{k}$ to any excited state $\underset{\sim}{k}'$ can be neglected, we can write equation (8) as

$$\left[\sum_a \frac{1}{2M_a} \underset{\sim}{P}_a^2 + V\right] \Phi_q(\underset{\sim}{x}) = E_q \Phi_q(\underset{\sim}{x}) \tag{9}$$

where the effective nuclear potential energy V is defined as

$$V = \tfrac{1}{2} \sum_{ab}' V_{nn}(\underset{\sim}{x}_a - \underset{\sim}{x}_b) + \mathcal{E}(\underset{\sim}{x}) \tag{10}$$

and where we have now dropped the explicite reference to the ground state $\underset{\sim}{k}$. Thus the adiabatic approximation allows us to define an effective nuclear potential energy which is the sum of the coulomb potential energy and the total energy of the electrons in the ground state - both, of course, evaluated for the particular nuclear configuration.

THE EXPANSION OF V.

We now wish to examine the variations in the effective nuclear potential energy V due to small displacements $\underset{\sim}{y}_a$, $\underset{\sim}{y}_b$ etc in the nuclear co-ordinates x_a, x_b.

We can write this as:

$$V - V^{(o)} = \sum_{m=1}^{\infty} V^{(m)} = \sum_{m=1}^{\infty} (V_{nn}^{(m)} + \mathcal{E}^{(n)}) \tag{11}$$

where the superscript indicates the order of each term in the nuclear displacements. Now the coulomb part can be written down immediately as:

$$V_{nn}^{(m)} = \frac{1}{2} \frac{1}{m!} \sum_{ab} (\underset{\sim}{y}_a \cdot \nabla_a + \underset{\sim}{y}_b \cdot \nabla_b)^m V_{nn} (\underset{\sim}{x}_a - \underset{\sim}{x}_b) \tag{12}$$

and this can be further simplified since ∇_b is equivalent to $-\nabla_a$ and thus we can write

$$V_{nn}^{(m)} = \frac{1}{2m!} \sum_{ab} (\underset{\sim}{y}_a \cdot \nabla_a - \underset{\sim}{y}_b \cdot \nabla_a)^m V_{nn} (\underset{\sim}{x}_a - \underset{\sim}{x}_b) \tag{13}$$

we note too that we can drop the prime on the summation since self energy terms will vanish on differentiation.

We now turn to the more difficult problem of the expansion of $\mathcal{E}(\underset{\sim}{x})$. It follows from equation (3) that the variations ΔH_e in the Hamiltonian for the electrons due to displacements $\lambda \underset{\sim}{y}_a$ in the nuclear-co-ordinates is given by

$$\begin{aligned} \Delta H_e &= \sum_{ja} V_{en} (\underset{\sim}{r}_j - \underset{\sim}{x}_a - \lambda \underset{\sim}{y}_a) \\ &= \sum_{j} \left\{ \lambda V_{en}^{(1)} (\underset{\sim}{r}_j, \underset{\sim}{x}) + \lambda^2 V_{en}^{(2)} (\underset{\sim}{r}_j, \underset{\sim}{x}) + \ldots \right\} \end{aligned} \tag{14}$$

where

$$V_{en}^{(m)} (\underset{\sim}{r}_j, \underset{\sim}{x}) = \frac{1}{m!} \sum_{a} (\underset{\sim}{y}_a \cdot \nabla_a)^m V_{en} (\underset{\sim}{r}_j - \underset{\sim}{x}_a) \tag{15}$$

Now the probability of finding electron j and r_j is

$$\int \psi^* (\underset{\sim}{r}, \underset{\sim}{x}) \psi (\underset{\sim}{r}, \underset{\sim}{x}) d^3 \underset{\sim}{r}_1 \ldots d^3 \underset{\sim}{r}_{j-1} d^3 \underset{\sim}{r}_{j+1} \ldots d^3 \underset{\sim}{r}_N$$

where N is the total number of electrons ie. we integrate the wave functions over all but the j'th electronic co-ordinate. However, since all electrons are identical the probability of finding any other electron at $\underset{\sim}{r}_j$ is the same thus the probability of finding any of the N electrons at $\underset{\sim}{r}_j$ - the charge density $\rho(\underset{\sim}{r}, \underset{\sim}{x})$ - is given by

$$\rho(\underset{\sim}{r}, \underset{\sim}{x}) = N \int \psi^* (\underset{\sim}{r}, \underset{\sim}{x}) \psi (\underset{\sim}{r}, \underset{\sim}{x}) d^3 \underset{\sim}{r}_1 \ldots d^3 \underset{\sim}{r}_{j-1} d^3 \underset{\sim}{r}_{j+1} \ldots d^3 \underset{\sim}{r}_N \tag{16}$$

and $\rho(r,x)$ is independent of which co-ordinate $\underset{\sim}{r}_j$ is excluded. We can now combine equations (14) and (16) and write

$$\int \psi^* (\underset{\sim}{r},\underset{\sim}{x}) \frac{\partial H_e}{\partial \lambda} \psi (\underset{\sim}{r},\underset{\sim}{x})\, d^3\underset{\sim}{r}_1 \dots d^3\underset{\sim}{r}_N$$

$$= \sum_j \int \psi^* (\underset{\sim}{r},\underset{\sim}{x}) \left| V_{en}^{(1)}(\underset{\sim}{r}_j,\underset{\sim}{x}) + 2\lambda\, V_{en}^{(2)}(\underset{\sim}{r}_j,\underset{\sim}{x}) + \dots \right| \psi(\underset{\sim}{r},\underset{\sim}{x})\, d^3\underset{\sim}{r}_1 \dots d^3\underset{\sim}{r}_N$$

$$= N^{-1} \sum_j \int \rho(\underset{\sim}{r}_j,\underset{\sim}{x}) \left| V_{en}^{(1)}(\underset{\sim}{r}_j,\underset{\sim}{x}) + 2\lambda\, V_{en}^{(2)}(\underset{\sim}{r}_j,\underset{\sim}{x}) + \dots \right| d^3\underset{\sim}{r}_j$$

$$= \int \rho (\underset{\sim}{r},\underset{\sim}{x}) \left| V_{en}^{(1)} (\underset{\sim}{r},\underset{\sim}{x}) + 2\lambda V_{en}^{(2)} (\underset{\sim}{r},\underset{\sim}{x}) + \dots \right| d^3\underset{\sim}{r} \tag{17}$$

since the integral over $\underset{\sim}{r}_j$ is clearly independent of j.

We can now make use of Fegnmons theorem (Phys. Rev. <u>56</u>, 340, 1939) which states that

$$\frac{\partial \mathcal{E}(\underset{\sim}{x})}{\partial \lambda} = \int \psi^* (\underset{\sim}{r},\underset{\sim}{x}) \frac{\partial H_e}{\partial \lambda} \psi (\underset{\sim}{r},\underset{\sim}{x})\, d^3\underset{\sim}{r}_1 \dots d^3\underset{\sim}{r}_N \tag{18}$$

thus we can write:

$$\frac{\partial \mathcal{E}(x)}{\partial \lambda} = \int \rho(\underset{\sim}{r},\underset{\sim}{x}) \left| V_{en}^{(1)}(\underset{\sim}{r},\underset{\sim}{x}) + 2\lambda\, V_{en}^{(2)}(\underset{\sim}{r},\underset{\sim}{x}) + 3\lambda^2 V_{en}^{(3)}(\underset{\sim}{r},\underset{\sim}{x}) + \dots \right| d^3\underset{\sim}{r} \tag{19}$$

(Note that the $\underset{\sim}{r}$ is a single co-ordinate but the $\underset{\sim}{x}$ represents all the nuclear co-ordinates). Now as we displace the nuclei the wave functions $\psi(\underset{\sim}{r},\underset{\sim}{x})$ change and thus $\rho(\underset{\sim}{r},\underset{\sim}{x})$ also change. We can write this as:

$$\rho(\underset{\sim}{r},\underset{\sim}{x}) = \rho^{(0)} (\underset{\sim}{r},\underset{\sim}{x}) + \lambda\rho^{(1)}(\underset{\sim}{r},\underset{\sim}{x}) + \lambda^2\rho^{(2)}(\underset{\sim}{r},\underset{\sim}{x}) + \dots \tag{20}$$

and substituting this into equation (19) and collecting terms of the same order we obtain:

$$\frac{\partial \mathcal{E}(\underset{\sim}{x})}{\partial\lambda} = \int \rho^{(0)}(\underset{\sim}{r},\underset{\sim}{x})\, V_{en}^{(1)}(\underset{\sim}{r},\underset{\sim}{x})\, d^3\underset{\sim}{r}$$

$$+ \lambda \int \left\{ 2\rho^{(0)}(\underset{\sim}{r},\underset{\sim}{x})\, V_{en}^{(2)}(\underset{\sim}{r},\underset{\sim}{x}) + \rho^{(1)}(\underset{\sim}{r},\underset{\sim}{x})\, V_{en}^{(1)}(\underset{\sim}{r},\underset{\sim}{x}) \right\} d^3\underset{\sim}{r}$$

$$+ \cdots + \lambda^n \sum_{m=1}^{n+1} \int m\rho^{(n+1-m)}(\underset{\sim}{r},\underset{\sim}{x})\, V_{en}^{(m)}(\underset{\sim}{r},\underset{\sim}{x})\, d^3\underset{\sim}{r} + \cdots \tag{21}$$

We can now write:

$$\mathcal{E}(\underset{\sim}{x}) - \mathcal{E}^{(0)}(\underset{\sim}{x}) = \int_0^1 \frac{\partial \mathcal{E}(\underset{\sim}{x})}{\partial\lambda}\, d\lambda$$

$$= \int \rho^{(0)}(\underset{\sim}{r},\underset{\sim}{x})\, V_{en}^{(1)}(\underset{\sim}{r},\underset{\sim}{x})\, d^3\underset{\sim}{r}$$

$$+ \int \left\{ \rho^{(0)}(\underset{\sim}{r},\underset{\sim}{x})\, V_{en}^{(2)}(\underset{\sim}{r},\underset{\sim}{x}) + \tfrac{1}{2}\rho^{(1)}(\underset{\sim}{r},\underset{\sim}{x})\, V_{en}^{(1)}(\underset{\sim}{r},\underset{\sim}{x}) \right\} d^3\underset{\sim}{r}$$

$$+ \cdots + \sum_{m=1}^{n} \frac{m}{n} \int \rho^{(n-m)}(\underset{\sim}{r},\underset{\sim}{x})\, V_{en}^{(m)}(\underset{\sim}{r},\underset{\sim}{x})\, d^3\underset{\sim}{r} + \cdots \tag{22}$$

and we have the required result namely that:

$$\mathcal{E}^{(n)}(\underset{\sim}{x}) = \sum_{m=1}^{n} \frac{m}{n} \int \rho^{(n-m)}(\underset{\sim}{r},\underset{\sim}{x})\, V_{en}^{(m)}(\underset{\sim}{r},\underset{\sim}{x})\, d^3\underset{\sim}{r} \tag{23}$$

PHONONS

Now if the initial nuclear configuration was in equilibrium then $V^{(1)}$ must be zero and it is well known that provided $V^{(3)} \ll V^{(2)}$ then the **natural** vibrational frequencies of the nuclei are determined by the mathematical form of $V^{(2)}$. We can write this as:

$$V^{(2)} = \tfrac{1}{4} \sum_{ab} (\underset{\sim}{y}_a \nabla_a - \underset{\sim}{y}_b \nabla_a)^2\, V_{nn}(\underset{\sim}{x}_a - \underset{\sim}{x}_b) + \mathcal{E}^{(2)}(x)$$

$$= \tfrac{1}{2} \sum_{ab} \underset{\sim}{y}_a \nabla_a (\underset{\sim}{y}_a \nabla_a - \underset{\sim}{y}_b \nabla_a)\, V_{nn}(\underset{\sim}{x}_a - \underset{\sim}{x}_b)$$

$$+ \int \rho^{(0)}(\underset{\sim}{r},\underset{\sim}{x})\, V_{en}^{(2)}(\underset{\sim}{r},\underset{\sim}{x})\, d^3\underset{\sim}{r}$$

$$+ \tfrac{1}{2} \int \rho^{(1)}(\underset{\sim}{r},\underset{\sim}{x})\, V_{en}^{(1)}(\underset{\sim}{r},\underset{\sim}{x})\, d^3\underset{\sim}{r} \tag{24}$$

This equation sets out the physics of the problem. The first term is due to the nuclear-nuclear coulomb interaction and is clearly translationally invariant since if all the $\underset{\sim}{y}$'s are the same this term is zero. The second term describes the direct electrostatic interaction between pairs of nuclear displacements and the unperturbed electronic charge distribution. The last term describes a more indirect interaction: $\rho^{(1)}(\underset{\sim}{r},\underset{\sim}{x})$ is the linear response of the electron distribution to a nuclear displacement and this, in turn, interacts with another nuclear displacement. This is often called the ion-electron-ion a interaction. Physically the effect of the electrons is to screen out most of the bare nuclear-nuclear interactions which results in phonon energies being so much smaller than electronic excitation energies thus justifying the adiabatic approximation.

The important point to note, however, is that the electronic contribution to $V^{(2)}$ comes from the electronic ground state $\rho^{(o)}(\underset{\sim}{r},\underset{\sim}{x})$ and its variation $\rho^{(1)}(\underset{\sim}{r},\underset{\sim}{x})$ when the nuclei are displaced. These two quantities can in principle, be measured by X-ray scattering techniques although to measure $\rho^{(1)}(\underset{\sim}{r},\underset{\sim}{x})$ would probably require a modulation technique similar to that of piezoreflectance measurements. It can readily be shown that $\rho^{(1)}(\underset{\sim}{r},\underset{\sim}{x})$ has non-zero fourier coefficients at $\underset{\sim}{g} + \underset{\sim}{q}$ only where $\underset{\sim}{g}$ is any vector of the reciprocal lattice and $\underset{\sim}{q}$ is the wave-vector of the phonon. It is difficult to calculate these coefficients in general because of the loss of the simple translational symmetry when the phonon is present but in the special case of optic phonons of zero wave-vector the calculation reduces to the change in the electronic charge density with changes in the basis vector and as the translational symmetry is not altered this is no more difficult than a normal band structure calculation.

Bibliography

M. Born and K. Huang, Dynamical Theory of Crystal Lattices, Oxford University Press, 1954.

P. D. DeCicco and F. A. Johnson, Proc. Roy. Soc. A310, 111, 1969.

XIV

OPTICAL PROPERTIES OF ATOMIC VIBRATION

F. A. Johnson

INTRODUCTION

In the previous lecture we discussed the changes produced in the ground state wave functions when we displaced atomic nuclei from their equilibrium configuration and the role this played in determining the frequencies of these atomic vibrations. In this lecture we shall consider the corresponding changes in the optical properties which depend on variations to both the ground and excited state wave functions.

OPTICAL PROPERTIES DUE TO ATOMIC VIBRATIONS

We start by writing down Maxwells equations for the field as follows:

$$\nabla \cdot \underset{\sim}{\epsilon} \underset{\sim}{E} = 0 \qquad \nabla \times \underset{\sim}{H} - \frac{\partial}{\partial t} \underset{\sim}{\epsilon} \underset{\sim}{E} = 0$$
$$\nabla \cdot \underset{\sim}{B} = 0 \qquad \nabla \times \underset{\sim}{E} + \frac{\partial}{\partial t} \underset{\sim}{B} = 0 \tag{1}$$

where $\underset{\sim}{\epsilon}$ is the dielectric permittivity tensor. Now when we displace nuclei we modify the ground state wave functions and the excited state wave functions. Thus we must in general modify $\underset{\sim}{\epsilon}$ and we can expand this im powers of nuclear displacements as follows:

$$\underset{\sim}{\epsilon} = \underset{\sim}{\epsilon}^{(o)} + \lambda \underset{\sim}{\epsilon}^{(1)} + \lambda^2 \underset{\sim}{\epsilon}^{(2)} + \dots \tag{2}$$

Associated with these variations in $\underset{\sim}{\epsilon}$ we shall, in general, have corresponding variations in the electric and magnetic fields which we can write as:

$$\underset{\sim}{E} = \underset{\sim}{E}^{(o)} + \lambda \underset{\sim}{E}^{(1)} + \lambda^2 \underset{\sim}{E}^{(2)} + \ldots$$

$$\underset{\sim}{H} = \underset{\sim}{H}^{(o)} + \lambda \underset{\sim}{H}^{(1)} + \lambda^2 \underset{\sim}{H}^{(2)} + \ldots$$

$$\underset{\sim}{B} = \underset{\sim}{B}^{(o)} + \lambda \underset{\sim}{B}^{(1)} + \lambda^2 \underset{\sim}{B}^{(2)} + \ldots \tag{3}$$

If we substitute equations (2) and (3) into equations (1) and look for the terms in λ^n we obtain

$$\nabla \cdot [\underset{\sim}{\epsilon}^{(o)} \underset{\sim}{E}^{(n)} + \underset{\sim}{\epsilon}^{(1)} \underset{\sim}{E}^{(n-1)} + \ldots + \underset{\sim}{\epsilon}^{(n)} \underset{\sim}{E}^{(o)}] = 0$$

$$\nabla \times \underset{\sim}{H}^{(n)} - \frac{\partial}{\partial t} [\underset{\sim}{\epsilon}^{(o)} \underset{\sim}{E}^{(n)} + \ldots + \underset{\sim}{\epsilon}^{(n)} \underset{\sim}{E}^{(o)}] = 0$$

$$\nabla \cdot \underset{\sim}{B}^{(n)} = 0$$

$$\nabla \times \underset{\sim}{E}^{(n)} + \frac{\partial}{\partial t} \underset{\sim}{B}^{(n)} = 0 \tag{4}$$

We can rewrite these equations as:

$$\nabla \cdot \underset{\sim}{\epsilon}^{(o)} \underset{\sim}{E}^{(n)} = - \nabla \cdot \underset{\sim}{P}^{(n)}$$

$$\nabla \times \underset{\sim}{H}^{(n)} - \frac{\partial}{\partial t} \underset{\sim}{\epsilon}^{(o)} \underset{\sim}{E}^{(n)} = \frac{\partial}{\partial t} \underset{\sim}{P}^{(n)}$$

$$\nabla \cdot \underset{\sim}{B}^{(n)} = 0$$

$$\nabla \times \underset{\sim}{E}^{(n)} + \frac{\partial}{\partial t} \underset{\sim}{B}^{(n)} = 0 \tag{5}$$

where

$$P^{(n)} = \sum_{m=1}^{n} \underset{\sim}{\epsilon}^{(m)} \underset{\sim}{E}^{(n-m)} \quad \text{for } n \geq 1$$

$$= \quad 0 \quad \text{for } n = 0 \tag{6}$$

Thus we see that the radiation associated with the n'th order nuclear displacements obeys a set of Maxwells equations with an effective charge density $-\nabla . \underset{\sim}{P}^{(n)}$ and an effective current density $\frac{\partial}{\partial t} \underset{\sim}{P}^{(n)}$; in fact $\underset{\sim}{P}^{(n)}$ is the driving polarisation which produces these field.

We must recall that the various terms in equation (2) are in general complex and that the real and imaginary parts vary with frequency thus we can write:

$$\underset{\sim}{\epsilon}^{(n)}(\omega_o) = \underset{\sim}{\epsilon}_1^{(n)}(\omega_o) + i\, \underset{\sim}{\epsilon}_2^{(n)}(\omega_o) \tag{7}$$

The Kramers-Kronig relations further require $\underset{\sim}{\epsilon}_1^{(n)}$ and $\underset{\sim}{\epsilon}_2^{(n)}$ to be related in the following way:

$$\underset{\sim}{\epsilon}_1^{(o)}(\omega_o) - 1 = \frac{2}{\pi} \int_0^{\infty} \frac{\omega\, \underset{\sim}{\epsilon}_2^{(o)}(\omega)}{\omega^2 - \omega_o^2}\, d\omega$$

$$\underset{\sim}{\epsilon}_1^{(n)}(\omega_o) = \frac{2}{\pi} \int_0^{\infty} \frac{\omega\, \underset{\sim}{\epsilon}_2^{(n)}(\omega)}{\omega^2 - \omega_o^2}\, d\omega \quad \text{for } n \geq 1 \tag{8}$$

Finally we note that $\underset{\sim}{\epsilon}_2^{(n)}(\omega_o)$ corresponds to n'th order absorption processes while $\underset{\sim}{\epsilon}_1^{(n)}(\omega_o)$ corresponds to n'th order dispersive processes.

Two interesting conclusions can be deduced from the dispersion relations:

(i) for a uniform translation of the crystal $\underset{\sim}{\epsilon}_2^{(n)}(\omega) = 0$ for $n \geq 1$ thus $\underset{\sim}{\epsilon}_1^{(n)}(\omega) = 0$ for $n \geq 1$ and this means that for acoustic phonons we expect all the optical effects to tend to zero as $\underset{\sim}{q} \to 0$.

(ii) if $\underset{\sim}{\epsilon}_2^{(n)}(\omega)$ has a rapid onset at some frequency ω_o then the $\underset{\sim}{\epsilon}_1^{(n)}(\omega)$ will increase rapidly as we approach ω_o from the low frequency side and this is responsible for the phenomenon of resonant enhancement of Raman scattering.

To complete this section it is worth quoting a simple example of equation (8) for future reference. Consider a hypothetical case where $\underset{\sim}{\epsilon}_2^{(n)}(\omega)$ was zero except in the range ω_1 to ω_2 where it had the constant value πA. It is readily seen from equation (8) that

$$\underset{\sim}{\epsilon}_1^{(n)}(\omega) = A\, \ell n \left| \frac{\omega_2^2 - \omega^2}{\omega_1^2 - \omega^2} \right|$$

for $\underset{\sim}{\epsilon}_2^{(n)}(\omega) = \pi A$ for $\omega_1 \leq \omega \leq \omega_2$ otherwise

$$\underset{\sim}{\epsilon}_2^{(n)}(\omega) = 0$$

Note that for $\omega = 0$ $\underset{\sim}{\epsilon}_1^{(n)}(\omega) = A\, \ell n\, (\omega_2^2/\omega_1^2)$ and that $\underset{\sim}{\epsilon}_1^{(n)}(\omega)$ increases to $+\infty$ as $\omega \to \omega_1$ then falls to $-\infty$ as $\omega \to \omega_2$ and finally rises to zero for $\omega >> \omega_2$. Although the example is artifical it demonstrates the typical form of the dispersion relations.

ONE AND TWO PHONON PROCESSES

In the remainder of this lecture we shall be dealing exclusively with the optical effects produced by phonons in crystals in particular with one and two phonon processes. We note first from equation (6) that

$$\underset{\sim}{P}^{(1)} = \underset{\sim}{\epsilon}^{(1)}\, \underset{\sim}{E}^{(0)}$$

$$\underset{\sim}{P}^{(2)} = \underset{\sim}{\epsilon}^{(2)}\, \underset{\sim}{E}^{(0)} + \underset{\sim}{\epsilon}^{(1)}\, \underset{\sim}{E}^{(1)}$$

thus the optical properties of single phonon processes arise from the direct interaction of the incident field $\underset{\sim}{E}^{(o)}$ with the first order dielectric permittivity $\underset{\sim}{\epsilon}^{(1)}$. Two-phonon effects have two contributions namely the direct interaction of the incident field $\underset{\sim}{E}^{(o)}$ with the second order permittivity $\underset{\sim}{\epsilon}^{(2)}$ and the repeated single phonon interaction $\underset{\sim}{\epsilon}^{(1)}\ \underset{\sim}{E}^{(1)}$.

TWO-PHONON EFFECTS

As we have already stated two-phonon effects arise both from direct second-order processes contributing to $\underset{\sim}{\epsilon}_2^{(2)}(\omega)$ and also from repeated single-phonon effects. In the repeated single-phonon processes it is not necessary for the intermediate state to be real and energy and wavevector need only be conserved between the initial and final states. Sometimes in Raman scattering experiments one does see a repeated single-phonon process in which the intermediate state is real but it is probably more appropriate to call this a two stage scattering process. This effect is observed in diamond.

The main two phonon effects are the direct infra-red absorption by two phonons extending from twice the maximum phonon frequency to longer wavelengths. These are known as lattice bands and have been studied fairly extensively. The second is the two-phonon effects on the interband transitions which, through their dispersive part, lead to two phonon Raman scattering which has also been measured fairly extensively. Although these spectra display many of the more obvious features of the combined density of states for phonons-including critical points - their detailed interpretation has seldom been entirely successful and we shall not discuss them further.

SINGLE PHONON EFFECTS

We now ask the question where do we expect $\underset{\sim}{\epsilon}_2^{(1)}(\omega)$ to be non-zero?

In polar crystals one has the strong and sharp peak in $\underset{\sim}{\epsilon}_2^{(1)}(\omega)$ at the transverse optic phonon frequencies for $\underset{\sim}{q} = 0$ optic phonons. This has been dealt with in the session on polaritons. The corresponding variations in $\underset{\sim}{\epsilon}_1^{(1)}(\omega)$ can contribute to single phonon Raman scattering and, as is well known, the reststrahlen reflection bands are due to these variations in $\underset{\sim}{\epsilon}^{(1)}(\omega)$.

The next strongest interaction is due to the changes in the direct interband transistions. These contributions to $\underset{\sim}{\epsilon}_2^{(1)}(\omega)$ can sometimes be measured by piezo-reflectance techniques and the corresponding real part $\underset{\sim}{\epsilon}_1^{(1)}(\omega)$ is primarily responsible for single phonon Raman scattering. Note that $\underset{\sim}{\epsilon}_1^{(1)}(\omega)$ extends down to zero frequencies but falls to zero on the high frequency side of $\underset{\sim}{\epsilon}_2^{(1)}(\omega)$ thus in a polar crystal the relative contributions of the inter-band and transverse-optic phonon to $\underset{\sim}{\epsilon}_1^{(1)}(\omega)$ will depend on the frequency. Usually one carries out Raman scattering experiments as near to the band gap as one can and this means that the interband contributions dominate-in fact one often exploits the resonant increase in $\underset{\sim}{\epsilon}_1^{(1)}(\omega)$ as the frequency approaches the first onset of $\underset{\sim}{\epsilon}_2^{(1)}(\omega)$.

In some crystals exciton transistions are important and these contribute to the dielectric permittivity. In these cases one must also include the effects of nuclear displacement on the exciton to obtain its contribution to $\underset{\sim}{\epsilon}_2^{(1)}(\omega)$.

THE CALCULATION OF $\underset{\sim}{\epsilon}_2^{(1)}(\omega)$

If we wish to calculate the optical properties of single phonons we must start by considering how we could compute the change in the inter-band absorption when a phonon is frozen into the lattice. For a phonon of arbitrary wave-vector this will be difficult since the fundamental translation symmetry of the unperturbed lattice is modified. However, if we recall that the single phonon Raman scattering comes from optical phonons of $\underset{\sim}{q} \approx 0$ then we immediately see that this is simply equivalent to changing the basis vector of the crystal and the basic translational symmetry is not altered. This problem is now no more difficult than any band structure calculation and I understand that preliminary calculations of this type are showing reasonable success. (A. Maradudin private communication 1969).

XV

DYNAMICS OF INTERACTING POLARITONS IN DIELECTRIC CRYSTALS

C. Mavroyannis

National Research Council of Canada
Ottawa 2, Canada

I. INTRODUCTION

In this lecture we shall discuss the microscopic view of the dynamical behavior of interacting polaritons in dielectric crystals. In the infrared range of frequencies of dielectric crystals, polaritons are defined as quasi-particles which consist of long wavelength transverse photons dressed by the harmonic field of optical phonons and propagate in the medium with energy and wavevector that satisfy the linear Maxwell equations[1]. The existence of polaritons has been demonstrated by Henry and Hopfield[2] by Raman scattering experiments on GaP.

The work to be presented here was performed in collaboration with K.N. Pathak[3]. This work was stimulated by the fact that though polaritons have been observed[2], the only existing theory is that presented by Burstein et al.[4] on Ram scattering by polaritons. However, this is a phenomenological approach and does not reveal the nature of the interactions involved.

We shall first derive the total polariton Hamiltonian consisting of the harmonic and anharmonic parts with respect to the polariton operators which describe polariton-polariton interactions. Then using the polariton Hamiltonian the Dyson equation for the polariton field is developed. The polarization operator and, hence, the polariton Green's function are evaluated in two successive approximations by means of constructed equivalent renormalized Hamiltonians which are correct in the lowest and first approximation respectively. The derived results for the phonon spectrum are compared with those obtained by Wehner[5] and Kwok[6]. Finally the polariton excitation spectrum is considered in detail and the corresponding bare phonon excitation spectrum is compared with those results derived by means of diagram techniques[7].

II. THE POLARITON HAMILTONIAN

The Hamiltonian of the coupled system, crystal plus radiation, can be written in the form

$$\mathcal{H} = \mathcal{H}_{\ell} + \mathcal{H}_{\gamma} + \mathcal{H}_{\ell\gamma}, \tag{1}$$

where the lattice Hamiltonian, in terms of phonon annihilation and creation operators is known to be[7]

$$\mathcal{H}_{\ell} = \sum_{\underline{k}j} \omega^{o}_{\underline{k}j} (a^{\dagger}_{\underline{k}j} a_{\underline{k}j} + \tfrac{1}{2}) + \sum_{n=3}^{\infty} \sum_{\underline{k}_1 j_1, \underline{k}_2 j_2, \ldots, \underline{k}_n j_n}$$

$$V_n(\underline{k}_1 j_1, \underline{k}_2 j_2, \ldots, \underline{k}_n j_n) \times (a_{\underline{k}_1 j_1} + a^{\dagger}_{-\underline{k}_1 j_1}) \ldots (a_{\underline{k}_n j_n} + a^{\dagger}_{-\underline{k}_n j_n}). \tag{2}$$

In (2) the first term describes the Hamiltonian of the free phonon assembly and the remaining terms describe the interactions between them. ω^{o}_{k} is the energy of the bare phonon of wavevector $\underline{k}$ and polarization j. We use $\hbar = 1$ throughout. The Hamiltonian for the radiation field is[8]

$$\mathcal{H}_{\gamma} = \sum_{\underline{k}\lambda} ck\, b^{\dagger}_{\underline{k}\lambda} b_{\underline{k}\lambda}, \tag{3}$$

where $b_{\underline{k}\lambda}$ and $b^{\dagger}_{\underline{k}\lambda}$ are the annihilation and creation operators for the bare photon of momentum $\underline{k}$ and transverse polarization λ (=1,2); c denotes velocity of light in vacuum. The interaction Hamiltonian due to the electromagnetic field can be represented as

$$\mathcal{H}_{\ell\gamma} = \sum_{\ell\kappa} \frac{Z_{\kappa}}{M_{\kappa} c} \underline{A}\, (\underline{R}(^{\ell}_{\kappa})) \cdot \underline{p}(^{\ell}_{\kappa}) + \sum_{\ell\kappa} \frac{Z^2_{\kappa}}{2M_{\kappa} c^2} A^2\, (\underline{R}(^{\ell}_{\kappa})), \tag{4}$$

where $\underline{A}\, (\underline{R}(^{\ell}_{\kappa}))$ is the vector potential of the electromagnetic field evaluated at the instantaneous position

$$\underline{R}\binom{\ell}{\kappa} = \underline{X}\binom{\ell}{\kappa} + \underline{U}\binom{\ell}{\kappa} = \underline{X}(\kappa) + \underline{X}(\ell) + \underline{U}\binom{\ell}{\kappa}. \qquad (5)$$

Here $\underline{X}\binom{\ell}{\kappa}$ is the equilibrium position of κth ion in ℓ^{th} unit cell, $\underline{U}\binom{\ell}{\kappa}$ is the displacement from the equilibrium position and Z_κ is the effective charge of the κth ion. We consider N unit cells in volume V and there are r atoms per unit cell. The displacement, momentum amplitude, and vector potential can be expressed in the second quantized representation respectively as [3]

$$\underline{U}\binom{\ell}{\kappa} = \left(\frac{1}{2M_\kappa N}\right)^{\frac{1}{2}} \sum_{\underline{k}j} \frac{\underline{e}(\kappa|\underline{k}j)}{\sqrt{\omega^o_{\underline{k}j}}} (a_{\underline{k}j}+a^\dagger_{-\underline{k}j})\, e^{i\underline{k}\cdot\underline{X}(\ell)}, \qquad (6a)$$

$$\underline{p}\binom{\ell}{\kappa} = \frac{1}{i}\left(\frac{M_\kappa}{2N}\right)^{\frac{1}{2}} \sum_{\underline{k}j} \underline{e}(\kappa|\underline{k}j) \sqrt{\omega^o_{\underline{k}j}}\, (a_{\underline{k}j}-a^\dagger_{-\underline{k}j})\, e^{i\underline{k}\cdot\underline{X}(\ell)}, \qquad (6b)$$

and

$$\underline{A}(\underline{R}\binom{\ell}{\kappa}) = \sum_{\underline{k}\lambda} \left(\frac{2\pi c^2}{Vck}\right)^{\frac{1}{2}} \underline{\xi}(\underline{k}\lambda)\, (b_{\underline{k}\lambda}+b^\dagger_{-\underline{k}\lambda})\, e^{i\underline{k}\cdot\underline{R}\binom{\ell}{\kappa}}. \qquad (6c)$$

In Eq. (6) $\underline{e}(\kappa|\underline{k}j)$ is the polarization vector for the phonons and satisfies the usual orthonormality and closure relations; $\underline{\xi}(\underline{k}\lambda)$ denotes the unit photon polarization vector. We substitute the expression for $\underline{R}\binom{\ell}{\kappa}$ from Eq. (5) into (6c) and expand the exponential in powers of the displacement amplitude $\underline{U}\binom{\ell}{\kappa}$ from the equilibrium position $\underline{X}\binom{\ell}{\kappa}$. The first term in this expansion corresponds to the usual long wavelength approximation, i.e., the vector potential is evaluated at the equilibrium position. In this approximation, the interaction Hamiltonian (4) can be easily expressed in the form

$$\mathcal{H}^o_{\ell\gamma} = \frac{\omega_p^2}{4} \sum_{\underline{k}\lambda} \frac{1}{ck} (b_{\underline{k}\lambda}+b^\dagger_{-\underline{k}\lambda})(b^\dagger_{\underline{k}\lambda}+b_{-\underline{k}\lambda})$$

$$-\frac{i}{2}\sum_{\underline{k},j,\lambda}\sqrt{\frac{\omega^{o}_{\underline{k}j}}{kc}}\; x_j(\underline{k}\lambda)\;(b^{\dagger}_{\underline{k}\lambda}+b_{\underline{k}\lambda})(a_{\underline{k}j}-a^{\dagger}_{\underline{k}j}), \tag{7}$$

where the coupling constant $x_j(\underline{k}\lambda)$ is expressed as

$$x_j(\underline{k}\lambda) = \left(\frac{4\pi N}{V}\right)^{\frac{1}{2}} \sum_{\kappa=1}^{r} \frac{Z_\kappa}{\sqrt{M_\kappa}}\, e^{-i\underline{k}\cdot\underline{X}(\kappa)} \quad [\underline{e}(\kappa|\underline{k}j)\cdot\underline{\xi}(\underline{k}\lambda)], \tag{8}$$

from which it follows that photons interact only with the transverse optical phonons. It is easy to see that the plasma frequency of the ions satisfies the relation

$$\omega_p^2 = \frac{4\pi N}{V}\sum_{\kappa}\frac{Z_\kappa^{\;2}}{M_\kappa} = \sum_j |x_j(\underline{k}\lambda)|^2. \tag{9}$$

The higher order terms arising from the interaction Hamiltonian (4) can be written in the general form

$$\mathcal{H}'_{\ell\gamma} = -\sum_{n=3}^{\infty}\;\sum_{\underline{q}\lambda,\underline{k}_1 j_1,\ldots,\underline{k}_{n-1}j_{n-1}} \phi_n(\underline{q}\lambda,\underline{k}_1 j_1,\underline{k}_2 j_2,\ldots,\underline{k}_{n-1}j_{n-1})$$

$$\times\;(b_{\underline{q}\lambda}+b^{\dagger}_{-\underline{q}\lambda})(a_{\underline{k}_1 j_1}-a^{\dagger}_{\underline{k}_1 j_1})(a_{\underline{k}_2 j_2}+a^{\dagger}_{-\underline{k}_2 j_2})\ldots(a_{\underline{k}_{n-1}j_{n-1}}+a^{\dagger}_{-\underline{k}_{n-1}j_{n-1}})$$

$$+\sum_{n=3}^{\infty}\;\sum_{\substack{\underline{q}_1\lambda_1,\underline{q}_2\lambda_2\\ \underline{k}_3 j_3,\ldots,\underline{k}_{n-2}j_{n-2}}} g_n(\underline{q}_1\lambda_1,\underline{q}_2\lambda_2,\underline{k}_3 j_3,\ldots,\underline{k}_{n-2}j_{n-2})$$

$$\times\;(b_{\underline{q}_1\lambda_1}+b^{\dagger}_{-\underline{q}_1\lambda_1})(b_{\underline{q}_2\lambda_2}+b^{\dagger}_{-\underline{q}_2\lambda_2})(a_{\underline{k}_3 j_3}+a^{\dagger}_{-\underline{k}_3 j_3})\ldots(a_{\underline{k}_{n-2}j_{n-2}}+a^{\dagger}_{-\underline{k}_{n-2}j_{n-2}}) \tag{10a}$$

where the coupling functions $\phi_n(\underline{q}\lambda,\underline{k}_1 j_1,\underline{k}_2 j_2,\ldots,\underline{k}_{n-1}j_{n-1})$ and $g_n(\underline{q}_1\lambda_1,\underline{q}_2\lambda_2,\underline{k}_3 j_3,\ldots,\underline{k}_{n-2}j_{n-2})$ result from terms in the expression (4) which are linear and quadratic with respect to the vector potential respectively, and are equal to

$$\phi_n(\underline{q}\lambda,\underline{k}_1 j_1,\underline{k}_2 j_2,\ldots,\underline{k}_{n-1}j_{n-1}) = i\sum_{\kappa}\left(\frac{\pi Z_\kappa^2 N}{M_\kappa V}\right)^{\frac{1}{2}} e^{i\underline{q}\cdot\underline{X}(\kappa)}$$

$$\times\ [\underline{\xi}(\underline{q}\lambda)\cdot\underline{e}(\kappa|\underline{k}_1 j_1)][\underline{q}\cdot\underline{e}(\kappa|\underline{k}_1 j_1)]\ldots[\underline{q}\cdot\underline{e}(\kappa|\underline{k}_{n-1} j_{n-1})]$$

$$\times\ \frac{(i)^{n-2}}{(n-2)!}\,(2M_\kappa N)^{-(n-2)/2}\ \omega^{o\ \frac{1}{2}}_{\underline{k}_1 j_1}(cq\omega^o_{\underline{k}_2 j_2}\,\omega^o_{k_3 j_3}\cdots\omega^o_{\underline{k}_{n-1} j_{n-1}})^{-\frac{1}{2}}$$

$$\times\ \underline{\Delta}(\underline{q}+\underline{k}_1+\underline{k}_2+\ \cdots\ +\underline{k}_{n-1}), \qquad (10b)$$

$$g_n(\underline{q}_1\lambda_1,\underline{q}_2\lambda_2,\underline{k}_3 j_3,\ldots,\underline{k}_{n-2} j_{n-2}) = \sum_\kappa \left(\frac{Z_\kappa^2 N}{2M_\kappa c}\right) e^{i(\underline{q}_1+\underline{q}_2)\cdot\underline{X}(\kappa)}$$

$$\times\ [\underline{\xi}(\underline{q}_1\lambda_1)\cdot\underline{\xi}(\underline{q}_2\lambda_2)]\left(\frac{4\pi^2 Z_\kappa^2}{V^2 \underline{q}_1\underline{q}_2}\right)^{\frac{1}{2}} \times \frac{(i)^{n-2}}{(n-2)!}\,(2M_\kappa N)^{-(n-2)/2}$$

$$\times\ (\omega^o_{\underline{k}_3 j_3}\,\omega^o_{\underline{k}_4 j_4}\cdots\omega^o_{\underline{k}_{n-2} j_{n-2}})^{-\frac{1}{2}}[(\underline{q}_1+\underline{q}_2)\cdot\underline{e}(\kappa|k_3 j_3)]\ldots$$

$$[(\underline{q}_1+\underline{q}_2)\cdot\underline{e}(\kappa|\underline{k}_{n-2} j_{n-2})] \times \underline{\Delta}(\underline{q}_1+\underline{q}_2+\underline{k}_3+\underline{k}_4+\cdots+\underline{k}_{n-2}). \qquad (10c)$$

The interaction Hamiltonian (10a) describes all the possible photon-phonon scattering processes. The function $\underline{\Delta}(\underline{k})$ appearing in (10) is equal to unity when $\underline{k}$ is equal to a reciprocal lattice vector, and vanishes otherwise.

It is now well known that the quadratic parts of the Hamiltonian can be diagonalized with a canonical transformation to exhibit the independent polariton modes. Hence, we diagonalize the Hamiltonian consisting of the harmonic parts of (2) and (3) for the lattice and the radiation field respectively plus the interaction between them, (7), by using the canonical transformation[8]

$$a_{\underline{k}j} = \sum_\rho [u_j(\underline{k}\rho)\alpha_{\underline{k}\rho} + v_j^*(-\underline{k}\rho)\alpha^\dagger_{-\underline{k}\rho}], \qquad (11a)$$

and

$$b_{\underline{k}\lambda} = \sum_\rho [u_\lambda(\underline{k}\rho)\alpha_{\underline{k}\rho} + v_\lambda^*(-\underline{k}\rho)\alpha^\dagger_{-\underline{k}\rho}], \qquad (11b)$$

where u_j, v_j and u_λ, v_λ are the transformation amplitudes for phonons and photons respectively and are determined by solving a set of four simultaneous equations; ρ designates the polariton bands. In the long wavelength approximation the Hamiltonian is now expressed in the polariton representation as

$$\mathcal{H}_o = \text{const.} + \sum_{\underline{k}\rho} \omega_{\underline{k}\rho}\, \alpha^\dagger_{\underline{k}\rho}\, \alpha_{\underline{k}\rho} \tag{12}$$

where $\omega_{\underline{k}\rho}$ are the energies of excitation derived from the solutions of Maxwell's equations

$$c^2k^2 = \omega^2_{\underline{k}\rho}\ \tilde{\varepsilon}(\underline{k},\omega_{\underline{k}\rho}), \tag{13a}$$

$$\tilde{\varepsilon}(\underline{k},\omega_{\underline{k}\rho}) = 1 + \sum_j \frac{|x_j(\underline{k})|^2}{\omega^{o2}_{\underline{k}j} - \omega^2_{\underline{k}\rho}} \tag{13b}$$

The polariton operators $\alpha^\dagger_{\underline{k}\rho}$ and $\alpha_{\underline{k}\rho}$ satisfy Bose commutation relations while the summation in (13b) is over those phonon branches which interact with the photons of the electromagnetic field. Because of the presence of anharmonicity both in the expression for the lattice Hamiltonian as well as in that for the coupled photon-phonon field, polaritons are no longer independent but interact with each other. To derive the expression for the polariton interaction Hamiltonian, we express both the anharmonic lattice Hamiltonian and the photon-phonon interaction Hamiltonian in the polariton representation with the result

$$\mathcal{H}_{int.} = \sum_{n=3}^{\infty} \sum_{k_1,k_2,\ldots,k_n} \Big\{ \tilde{V}_n(k_1,k_2,\ldots,k_n) A_{k_1} A_{k_2} \ldots A_{k_n} - \tilde{\phi}_n(k_1,k_2,\ldots,k_n) B_{k_1} A_{k_2} A_{k_3} \ldots A_{k_n} \Big\}, \tag{14}$$

where

$$A_k \equiv \alpha_k + \alpha^{\dagger}_{-k} \quad , \quad B_k \equiv \alpha_k - \alpha^{\dagger}_{-k} \quad , \quad k \equiv (\underline{k}, \rho).$$

In (14), $\tilde{V}_n(k_1,k_2,\ldots,k_n)$ and $\tilde{\phi}_n(k_1,k_2,\ldots k_n)$ are the anharmonic coupling functions for the polariton field; their explicit expressions are given elsewhere[3]. In the range of wavevectors where retardation effects can be neglected, i.e., in the limit when $\omega_k \to \omega^o_{kj}$ and $\tilde{\varepsilon}(k,\omega_k) \to 1$, then $\tilde{\phi}_n(k_1,k_2,\ldots,k_n) \to 0$, $\tilde{V}_n(k_1,k_2,\ldots,k_n) \to V_n(k_1j_1,k_2j_2,\ldots,k_nj_n)$ and the interaction Hamiltonian (14) describes the anharmonic phonon field.

III. DYSON'S EQUATION

We shall now derive the Dyson equation and, consequently, the polarization operator describing the polariton spectrum. The retarded polariton Green's function is defined as[9]

$$G(k,t-t') \equiv \langle\langle A_k;A^\dagger_{k'}\rangle\rangle = -i\theta(t-t')\langle[A_k(t),A^\dagger_{k'}(t')]_-\rangle, \quad (15)$$

where the angular brackets denote the average over the canonical ensemble appropriate to the total Hamiltonian $\mathcal{H}$, $\theta(t)$ is the usual step function and the operators A_k and $A^\dagger_{k'}$ are in the Heisenberg representation. In (15) and in what follows, the time arguments of the operators have been suppressed for convenience. The Fourier transform of the polariton Green's function with respect to the argument t satisfies the equation of motion

$$\omega G(k,\omega) = \frac{1}{2\pi}\langle[A_k,A^\dagger_{k'}]_-\rangle_{t=t'} + \langle\langle[A_k,\mathcal{H}]_-;A^\dagger_{k'}\rangle\rangle. \quad (16)$$

Using (12), (14) and (16) the equation of motion for the polariton Green's function is found to be

$$G^{-1}_\infty(k,\omega)G(k,\omega) = 1 + \langle\langle F(k);A^\dagger_{k'}\rangle\rangle, \quad (17)$$

where

$$G_\infty(k,\omega) = \frac{\omega_k}{\pi}[\omega^2-\omega^2_k]^{-1}\delta_{kk'}, \quad (18)$$

is the unperturbed polariton Green's function and

$$F(k) = 2\pi\sum_{n=3}^{\infty}\sum_{k_1,k_2,\ldots,k_{n-1}}[Q_n(k_1,k_2,\ldots,k_{n-1},-k;\omega)A_{k_1}A_{k_2}\ldots A_{k_{n-1}}$$

$$-(n-1)\Phi_n(k_1,k_2,\ldots,k_{n-2},-k)B_{k_1}A_{k_2}\ldots A_{k_{n-2}}], \quad (19)$$

$$Q_n(k_1,k_2,\ldots,k_{n-1},-k;\omega) \equiv n\,\tilde{V}_n(k_1,k_2,\ldots,k_{n-1},-k)$$
$$+ \left(\frac{\omega}{\omega_k}\right) \tilde{\phi}_n(k_1,k_2,\ldots,k_{n-1},-k). \tag{20}$$

Considering the equation of motion for the Green's function that appears on the right-hand side of (17) with respect to the argument t' and substituting the resulting expression into (17) we derive Dyson's equation

$$G(k,\omega) = G_{oo}(k,\omega) + G_{oo}(k,\omega)\tilde{P}(k,\omega)G_{oo}(k,\omega)$$
$$= G_{oo}(k,\omega) + G_{oo}(k,\omega)\Pi(k,\omega)G(k,\omega), \tag{21}$$

where the polarization operator, $\Pi(k,\omega)$, is equal to

$$\Pi(k,\omega) = \tilde{P}(k,\omega)[1 + G_{oo}(k,\omega)\tilde{P}(k,\omega)]^{-1}, \tag{22}$$

and

$$\tilde{P}(k,\omega) = \tfrac{1}{2}\left[\langle [F(k),B_k^\dagger]\rangle + \left(\frac{\omega}{\omega_k}\right)\langle [F(k),A_k^\dagger]\rangle\right]_{t=t'} + \langle\langle F(k);F^\dagger(k)\rangle\rangle. \tag{23}$$

In the range of frequencies ω far from the zeros of the denominator appearing in the expression for the polarization operator (22), we may expand the denominator of (22) in a power series of $\tilde{P}(k,\omega)$ as

$$\Pi(k,\omega) \doteq \tilde{P}(k,\omega)[1 - G_{oo}(k,\omega)\tilde{P}(k,\omega) + \cdots\cdots]. \tag{24}$$

In what follows we shall retain only the first term in the expression (24). Then Dyson's equation (21) may be written as

$$G(k,\omega) = \omega_k/\pi\ \delta_{kk'}\ [\omega^2-\tilde{\omega}_k^2-2\omega_k P(k,\omega)]^{-1}, \tag{25}$$

where the energies $\tilde{\omega}_k$ are determined from the solutions of the equation

$$\tilde{\omega}_k^2 = \omega_k^2 + \left(\frac{\omega_k}{2\pi}\right)\left[\left\langle [F(k),B_k^{\dagger}]_- \right\rangle^{(o)} + \left(\frac{\omega}{\omega_k}\right)\left\langle [F(k),A_k^{\dagger}]_- \right\rangle^{(o)}\right]_{\omega=\tilde{\omega}_k, t=t'}, \tag{26}$$

and $P(k,\omega)$ is given by

$$P(k,\omega) = \frac{1}{2\pi} \langle\langle F(k);F^{\dagger}(k)\rangle\rangle. \tag{27}$$

The solutions of the equation (26) determine the frequencies, $\tilde{\omega}_k$, for the renormalized mode k in the lowest approximation of perturbation theory and the renormalization is caused by the terms arising from the evaluation of the commutators. In (26), $\omega=\tilde{\omega}_k$ and the superscript (o) indicate that the polariton occupation numbers, $N_q = \langle A_q^{\dagger}A_q\rangle$, resulting from the calculation of the commutators have to be evaluated by means of the zeroth order Green's function (30) or the renormalized Hamiltonian (31); for example, $N_q^{(o)} = \left(\frac{\omega_q}{\tilde{\omega}_q}\right) \coth \tfrac{1}{2}\beta\tilde{\omega}_q$. In this approximation the correlation functions with $n>4$ have to be decoupled in the usual way into products of polariton occupation numbers, $N_q^{(o)}$, and hence only correlation functions with even number of operators give nonzero contributions. In this manner, the second term on the r.h.s. of (26) can be easily expressed as a function of the polariton occupation numbers and $\tilde{\omega}_k$ including terms up to any desired value of n, then the final solution for the frequencies $\tilde{\omega}_k$ has to be derived from eq. (26) by computation. The polarization operator, $P(k,\omega)$, will be calculated in successive approximations, i.e., it may be written as

$$P(k,\omega) \approx P^{(o)}(k,\omega) + P^{(1)}(k,\omega) + \cdots, \tag{28}$$

where the superscripts (o) and (1) indicate that the various Green's functions that appear in (27) must be evaluated in the lowest and first approximation respectively in the manner which shall be described below.

i) Lowest Approximation for $P(k,\omega)$:

In the first approximation the Green's function (25) may be written as

$$G^{(1)}(k,\omega) = \frac{\omega_k}{\pi} \delta_{kk'} [\omega^2 - \tilde{\omega}_k^2 - 2\omega_k P^{(o)}(k,\omega)]^{-1}. \tag{29}$$

From (29) we observe that when $P^{(o)}(k,\omega)$ is taken equal to zero then the zeroth order renormalized polariton Green's function is given by

$$G^{(o)}(k,\omega) = \frac{\omega_k}{\pi} \delta_{kk'} [\omega^2 - \tilde{\omega}_k^2]^{-1}. \tag{30}$$

The excitation spectrum described by (30) results from an equivalent zeroth order renormalized Hamiltonian which has the form

$$\mathcal{H}^{(o)}_{ren.} = \text{const.} + \tfrac{1}{4} \sum_k \left[\left(\frac{\tilde{\omega}_k^2}{\omega_k}\right) A_k^\dagger A_k + \omega_k B_k^\dagger B_k\right]. \tag{31}$$

Therefore, the Hamiltonian (31) must be used for the evaluation of $P^{(o)}(k,\omega)$. Because of the form of the Hamiltonian (31), only the Green's functions appearing in (27) with $n=n'$ give contributions while terms with $n \neq n'$ disappear. In the Appendix, an example is carried out where the various two- and three-polariton Green's functions are evaluated via the Hamiltonian (31). Substituting (A.5) and (A.9) into (27) we obtain

$$P^{(o)}(k,\omega) = 2 \sum_{k_1,k_2} \Big\{ [\tilde{Q}_3^{(+)}(k_1,k_2,-k;\omega)(\tilde{\omega}_{k_1}+\tilde{\omega}_{k_2})$$
$$- 2\omega\hat{Q}_3^{(+)}(k_1,k_2,-k;\omega)] \times (\tilde{n}_{k_1}+\tilde{n}_{k_2})[\omega^2-(\tilde{\omega}_{k_1}+\tilde{\omega}_{k_2})^2]^{-1}$$
$$+ [\tilde{Q}_3^{(-)}(k_1,k_2,-k;\omega)(\tilde{\omega}_{k_1}-\tilde{\omega}_{k_2})-2\omega\hat{Q}_3^{(-)}(k_1,k_2,-k;\omega)]$$

$$\times (\tilde{n}_{k_2}-\tilde{n}_{k_1})[\omega^2-(\tilde{\omega}_{k_1}-\tilde{\omega}_{k_2})^2]^{-1}\Big\}$$

$$+ 2 \sum_{k_1,k_2,k_3}\Big\{[\tilde{Q}_4^{(+)}(k_1,k_2,k_3,-k;\omega)(\tilde{\omega}_{k_1}+\tilde{\omega}_{k_2}+\tilde{\omega}_{k_3})$$

$$- 2\omega\hat{Q}_4(k_1,k_2,k_3,-k;\omega)] \times (1+\tilde{n}_{k_1}\tilde{n}_{k_2}+\tilde{n}_{k_2}\tilde{n}_{k_3}+\tilde{n}_{k_1}\tilde{n}_{k_3})$$

$$[\omega^2-(\tilde{\omega}_{k_1}+\tilde{\omega}_{k_2}+\tilde{\omega}_{k_3})^2]^{-1} + [\tilde{Q}_4^{(-)}(k_1,k_2,k_3,-k;\omega)(\tilde{\omega}_{k_1}-\tilde{\omega}_{k_2}-\tilde{\omega}_{k_3})$$

$$-2\omega\hat{Q}_4(k_1,k_2,k_3,-k;\omega)] \times (1+\tilde{n}_{k_2}\tilde{n}_{k_3}-\tilde{n}_{k_1}\tilde{n}_{k_3}-\tilde{n}_{k_1}\tilde{n}_{k_2})$$

$$[\omega^2-(\tilde{\omega}_{k_1}-\tilde{\omega}_{k_2}-\tilde{\omega}_{k_3})^2]^{-1} + (\tilde{\omega}_{k_2}\leftrightarrow-\tilde{\omega}_{k_2}) + (\tilde{\omega}_{k_3}\leftrightarrow-\tilde{\omega}_{k_3})\Big\}$$

$$+ \text{terms with } n > 4. \qquad (32)$$

The coupling functions $\tilde{Q}_3^{(\pm)}(k_1,k_2,-k;\omega)$, $\hat{Q}_3^{(\pm)}(k_1,k_2,-k;\omega)$, $\tilde{Q}_4^{(\pm)}(k_1,k_2,k_3,-k;\omega)$ and $\hat{Q}_4(k_1,k_2,k_3,-k;\omega)$ are given by the expressions (A.10a)-(A.10d) in the Appendix respectively. In (32) $(\tilde{\omega}_{k_2}\leftrightarrow-\tilde{\omega}_{k_2})$ and $(\tilde{\omega}_{k_3}\leftrightarrow-\tilde{\omega}_{k_3})$ for the quartic anharmonic terms indicate that there are two further terms obtained by changing the sign of $\tilde{\omega}_{k_2}$ and $\tilde{\omega}_{k_3}$ in the first term respectively. Hence, the excitation spectrum in the first approximation is described by the expression (29) with $P^{(0)}(k,\omega)$ given by (32) and is a function of the frequency modes, $\tilde{\omega}_k$'s, which are renormalized in the lowest order of perturbation theory. In this approximation (29) may be written as

$$G^{(1)}(k,\omega) = (\frac{\omega_k}{\pi})\,\delta_{kk'}\,[\omega^2-\tilde{\omega}_k^2-2\omega_k\mathrm{Re}P^{(0)}(k,\omega)-2\omega_k\mathrm{Im}P^{(0)}(k,\omega)]^{-1} \qquad (32a)$$

where $\mathrm{Re}P^{(0)}(k,\omega)$ and $\mathrm{Im}P^{(0)}(k,\omega)$ indicate the real and imaginary parts of (32) respectively. If $\mathrm{Im}P^{(0)}(k,\omega)$ is small but finite, $\mathrm{Im}P^{(0)}(k,\omega)<<\omega_k$ and varies slowly with ω in the vicinity of frequencies $\omega\sim\omega_k$, then the energy spectrum described by (32a) is a Lorentzian line peaked at $\omega\sim\tilde{\omega}_k$ with a width of the order of $\mathrm{Im}P^{(0)}(k,\tilde{\omega}_k)$ in energy units.

An alternative approach of calculating the Green's function

$$G(k,\omega) = (\frac{\omega_k}{\pi})\, \delta_{kk'}\, [\omega^2-\omega_k^2-(\frac{\omega_k}{\pi})\tilde{P}(k,\omega)]^{-1} \tag{32b}$$

is the following: If the frequency $\tilde{\omega}_k$ is replaced by Ω_k in the expression for the Hamiltonian (31), where Ω_k is the polariton energy of excitation which has yet to be determined self-consistently, then using this Hamiltonian to evaluate (32b) we find

$$G(k,\omega) = (\frac{\omega_k}{\pi})\, \delta_{kk'}\, [\omega^2-\Omega_k^2-2\omega_k \mathrm{Im}P(k,\omega)]^{-1} \tag{32c}$$

where Ω_k is determined by the solutions of the equation

$$\Omega_k^2-\omega_k^2-P_1(k,N_q;\omega=\Omega_k)-2\omega_k \mathrm{Re}P(k,\omega=\Omega_k) = 0. \tag{32d}$$

The function $P_1(k,N_q;\omega=\Omega_k)$ results from the evaluation of the last term on the r.h.s. of (26) in the manner described in the footnote (12) and is a function of the polariton occupation numbers, $N_q = (\frac{\omega_q}{\Omega_q}) \coth \tfrac{1}{2}\beta\Omega_q$, and the frequency Ω_k. The function $P(k,\omega=\Omega_k)$ is given by the expression (32) if all the frequencies $\tilde{\omega}_{k_1}$, $\tilde{\omega}_{k_2}$ and $\tilde{\omega}_{k_3}$ are replaced everywhere by Ω_{k_1}, Ω_{k_2} and Ω_{k_3} respectively. It is easily seen that when $\mathrm{Im}P(k,\omega)$ is small but finite then the excitation spectrum described by (32c) is reduced to that of (32a). In the limiting case when $\mathrm{Im}P(k,\omega)$ tends to zero for any value of $\omega = \Omega_k$, then the energies of excitation Ω_k, corresponding to a quasi-particle with infinite lifetime, are determined by the solutions of the equation

$$\Omega_k^2-\omega_k^2-P_1(k,N_q;\omega=\Omega_k)-2\omega_k P(k,\omega=\Omega_k) = 0. \tag{32e}$$

In principle, Eq. (32e) can be solved self-consistently by means of computing methods to obtain numerical results. In practice, not only the frequency Ω_k but also all the frequencies $\Omega_{k_1}, \Omega_{k_2}, \ldots, \Omega_{k_{n-1}}$ have to be determined self-consistently, therefore, for the general solution of (32e) one has to deal with infinite number of variables (frequencies). Hence, approximate methods have to be developed for the calculation of (32c).

ii) First Approximation for $P(k,\omega)$:

In the case when the imaginary part of $P^{(o)}(k,\omega)$ tends to zero for some frequency, say ε_k ($\neq\tilde{\omega}_k$), then the expression (29a) takes the form

$$G^{(1)}(k,\omega) = (\frac{\omega_k}{\pi})\ \delta_{kk'}\ [\omega^2-\varepsilon_k^2]^{-1} \tag{33}$$

where the energies of excitation ε_k are determined by the roots of the equation

$$\varepsilon_k^2-\tilde{\omega}_k^2-2\omega_k P^{(o)}(k,\varepsilon_k) = 0. \tag{34a}$$

Considering that when the system resonates in the neighborhood of frequencies $\omega\sim\tilde{\omega}_k$ the expressions (32a) and (32c) coincide, then far from resonance the inclusion in (34a) of the term $2\omega_k P^{(o)}(k,\varepsilon_k)$, in which all the frequencies $\Omega_{k'}$ have been replaced by their corresponding values at resonance $\tilde{\omega}_{k'}$, may give a reasonably good value for ε_k. In view of (32e), one may argue that the frequency ε_k determined by the solution of the equation

$$\varepsilon_k^2-\omega_k^2-P_1(k,N_q(\varepsilon_q);\omega=\varepsilon_k)-2\omega_k P^{(o)}(k,\varepsilon_k) = 0 \tag{34b}$$

is a better approximation to Ω_k than that obtained by the solution of (34a). In fact, this is generally true, the only disadvantage of (34b) is that much more effort is required for the computation of ε_k than that for the corresponding solution of (34a). Although the last term in (34b) is only an approximation to the corresponding term in (32e), it will be improved in the next approximation. The solutions of (34) corresponding to the frequencies ε_k imply that the bare polariton is dressed by the anharmonic interaction of the others and the resultant dressed quasiparticle with energy ε_k may be called "physical polariton". The excitation spectrum described by (33) results from an equivalent renormalized Hamiltonian which is correct in the first approximation, i.e.,

$$\mathcal{H}^{(1)}_{\text{ren.}} = \text{constant} + \tfrac{1}{4}\sum_k [(\varepsilon_k^2/\omega_k)A_k^\dagger A_k + \omega_k B_k^\dagger B_k]. \qquad (35)$$

Therefore, the polariton Green's function in the second approximation turns out to be

$$G^{(2)}(k,\omega) = \frac{\omega_k}{\pi}\,\delta_{kk'}\,[\omega^2 - \varepsilon_k^2 - 2\omega_k P^{(1)}(k,\omega)]^{-1}, \qquad (36)$$

and the polarization operator $P^{(1)}(k,\omega)$ has to be evaluated via the first order Hamiltonian (35). A comparison between (31) and (35) shows that the expression for $P^{(1)}(k,\omega)$ can be derived from that of $P^{(0)}(k,\omega)$, equation (32), by replacing the renormalized frequencies correct to the lowest order, viz., $\tilde{\omega}_{k_1}$, $\tilde{\omega}_{k_2}$ and $\tilde{\omega}_{k_3}$, by ε_{k_1}, ε_{k_2} and ε_{k_3} respectively, where the energies of excitation ε_{k_1}, ε_{k_2} and ε_{k_3} are obtained from the solutions of the equation (34) for each particular mode. Then the expression for $P^{(1)}(k,\omega)$ is found to be

$$P^{(1)}(k,\omega) = 2\sum_{k_1,k_2}\Big\{[\tilde{Q}_3^{(+)}(k_1,k_2,-k;\omega)(\varepsilon_{k_1}+\varepsilon_{k_2}) - 2\omega\hat{Q}_3^{(+)}(k_1,k_2,-k;\omega)]$$

$$\times\,(\eta_{k_1}+\eta_{k_2})[\omega^2-(\varepsilon_{k_1}+\varepsilon_{k_2})^2]^{-1} + [\tilde{Q}_3^{(-)}(k_1,k_2,-k;\omega)(\varepsilon_{k_1}-\varepsilon_{k_2})$$

$$-\,2\omega\hat{Q}_3^{(-)}(k_1,k_2,-k;\omega)]\times(\eta_{k_1}-\eta_{k_2})[\omega^2-(\varepsilon_{k_1}-\varepsilon_{k_2})^2]^{-1}\Big\}$$

$$+\,2\sum_{k_1,k_2,k_3}\Big\{[\tilde{Q}_4^{(+)}(k_1,k_2,k_3,-k;\omega)(\varepsilon_{k_1}+\varepsilon_{k_2}+\varepsilon_{k_3})$$

$$- 2\omega\hat{Q}_4(k_1,k_2,k_3,-k;\omega)] \times (1+\eta_{k_1}\eta_{k_2}+\eta_{k_2}\eta_{k_3}+\eta_{k_1}\eta_{k_3})$$

$$[\omega^2-(\varepsilon_{k_1}+\varepsilon_{k_2}+\varepsilon_{k_3})^2]^{-1} + [\tilde{Q}_4^{(-)}(k_1,k_2,k_3,-k;\omega)(\varepsilon_{k_1}-\varepsilon_{k_2}-\varepsilon_{k_3})$$

$$- 2\omega\hat{Q}_4(k_1,k_2,k_3,-k;\omega)] \times (1+\eta_{k_2}\eta_{k_3}-\eta_{k_1}\eta_{k_3}-\eta_{k_1}\eta_{k_2})$$

$$[\omega^2-(\varepsilon_{k_1}-\varepsilon_{k_2}-\varepsilon_{k_3})^2]^{-1} + (\varepsilon_{k_2}\leftrightarrow-\varepsilon_{k_2}) + (\varepsilon_{k_3}\leftrightarrow-\varepsilon_{k_3})$$

$$+ \text{ terms with } n > 4. \tag{37a}$$

Where $\eta_{k_1} = \coth \tfrac{1}{2}\beta\varepsilon_{k_1}$, the coupling functions are given by the expressions

$$\tilde{Q}_3^{(\pm)}(k_1,k_2,-k;\omega) = \tfrac{1}{2} \sum_{k_1',k_2'} \Big[Q_3(k_1,k_2,-k;\omega)Q_3^*(k_1',k_2',-k;\omega)$$

$$(\delta_{k_1'k_1}\delta_{k_2'k_2}+\delta_{k_1'k_2}\delta_{k_2'k_1}) + 4\left(\frac{\varepsilon_{k_1}}{\omega_{k_1}}\right)\tilde{\Phi}_3(k_1,k_2,-k)\tilde{\Phi}_3^*(k_1',k_2',-k)$$

$$\left(\frac{\varepsilon_{k_1}}{\omega_{k_1}}\delta_{k_1'k_1}\delta_{k_2'k_2} \pm \frac{\varepsilon_{k_2}}{\omega_{k_2}}\delta_{k_1'k_2}\delta_{k_2'k_1}\right)\Bigg]\left(\frac{\omega_{k_1}\omega_{k_2}}{\varepsilon_{k_1}\varepsilon_{k_2}}\right), \tag{37b}$$

$$\hat{Q}_3^{(\pm)}(k_1,k_2,-k;\omega) = 2 \sum_{k_1',k_2'} Q_3(k_1,k_2,-k;\omega)\tilde{\Phi}_3^*(k_1',k_2',-k)$$

$$\times \left(\frac{\varepsilon_{k_1}}{\omega_{k_1}}\delta_{k_1'k_1}\delta_{k_2'k_2} \pm \frac{\varepsilon_{k_2}}{\omega_{k_2}}\delta_{k_1'k_2}\delta_{k_2'k_1}\right)\left(\frac{\omega_{k_1}\omega_{k_2}}{\varepsilon_{k_1}\varepsilon_{k_2}}\right), \tag{37c}$$

$$\tilde{Q}_4^{(\pm)}(k_1,k_2,k_3,-k;\omega) = \tfrac{1}{4} \sum_{k_1',k_2',k_3'} \Big[Q_4(k_1,k_2,k_3,-k;\omega)$$

$$Q_4^*(k_1',k_2',k_3',-k;\omega)\delta + 9\left(\frac{\varepsilon_{k_1}}{\omega_{k_1}}\right)\tilde{\Phi}_4(k_1,k_2,k_3,-k)\tilde{\Phi}_4^*(k_1',k_2',k_3',-k)$$

$$\times \left(\frac{\varepsilon_{k_1}}{\omega_{k_1}}\delta_{123} \pm \frac{\varepsilon_{k_2}}{\omega_{k_2}}\delta_{213} \pm \frac{\varepsilon_{k_3}}{\omega_{k_3}}\delta_{321}\right)\Bigg]\left(\frac{\omega_{k_1}\omega_{k_2}\omega_{k_3}}{\varepsilon_{k_1}\varepsilon_{k_2}\varepsilon_{k_3}}\right), \tag{37d}$$

$$\hat{Q}_4(k_1,k_2,k_3,-k;\omega) = \tfrac{3}{4} \sum_{k_1',k_2',k_3'} \tilde{\Phi}_4(k_1,k_2,k_3,-k)$$

$$Q_4^*(k_1',k_2',k_3',-k;\omega)\left(\frac{\omega_{k_2}\omega_{k_3}}{\varepsilon_{k_2}\varepsilon_{k_3}}\right)\delta, \tag{37e}$$

and the Kronecker symbols δ, δ_{123}, δ_{213} and δ_{321} have been defined by (A.9d) in the Appendix. The expression (37) for the polarization operator $P^{(1)}(k,\omega)$ is a function of the frequencies of the dressed polariton modes (physical polaritons) and, therefore, the Green's function $G^{(2)}(k,\omega)$ describes the excitation spectrum of dressed interacting polariton modes. We observed that in (37), and similarly in (32), the square of the anharmonic coupling functions Q_n and $\tilde{\phi}_n$ as well as their products $Q_n\tilde{\phi}_n^*$ are multiplied by factors having the general form $\frac{\omega_{k_1}\omega_{k_2}\cdots\omega_{k_{n-1}}}{\varepsilon_{k_1}\varepsilon_{k_2}\cdots\varepsilon_{k_{n-1}}}$ and $\frac{\varepsilon_{k_1}\omega_{k_2}\omega_{k_3}\cdots\omega_{k_{n-1}}}{\omega_{k_1}\varepsilon_{k_2}\varepsilon_{k_3}\cdots\varepsilon_{k_{n-1}}}$ respectively, where n is the number of modes contained in each particular term. They result from the evaluation of the Green's functions having the form of $\langle\langle A_{k_1}A_{k_2}\cdots A_{k_{n-1}};A^{\dagger}_{k'_1}A^{\dagger}_{k'_2}\cdots A^{\dagger}_{k'_{n-1}}\rangle\rangle$, $\langle\langle B_{k_1}A_{k_2}A_{k_3}\cdots A_{k_{n-1}};A^{\dagger}_{k'_{n-1}}\cdots A^{\dagger}_{k'_3}A^{\dagger}_{k'_2}B^{\dagger}_{k'_1}\rangle\rangle$, $\langle\langle B_{k_1}A_{k_2}A_{k_3}\cdots A_{k_{n-1}};A^{\dagger}_{k'_1}A^{\dagger}_{k'_2}\cdots A^{\dagger}_{k'_{n-1}}\rangle\rangle$ and give an account of the nondiagonal contributions arising from the renormalized first-order Hamiltonian (35). Considering that the ratio (ω_k/ε_k) is either greater than or less than unity depending on whether the energy shift is negative or positive respectively, we conclude that both factors result in increasing or in reducing the square of the amplitudes of the anharmonic coupling functions that appear in the expression for the polarization operator. Another important effect arising from the aforementioned factors is that they are temperature dependent. We remark that the expression (36) for $G^{(2)}(k,\omega)$ can be obtained directly from (32b) if use is made of the same approximations that have been done for the derivation of (36). The poles of the Green's function $G^{(2)}(k,\omega)$ give the energies of excitation for the polariton spectrum correct in the second approximation. The obtained results can be improved by going into higher order approximations but, as far as our problem is concerned, the second approximation is sufficient to describe the polariton excitation spectrum because the interactions appearing in the expression (14) for $\mathcal{H}_{int.}$ are weak.

It should be pointed out that in our calculation for $P(k,\omega)$ given by (27) terms with $n \neq n'$ vanish; this is due to the Hamiltonian representation employed at each successive approximation. The derived results can be improved if use is made of the complete expression for the polarization operator as given by (23) at each successive step. For example, we may write $G^{(2)}(k,\omega)$ as

$$G^{(2)}(k,\omega) = \frac{\omega_k}{\pi}\,\delta_{kk'}\,[\omega^2-\varepsilon_k^2-2\omega_k\Pi^{(1)}(k,\omega)]^{-1} \tag{38}$$

with

$$\Pi^{(1)}(k,\omega) \approx P^{(1)}(k,\omega)[1-G_{oo}(k,\omega)P^{(1)}(k,\omega) + \ldots.], \tag{39}$$

and $P^{(1)}(k,\omega)$ is expressed by (37). The second, third, ..., terms in the expansion (39) will contribute small corrections to the dominant first term. The small corrections obtained by this procedure at each successive step will compensate the terms appearing in (27) with $n' \neq n$ because both corrections are of the same order of magnitude.

In the range of wavevectors where retardation effects are not important and may be neglected, the expressions (36) and (37) can be reduced to those corresponding to the bare phonon spectrum by taking the proper limits when $\omega_k \to \omega_k^o$ and $\tilde{\varepsilon}(k,\omega_k) \to 1$ then $\tilde{\Phi}_n(k_1,k_2,\ldots,k_n) \to o$ and $\tilde{V}_n(k_1,k_2,\ldots,k_n) \to V_n(k_1,k_2,\ldots,k_n)$. In this case for the bare phonon spectrum we have

$$G_b^{(2)}(k,\omega) = \frac{\omega_k^o}{\pi}\,\delta_{kk'}\,[\omega^2-\varepsilon_k^{o2}-2\omega_k^o P_b^{(1)}(k,\omega)]^{-1}, \tag{40a}$$

where

$$P_b^{(1)}(k,\omega) = 18\sum_{k_1,k_2}|V_3(k_1,k_2,-k)|^2\left(\frac{\omega_{k_1}^o\omega_{k_2}^o}{\varepsilon_{k_1}^o\varepsilon_{k_2}^o}\right)$$

$$\times\left\{(n_{k_1}^o+n_{k_2}^o)\left[\frac{(\varepsilon_{k_1}^o+\varepsilon_{k_2}^o)}{\omega^2-(\varepsilon_{k_1}^o+\varepsilon_{k_2}^o)^2}\right]+(n_{k_2}^o-n_{k_1}^o)\left[\frac{(\varepsilon_{k_1}^o-\varepsilon_{k_2}^o)}{\omega^2-(\varepsilon_{k_1}^o-\varepsilon_{k_2}^o)^2}\right]\right\}$$

$$+\,48\sum_{k_1,k_2,k_3}|V_4(k_1,k_2,k_3,-k)|^2\left(\frac{\omega_{k_1}^o\omega_{k_2}^o\omega_{k_3}^o}{\varepsilon_{k_1}^o\varepsilon_{k_2}^o\varepsilon_{k_3}^o}\right)$$

$$\times\left\{(1+n_{k_1}^o n_{k_2}^o+n_{k_2}^o n_{k_3}^o+n_{k_1}^o n_{k_3}^o)\left[\frac{(\varepsilon_{k_1}^o+\varepsilon_{k_2}^o+\varepsilon_{k_3}^o)}{\omega^2-(\varepsilon_{k_1}^o+\varepsilon_{k_2}^o+\varepsilon_{k_3}^o)^2}\right]\right.$$

$$\left.+\,3(1+n_{k_2}^o n_{k_3}^o-n_{k_1}^o n_{k_3}^o-n_{k_1}^o n_{k_2}^o)\left[\frac{(\varepsilon_{k_1}^o-\varepsilon_{k_2}^o-\varepsilon_{k_3}^o)}{\omega^2-(\varepsilon_{k_1}^o-\varepsilon_{k_2}^o-\varepsilon_{k_3}^o)^2}\right]\right\}$$

$$+\text{ terms with } n>4. \tag{40b}$$

$n_{k_1}^o = \coth \tfrac{1}{2}\beta\varepsilon_{k_1}^o$ and $k \equiv (\underline{k}j)$. The energies of excitation, ε_k^o, are determined from the solutions of the equation

$$\varepsilon_k^{o2}-\tilde{\omega}_k^{o2}-2\omega_k^o P_b^{(o)}(k,\varepsilon_k^o) = 0, \tag{40c}$$

and the expression for $P_b^{(o)}(k,\varepsilon_k^o)$ is obtained from (32) with $\omega = \varepsilon_k^o$, $\dot{\phi}_n(k_1,k_2,\ldots,k_n) = 0$, $\tilde{V}_n(k_1,k_2,\ldots,k_n) = V_n(k_1,k_2,\ldots,k_n)$ and $\tilde{\omega}_k$'s are replaced by $\tilde{\omega}_k^o$'s where

$$\tilde{\omega}_k^{o2} = \omega_k^{o2}+\omega_k^o\sum_{n=3}^{\infty} n\sum_{k_1,k_2,\ldots,k_{n-1}} V_n(k_1,k_2,\ldots,k_{n-1},-k)$$

$$\times\left\langle[A_{k_1}A_{k_2}\ldots.A_{k_{n-1}},B_k^\dagger]_-\right\rangle_{t=t',\omega=\tilde{\omega}_k^o}, \tag{40d}$$

The expression (40b) for $P_b^{(1)}(k,\omega)$ consists of terms which are proportional to two-, three- and many-phonon Green's functions having the form of $\langle\langle A_{k_1}A_{k_2}\ldots A_{k_{n-1}};A_{k_1'}^\dagger A_{k_2'}^\dagger\ldots A_{k_{n-1}'}^\dagger\rangle\rangle$

Wehner[5] and more recently Kwok[6] have developed methods for calculating the various many-phonon Green's functions by employing the functional derivative technique of Martin and Schwinger[10]. Although their expressions for the many-phonon Green's functions have a form similar to those appearing in (40b), they differ with respect to the expressions for the renormalized frequencies ε_k^o. Complete agreement with the results of Wehner[5] and Kwok[6] is achieved only if the equation (40c) is replaced by

$$\varepsilon_k^{o2} \sim \tilde{\omega}_k^{o2} + 2\omega_k^o P_b^{(o)}(k,\omega_k^o). \tag{41}$$

Also in the expression for $P_b^{(o)}(k,\omega_k^o)$ all the frequencies $\tilde{\omega}_k^o$ have been replaced by ω_k^o which are correct in the harmonic approximation. Having these approximations in mind, the one phonon Green's function (33) with $\varepsilon_k \to \varepsilon_k^o$ as well as the two- and three-phonon Green's functions appearing in (40b), which are correct in the first approximation, become identical to the correspondong ones derived by Wehner[5], but the expression (40a) for $G_b^{(2)}(k,\omega)$ is correct in the second approximation. In conclusion, the developed formalism based on successive approximations, apart from its simplicity, succeeds in the renormalization of the frequencies for the phonon modes.

In discussing the polariton spectrum we have dealt only with the diagonal part of the polariton Green's function with respect to the polariton branch index ρ $(=\rho')$. To include non-diagonal contributions one can easily derive the following equation

$$G_{\rho\rho'}(k,\omega) = G_{\rho\rho}(k,\omega)[\delta_{\rho\rho'} + \pi \sum_{\rho''\neq\rho'} \Pi_{\rho'\rho''}(k,\omega) G_{\rho''\rho'}(k,\omega)], \tag{42}$$

which can be solved by iteration.

IV. EXCITATION SPECTRUM

To study the excitation spectrum, we make use of the spectral function, $J_k(\omega)$, given by the relation[9]

$$J_k(\omega) = 2(1-e^{\beta\omega})^{-1}\ \mathrm{Im}G(k,\omega+i\eta), \tag{43}$$

where Im stands for the imaginary part and η is a small positive quantity, $\eta\to+o$. Substituting the imaginary part of (37) into (43) we obtain

$$J_k^{(2)}(\omega) = \left(\frac{2\omega_k}{\pi}\right) \frac{2\omega_k\ \Gamma_k^{(1)}(\omega)\ (e^{\beta\omega}-1)^{-1}}{[\omega^2-\nu_k^2]^2 + 4\omega_k^2\ \Gamma_k^{(1)2}(\omega)}\ , \tag{44}$$

where ν_k are the energies of excitation determined by the solutions of the equation

$$\nu_k^2 - \varepsilon_k^2 - 2\omega_k\ \Delta_k(\nu_k) = 0\ ; \tag{45}$$

$\Delta_k(\omega) = \mathrm{Re}P_k^{(1)}(\omega)$, is the real part of the polarization operator given by the expression (37) but now the principal value over the summations must be taken. In (44), $\Gamma_k^{(1)}(\omega) = \mathrm{Im}P^{(1)}(k,\omega)$ is the imaginary part of the polarization operator in the first approximation and is equal to

$$\Gamma^{(1)}(k,\omega) = \pi \sum_{k_1,k_2}\Big\{\Big[\tilde{Q}_3^{(+)}(k_1,k_2,-k;\omega)\Big(\delta(\omega+\varepsilon_{k_1}+\varepsilon_{k_2})-\delta(\omega-\varepsilon_{k_1}-\varepsilon_{k_2})\Big)$$

$$- 2\hat{Q}_3^{(+)}(k_1,k_2,-k;\omega)\Big(\delta(\omega+\varepsilon_{k_1}+\varepsilon_{k_2})+\delta(\omega-\varepsilon_{k_1}-\varepsilon_{k_2})\Big)\Big](\eta_{k_1}+\eta_{k_2})$$

$$+ \Big[\tilde{Q}_3^{(-)}(k_1,k_2,-k;\omega)\Big(\delta(\omega+\varepsilon_{k_1}-\varepsilon_{k_2})-\delta(\omega-\varepsilon_{k_1}+\varepsilon_{k_2})\Big)$$

$$- 2\hat{Q}_3^{(-)}(k_1,k_2,-k;\omega)\Big(\delta(\omega+\varepsilon_{k_1}-\varepsilon_{k_2})+\delta(\omega-\varepsilon_{k_1}+\varepsilon_{k_2})\Big)\Big](\eta_{k_1}-\eta_{k_2})\Big\}$$

$$+ \pi \sum_{k_1,k_2,k_3}\Big\{\Big[\tilde{Q}_4^{(+)}(k_1,k_2,k_3,-k;\omega)\Big(\delta(\omega+\varepsilon_{k_1}+\varepsilon_{k_2}+\varepsilon_{k_3})$$

$$- \delta(\omega-\varepsilon_{k_1}-\varepsilon_{k_2}-\varepsilon_{k_3})\Big) - 2\hat{Q}_4(k_1,k_2,k_3,-k;\omega)\Big(\delta(\omega+\varepsilon_{k_1}+\varepsilon_{k_2}+\varepsilon_{k_3})$$

$$+ \delta(\omega-\varepsilon_{k_1}-\varepsilon_{k_2}-\varepsilon_{k_3})\Big)\Big]\times(1+\eta_{k_1}\eta_{k_2}+\eta_{k_2}\eta_{k_3}+\eta_{k_1}\eta_{k_3})$$

$$+ \left[\tilde{Q}_4^{(-)}(k_1,k_2,k_3,-k;\omega)\left(\delta(\omega+\varepsilon_{k_1}-\varepsilon_{k_2}-\varepsilon_{k_3})-\delta(\omega-\varepsilon_{k_1}+\varepsilon_{k_2}+\varepsilon_{k_3})\right)\right.$$

$$\left. - 2\hat{Q}_4(k_1,k_2,k_3,-k;\omega)\left(\delta(\omega+\varepsilon_{k_1}-\varepsilon_{k_2}-\varepsilon_{k_3})+\delta(\omega-\varepsilon_{k_1}+\varepsilon_{k_2}+\varepsilon_{k_3})\right)\right]$$

$$\left.\times\ (1+\eta_{k_2}\eta_{k_3}-\eta_{k_1}\eta_{k_3}-\eta_{k_1}\eta_{k_2}) + (\varepsilon_{k_2}\leftrightarrow-\varepsilon_{k_2}) + (\varepsilon_{k_3}\leftrightarrow-\varepsilon_{k_3})\right\}$$

$$+ \text{ terms with } n > 4. \tag{46}$$

The spectral function (44) describes the polariton excitation spectrum in the whole range of frequencies ω.

As $\Gamma^{(1)}(k,\omega)$ goes to zero, the spectral function $J_k^{(2)}(\omega)$ tends to a delta-shape distribution, i.e.,

$$J_k^{(2)}(\omega) \approx 2\omega_k\ (e^{\beta\omega}-1)^{-1}\ \delta(\omega^2-\nu_k^2), \tag{47}$$

where ν_k are the energies of the elementary excitations determined by the solutions of (45) with $\Delta_k(\nu_k) = P^{(1)}(k,\nu_k)$. The excitation spectrum given by (47) corresponds to the transparent range of frequencies of the crystal and the expression for $P^{(1)}(k,\nu_k)$, as shown by (37) with $\omega=\nu_k$, is the scattering amplitude in energy units describing the scattering processes that occur among the dressed polariton modes far from resonances. Integration of (47) over ω at t=t' leads to the expression for the polariton distribution function correct in the second approximation as

$$\left\langle A_k^{\dagger}A_k\right\rangle^{(2)} = \left(\frac{\omega_k}{\nu_k}\right)\ \coth \tfrac{1}{2}\ \beta\nu_k.$$

In the limiting case where $\Gamma_k^{(1)}(\omega)$ may be considered to be very small but finite, $\Gamma_k^{(1)}(\omega) \ll \omega_k$, the spectral function (44) has a steep maximum at the frequencies $\omega^2=\nu_k^2$, provided that $\frac{\partial\Gamma_k^{(1)}(\omega)}{\partial\omega} \ll 1$ and $\frac{\partial}{\partial\omega}\Delta_k(\omega) \ll 1$. If it is further assumed that in the neighborhood of $\omega^2 \sim \nu_k^2$ the function $\Gamma_k^{(1)}(k,\omega)$ varies slowly with ω, i.e., $\Gamma_k^{(1)}(\omega) \approx \Gamma_k^{(1)}(\nu_k)$, then (44) reduces to that describing the spectrum of the fundamental absorption band

$$J_k^{(2)}(\omega) \approx \left(\frac{2\omega_k}{\pi}\right) \frac{2\omega_k \Gamma_k^{(1)}(\nu_k)(e^{\beta\omega}-1)^{-1}}{(\omega^2-\nu_k^2)^2+4\omega_k^2\Gamma_k^{(1)2}(\nu_k)} . \tag{48}$$

The function (48) is a Lorentzian line with a maximum at $\omega^2=\nu_k^2$, the energy shift is equal to $(\nu_k^2-\omega_k^2)^{\frac{1}{2}}$ and the width is of the order of $\Gamma_k^{(1)}(\nu_k)$ in energy units, provided that the frequency ν_k satisfies at least one of the arguments of the delta function appearing in (46). It should be pointed out that in evaluating (48) from (44) the ω dependence of the functions $\Gamma_k^{(1)}(\omega)$ and $\Delta_k(\omega)$ has been neglected. If the variation of $\Gamma_k^{(1)}(\omega)$ and $\Delta_k(\omega)$ with respect to ω is taken into consideration then, apart from the appearance of the combination bands which arise from the derivatives of the delta functions with respect to ω, there will appear new peaks and possibly the shape of the absorption band may change as a result of the ω-dependent terms in the coupling functions of (46) and $\Delta_k(\omega)$. Therefore, for an accurate calculation of the excitation spectrum (44) for a specific crystal the ω-dependence of $\Gamma_k^{(1)}(\omega)$ and $\Delta_k(\omega)$ must be examined with care. Further, the possibility of observing these new peaks depends on the numerical value of the quantity $(\tilde{\phi}_n/\omega_k)$.

For values of frequencies ω not near ν_k, that is, in the vicinity of the edges of the polariton absorption bands, we may discard $\Gamma_k^{(1)2}(\omega)$ from the denominator of (44) with the result

$$J_k^{(2)}(\omega) \approx \left(\frac{2\omega_k}{\pi}\right) \frac{2\omega_k \Gamma_k^{(1)}(\omega)(e^{\beta\omega}-1)^{-1}}{[\omega^2-\varepsilon_k^2-2\omega_k\Delta_k(\omega)]^2} . \tag{49}$$

If now the expression $\Gamma_k^{(1)}(\omega)$ given by (46) is written as

$$\Gamma_k^{(1)}(\omega) = \gamma_1(k,\omega) + \omega^2\gamma_2(k,\omega) + \omega\gamma_3(k,\omega), \tag{50}$$

where the functions $\gamma_1(k,\omega)$, $\gamma_2(k,\omega)$ and $\gamma_3(k,\omega)$ depend on ω only through the delta functions that appear in (46), then substitution of (50) into (49) leads to

$$J_k^{(2)}(\omega) \approx \left(\frac{2\omega_k}{\pi}\right) \frac{2\omega_k[\gamma_1(k,\omega)+\omega^2\gamma_2(k,\omega)+\omega\gamma_3(k,\omega)](e^{\beta\omega}-1)^{-1}}{[\omega^2-\varepsilon_k^2-2\omega_k\Delta_k(\omega)]^2} \tag{51a}$$

For large ω the expression (51a) may be written as

$$J_k^{(2)}(\omega) \sim \left(\frac{4\omega_k^2}{\pi}\right)[e^{\beta\omega}-1]^{-1}[\omega^{-2}\gamma_2(k,\omega)+\omega^{-3}\gamma_3(k,\omega)+\omega^{-4}\gamma_1(k,\omega)], \tag{51b}$$

where the functions $\gamma_2(k,\omega)$, $\gamma_3(k,\omega)$ and $\gamma_1(k,\omega)$ consist of the anharmonic coupling functions of the general form $|\tilde{\phi}_n|^2$, $\tilde{V}_n\tilde{\phi}_n$ and $|\tilde{V}_n|^2$ respectively. Therefore, $\omega^{-2}\gamma_2(k,\omega)$ is the dominant term in the neighborhood of the edges of the polariton fundamental absorption bands, where $\omega^2-c_k^2>>2\omega_k\Delta_k(\omega)$ indicating the importance of the function $|\tilde{\phi}_n|^2$; such behavior could be observed for large values of ω.

For the renormalized bare phonon spectrum, the spectral function (44) becomes

$$J_b^{(2)}(k,\omega) = \left(\frac{2\omega_k^o}{\pi}\right) \frac{2\omega_k^o\Gamma_b^{(1)}(k,\omega)(e^{\beta\omega}-1)^{-1}}{(\omega^2-\nu_k^{o2})^2+4\omega_k^{o2}\Gamma_b^{(1)2}(k,\omega)}, \tag{52}$$

where the energies of excitation ν_k^o are given by the solutions of the equation

$$\nu_k^{o2}-\varepsilon_k^{o2}-2\omega_k^o\Delta_k^o(\nu_k^o) = 0 , \tag{53}$$

with $\Delta_k^o(\omega) = \mathrm{Re}P_b^{(1)}(k,\omega)$ and the imaginary part of $P_b^{(1)}(k,\omega) = \Gamma_b^{(1)}(k,\omega)$ is given by

$$\Gamma_b^{(1)}(k,\omega) = 9\pi \sum_{k_1,k_2} |V_3(k_1,k_2,-k)|^2 \left[\frac{\omega_{k_1}^o\omega_{k_2}^o}{\varepsilon_{k_1}^o\varepsilon_{k_2}^o}\right] \Big\{ (n_{k_1}^o+n_{k_2}^o)$$

$$[\delta(\omega+\varepsilon_{k_1}^o+\varepsilon_{k_2}^o)-\delta(\omega-\varepsilon_{k_1}^o-\varepsilon_{k_2}^o)]+(n_{k_2}^o-n_{k_1}^o)[\delta(\omega+\varepsilon_{k_1}^o-\varepsilon_{k_2}^o)$$

$$- \delta(\omega-\varepsilon^{o}_{k_1}+\varepsilon^{o}_{k_2})]\Big\} + 24\pi \sum_{k_1,k_2,k_3} |V_4(k_1,k_2,k_3,-k)|^2 \left(\frac{\omega^{o}_{k_1}\omega^{o}_{k_2}\omega^{o}_{k_3}}{\varepsilon^{o}_{k_1}\varepsilon^{o}_{k_2}\varepsilon^{o}_{k_3}}\right)$$

$$\Big\{(1+n^{o}_{k_1}n^{o}_{k_2}+n^{o}_{k_2}n^{o}_{k_3}+n^{o}_{k_1}n^{o}_{k_3})[\delta(\omega+\varepsilon^{o}_{k_1}+\varepsilon^{o}_{k_2}+\varepsilon^{o}_{k_3})-\delta(\omega-\varepsilon^{o}_{k_1}-\varepsilon^{o}_{k_2}-\varepsilon^{o}_{k_3})]$$

$$+3(1+n^{o}_{k_2}n^{o}_{k_3}-n^{o}_{k_1}n^{o}_{k_3}-n^{o}_{k_1}n^{o}_{k_2})\,[\delta(\omega+\varepsilon^{o}_{k_1}-\varepsilon^{o}_{k_2}-\varepsilon^{o}_{k_3})-\delta(\omega-\varepsilon^{o}_{k_1}+\varepsilon^{o}_{k_2}+\varepsilon^{o}_{k_3})]\Big\}$$

$$+ \text{ terms with } n > 4. \qquad (54)$$

The renormalized phonon spectrum described by (52) is completely different from that derived by means of diagram techniques[7]. The main difference is due to the appearance of the complete renormalization of the frequencies for each particular mode in the expression (52). Agreement with the results of previous studies[7,11] is achieved only when all the renormalized effects are ignored, i.e., all the ε^{o}_{k}'s in the expressions appearing in (52) have to be replaced by the ω^{o}_{k}'s. Formulas (52)-(54) are reduced to those derived by Kashcheev[12] if the factors $\left(\frac{\omega^{o}_{k_1}\omega^{o}_{k_2}}{\varepsilon^{o}_{k_1}\varepsilon^{o}_{k_2}}\right)$ and $\left(\frac{\omega^{o}_{k_1}\omega^{o}_{k_2}\omega^{o}_{k_3}}{\varepsilon^{o}_{k_1}\varepsilon^{o}_{k_2}\varepsilon^{o}_{k_3}}\right)$ are taken equal to unity and the energies of excitation ε^{o}_{k}'s are replaced by $\tilde{\omega}^{o}_{k}$'s with the expression for $\tilde{\omega}^{o}_{k}$ including only quartic anharmonic contributions for each particular frequency mode.

As far as the temperature dependence for the excitation spectrum (44) or (52) is concerned, the renormalization effects are twofold: firstly, the renormalized frequencies for the polariton (or the bare phonon) modes are temperature dependent and, secondly, the factors of the general form $\left(\frac{\omega_{k_1}\omega_{k_2}\cdots\omega_{k}}{\varepsilon_{k_1}\varepsilon_{k_2}\cdots\varepsilon_{k_n}}\right)$ and $\left(\frac{\varepsilon_{k_1}\omega_{k_2}\omega_{k_3}\cdots\omega_{k_n}}{\omega_{k_1}\varepsilon_{k_2}\varepsilon_{k_3}\cdots\varepsilon_{k_n}}\right)$ that appear in the expressions for the real and imaginary parts of the polarization operator change the temperature dependence of the frequency shift and spectral width of the absorption bands.

Therefore, it is evident that the renormalization effects are important in connection with the temperature variations as well as the quantitative calculations[13] of the frequency shift and spectral width of the fundamental infrared absorption spectra of ionic crystals.

It is easy to see that one of the effects of taking into account the intrinsic coupling between the radiation field and the transverse optical phonons, when $k \sim 10^3\ cm^{-1}$, is to reduce the magnitudes of the anharmonic coupling functions $V_n(k_1,k_2,\ldots,k_n)$. A rough estimate, for the ratio of the square of the cubic anharmonic coefficient with and without retardation is of the order of

$$|\bar{V}_3/V_3|^2 \sim \left(\frac{\omega_T\omega_K\omega_p^2}{(\omega_T^2-\omega_k^2)^2+\omega_T^2\omega_p^2}\right) \tag{55}$$

where $\omega_T \sim \omega_k^0$ is the frequency for the transverse optical phonons. In the absence of retardation, $\omega_k \sim \omega_T$, the right-hand side of (55) is equal to unity. However, in the long wavelength region, $k \sim 10^3\ cm^{-1}$, ω_k for the two polariton branches may be taken to be of the order of $\frac{1}{2}\omega_T$ and ω_L respectively, where ω_L is the longitudinal frequency for optical phonons. Moreover, if we consider that for NaCl $\omega_L^2 \sim 2\omega_T^2$, then the right-hand side of (55) is roughly of the order of 0.12, that is, the magnitude of the square of the cubic anharmonic coefficient, $|V_3|^2$, is decreased by about 12% due to retardation. Similarly, the square of the quartic coefficient is decreased roughly by 6%. Of course, these are only rough estimates.

APPENDIX

We wish to carry out the evaluation of the two- and three-particle Green's functions by means of the zeroth order renormalized Hamiltonian (31). If we introduce the operator which is a four component row vector

$$\tilde{A}^{\dagger}(k_1',k_2') \equiv (A^{\dagger}_{k_1'}A^{\dagger}_{k_2'} \quad A^{\dagger}_{k_1'}B^{\dagger}_{k_2'} \quad B^{\dagger}_{k_1'}A^{\dagger}_{k_2'} \quad B^{\dagger}_{k_1'}B^{\dagger}_{k_2'}), \qquad (A.1)$$

then using (31) we derive the equation of motion for the Green's function $\langle\langle(\tilde{A}(k_1,k_2);\tilde{A}^{\dagger}(k_1',k_2')\rangle\rangle$, which is in a matrix form, as

$$L(k_1,k_2;\omega)\langle\langle\tilde{A}(k_1,k_2);\tilde{A}^{\dagger}(k_1',k_2')\rangle\rangle = \frac{1}{2\pi}\langle[\tilde{A}(k_1,k_2),\tilde{A}^{\dagger}(k_1',k_2')]_-\rangle_{t=t'}, \qquad (A.2)$$

where $L(k_1,k_2;\omega)$ is a four by four matrix whose nonzero matrix elements are given by:

$$\left.\begin{aligned} &L_{11} = L_{22} = L_{33} = L_{44} = \omega;\ L_{12} = L_{34} = -\omega_{k_2};\ L_{13} = L_{24} = -\omega_{k_1};\\ &L_{21} = L_{43} = -\tilde{\omega}^2_{k_2}/\omega_{k_2};\ L_{31} = L_{42} = -\tilde{\omega}^2_{k_1}/\omega_{k_1}. \end{aligned}\right\} \qquad (A.3)$$

The determinant of the matrix $L(k_1,k_2;\omega)$ is found to be

$$\text{det. } L(k_1,k_2;\omega) = [\omega^2-(\tilde{\omega}_{k_1}+\tilde{\omega}_{k_2})^2][\omega^2-(\tilde{\omega}_{k_1}-\tilde{\omega}_{k_2})^2], \qquad (A.4)$$

which gives the poles of the Green's function $\langle\langle\tilde{A}(k_1,k_2);\tilde{A}^{\dagger}(k_1',k_2')\rangle\rangle$. Solving the set of equations (A.2) we obtain the following expression for the various two-particle Green's functions:

$$\langle\langle A_{k_1}A_{k_2};A^{\dagger}_{k_1'}A^{\dagger}_{k_2'}\rangle\rangle = \frac{1}{2\pi}\left(\frac{\omega_{k_1}\omega_{k_2}}{\tilde{\omega}_{k_1}\tilde{\omega}_{k_2}}\right)\left[(\tilde{n}_{k_1}+\tilde{n}_{k_2})\frac{(\tilde{\omega}_{k_1}+\tilde{\omega}_{k_2})}{\omega^2-(\tilde{\omega}_{k_1}+\tilde{\omega}_{k_2})^2}\right.$$

$$\left.+(\tilde{n}_{k_1}-\tilde{n}_{k_2})\frac{(\tilde{\omega}_{k_1}-\tilde{\omega}_{k_2})}{\omega^2-(\tilde{\omega}_{k_1}-\tilde{\omega}_{k_2})^2}\right]\left(\delta_{k_1k_1'}\delta_{k_2k_2'}+\delta_{k_1k_2'}\delta_{k_2k_1'}\right), \qquad (A.5a)$$

$$\langle\langle B_{k_1}A_{k_2};A^{\dagger}_{k_2'}B^{\dagger}_{k_1'}\rangle\rangle = \frac{1}{2\pi}\left(\frac{\tilde{\omega}_{k_1}\omega_{k_2}}{\omega_{k_1}\tilde{\omega}_{k_2}}\right)\left\{(\tilde{n}_{k_1}+\tilde{n}_{k_2})\frac{\tilde{\omega}_{k_1}+\tilde{\omega}_{k_2}}{\omega^2-(\tilde{\omega}_{k_1}+\tilde{\omega}_{k_2})^2}\right.$$

$$+ (\tilde{n}_{k_2}-\tilde{n}_{k_1}) \frac{\tilde{\omega}_{k_1}-\tilde{\omega}_{k_2}}{\omega^2-(\tilde{\omega}_{k_1}-\tilde{\omega}_{k_2})^2} \Big\} \delta_{k_1 k_1'}\delta_{k_2 k_2'} + \frac{1}{2\pi}\Big\{(\tilde{n}_{k_1}+\tilde{n}_{k_2})$$

$$\times \frac{\tilde{\omega}_{k_1}+\tilde{\omega}_{k_2}}{\omega^2-(\tilde{\omega}_{k_1}-\tilde{\omega}_{k_2})} - (\tilde{n}_{k_2}-\tilde{n}_{k_1}) \frac{\tilde{\omega}_{k_1}-\tilde{\omega}_{k_2}}{\omega^2-(\tilde{\omega}_{k_1}-\tilde{\omega}_{k_2})^2}\Big\}\delta_{k_2' k_1}\delta_{k_1' k_2}, \qquad (A.5b)$$

$$\langle\langle A_{k_1} A_{k_2}; A^{\dagger}_{k_2'} B^{\dagger}_{k_1'}\rangle\rangle = \frac{1}{2\pi}\left(\frac{\omega_{k_2}\omega}{\tilde{\omega}_{k_2}}\right)\left[\frac{(\tilde{n}_{k_1}+\tilde{n}_{k_2})}{\omega^2-(\tilde{\omega}_{k_1}+\tilde{\omega}_{k_2})^2}+\frac{(\tilde{n}_{k_2}-\tilde{n}_{k_1})}{\omega^2-(\tilde{\omega}_{k_1}-\tilde{\omega}_{k_2})^2}\right]$$

$$\times\, \delta_{k_1' k_1}\delta_{k_2' k_2} + \frac{1}{2\pi}\left(\frac{\omega_{k_1}\omega}{\tilde{\omega}_{k_1}}\right)\left[\frac{\tilde{n}_{k_1}+\tilde{n}_{k_2}}{\omega^2-(\tilde{\omega}_{k_1}+\tilde{\omega}_{k_2})^2}-\frac{\tilde{n}_{k_2}-\tilde{n}_{k_1}}{\omega^2-(\tilde{\omega}_{k_1}-\tilde{\omega}_{k_2})^2}\right]\delta_{k_1' k_2}\delta_{k_2' k_1}, \qquad (A.5c)$$

where, $\tilde{n}_{k_1} = \coth \tfrac{1}{2}\beta\tilde{\omega}_{k_1}$ and $\beta = (\kappa_B T)^{-1}$; κ_B being the Boltzmann constant and T the absolute temperature. In evaluating (A.5) from (A.2) we have made use of $\left\langle A^{\dagger}_{k_1} A_{k_1}\right\rangle^{(o)} = \left(\frac{\omega_{k_1}}{\tilde{\omega}_{k_1}}\right) \coth \tfrac{1}{2}\beta\tilde{\omega}_{k_1}$, $\left\langle B^{\dagger}_{k_1} B_{k_1}\right\rangle^{(o)} = \left(\frac{\tilde{\omega}_{k_1}}{\omega_{k_1}}\right) \coth \tfrac{1}{2}\beta\tilde{\omega}_{k_1}$ and $\left\langle A^{\dagger}_{k_2} A_{k_2}\right\rangle^{(o)} = \left(\frac{\omega_{k_2}}{\tilde{\omega}_{k_2}}\right) \coth \tfrac{1}{2}\beta\tilde{\omega}_{k}$ for the distribution functions in the lowest approximation.

Similarly, if the operator $\tilde{A}^{\dagger}(k_1', k_2', k_3')$ is defined as the row vector

$$\tilde{A}^{\dagger}(k_1',k_2',k_3') \equiv \Big(A^{\dagger}_{k_1'}A^{\dagger}_{k_2'}A^{\dagger}_{k_3'} \quad B^{\dagger}_{k_1'}A^{\dagger}_{k_2'}A^{\dagger}_{k_3'} \quad A^{\dagger}_{k_1'}B^{\dagger}_{k_2'}A^{\dagger}_{k_3'} \quad A^{\dagger}_{k_1'}A^{\dagger}_{k_2'}B^{\dagger}_{k_3'}$$
$$B^{\dagger}_{k_1'}B^{\dagger}_{k_2'}A^{\dagger}_{k_3'} \quad A^{\dagger}_{k_1'}B^{\dagger}_{k_2'}B^{\dagger}_{k_3'} \quad B^{\dagger}_{k_1'}A^{\dagger}_{k_2'}B^{\dagger}_{k_3'} \quad B^{\dagger}_{k_1'}B^{\dagger}_{k_2'}B^{\dagger}_{k_3'}\Big) \qquad (A.6)$$

then by means of (31) we derive the equation of motion

$$L(k_1,k_2,k_3;\omega)\langle\langle\tilde{A}(k_1,k_2,k_3);\tilde{A}^{\dagger}(k_1',k_2',k_3')\rangle\rangle =$$
$$\frac{1}{2\pi}\left\langle[\tilde{A}(k_1,k_2,k_3),\tilde{A}^{\dagger}(k_1',k_2',k_3')]_{-}\right\rangle_{t=t'}, \qquad (A.7)$$

where $L(k_1,k_2,k_3;\omega)$ is an eight by eight matrix having all diagonal elements equal to ω while the nonzero off-diagonal elements are given by

$L_{12} = L_{35} = L_{47} = L_{68} = -\omega_{k_1}$; $L_{13} = L_{25} = L_{46} = L_{78} = -\omega_{k_2}$;

$L_{14} = L_{27} = L_{36} = L_{58} = -\omega_{k_3}$; $L_{21} = L_{53} = L_{74} = L_{86} = -\tilde{\omega}^2_{k_1}/\omega_{k_1}$;

$L_{31} = L_{52} = L_{64} = L_{87} = -\tilde{\omega}^2_{k_2}/\omega_{k_2}$; $L_{41} = L_{63} = L_{72} = L_{85} = -\tilde{\omega}^2_{k_3}/\omega_{k_3}$.

The determinant of the matrix $L(k_1,k_2,k_3;\omega)$ factorizes into

$$\text{det. } L(k_1,k_2,k_3;\omega) = [\omega^2-(\tilde{\omega}_{k_1}+\tilde{\omega}_{k_2}+\tilde{\omega}_{k_3})^2][\omega^2-(\tilde{\omega}_{k_1}+\tilde{\omega}_{k_2}-\tilde{\omega}_{k_3})^2]$$
$$[\omega^2-(\tilde{\omega}_{k_1}-\tilde{\omega}_{k_2}+\tilde{\omega}_{k_3})^2][\omega^2-(\tilde{\omega}_{k_2}+\tilde{\omega}_{k_3}-\tilde{\omega}_{k_1})^2], \tag{A.8}$$

and describes the poles of the Green's function $\langle\langle\tilde{A}(k_1,k_2,k_3);\tilde{A}^\dagger(k_1',k_2',k_3')\rangle\rangle$. Solving the system of equations (A.7) and after some lengthy algebra we derive the following expressions for the three-particle Green's functions:

$$\langle\langle A_{k_1}A_{k_2}A_{k_3};A^\dagger_{k_1'}A^\dagger_{k_2'}A^\dagger_{k_3'}\rangle\rangle = \left(\frac{\delta}{4\pi}\right)\left(\frac{\omega_{k_1}\omega_{k_2}\omega_{k_3}}{\tilde{\omega}_{k_1}\tilde{\omega}_{k_2}\tilde{\omega}_{k_3}}\right)$$
$$\times\Big\{(1+\tilde{n}_{k_1}\tilde{n}_{k_2}+\tilde{n}_{k_2}\tilde{n}_{k_3}+\tilde{n}_{k_1}\tilde{n}_{k_3})(\tilde{\omega}_{k_1}+\tilde{\omega}_{k_2}+\tilde{\omega}_{k_3})[\omega^2-(\tilde{\omega}_{k_1}+\tilde{\omega}_{k_2}+\tilde{\omega}_{k_3})^2]^{-1}$$
$$+(1+\tilde{n}_{k_2}\tilde{n}_{k_3}-\tilde{n}_{k_1}\tilde{n}_{k_3}-\tilde{n}_{k_1}\tilde{n}_{k_2})(\tilde{\omega}_{k_1}-\tilde{\omega}_{k_2}-\tilde{\omega}_{k_3})[\omega^2-(\tilde{\omega}_{k_1}-\tilde{\omega}_{k_2}-\tilde{\omega}_{k_3})^2]^{-1}$$
$$+(\tilde{\omega}_{k_2}\leftrightarrow-\tilde{\omega}_{k_2})+(\tilde{\omega}_{k_3}\leftrightarrow-\tilde{\omega}_{k_3})\Big\}, \tag{A.9a}$$

$$\langle\langle B_{k_1}A_{k_2}A_{k_3};A^\dagger_{k_1'}A^\dagger_{k_2'}A^\dagger_{k_3'}\rangle\rangle = \left(\frac{\omega\delta}{4\pi}\right)\left(\frac{\omega_{k_2}\omega_{k_3}}{\tilde{\omega}_{k_2}\tilde{\omega}_{k_3}}\right)$$
$$\times\Big\{(1+\tilde{n}_{k_1}\tilde{n}_{k_2}+\tilde{n}_{k_2}\tilde{n}_{k_3}+\tilde{n}_{k_1}\tilde{n}_{k_3})[\omega^2-(\tilde{\omega}_{k_1}+\tilde{\omega}_{k_2}+\tilde{\omega}_{k_3})^2]^{-1}$$
$$+(1+\tilde{n}_{k_2}\tilde{n}_{k_3}-\tilde{n}_{k_1}\tilde{n}_{k_2}-\tilde{n}_{k_1}\tilde{n}_{k_3})[\omega^2-(\tilde{\omega}_{k_1}-\tilde{\omega}_{k_2}-\tilde{\omega}_{k_3})^2]^{-1}$$
$$+(\tilde{\omega}_{k_2}\leftrightarrow-\tilde{\omega}_{k_2})+(\tilde{\omega}_{k_3}\leftrightarrow-\tilde{\omega}_{k_3})\Big\}, \tag{A.9b}$$

$$\langle\langle B_{k_1}A_{k_2}A_{k_3};A^\dagger_{k_3'}A^\dagger_{k_2'}B^\dagger_{k_1'}\rangle\rangle = \frac{1}{4\pi}\left(\frac{\omega_{k_2}\omega_{k_3}}{\tilde{\omega}_{k_2}\tilde{\omega}_{k_3}}\right)\Big\{(1+\tilde{n}_{k_1}\tilde{n}_{k_2}+\tilde{n}_{k_2}\tilde{n}_{k_3}+\tilde{n}_{k_3}\tilde{n}_{k_1})$$
$$\times\left(\frac{\tilde{\omega}_{k_1}}{\omega_{k_1}}\delta_{123}+\frac{\tilde{\omega}_{k_2}}{\omega_{k_2}}\delta_{213}+\frac{\tilde{\omega}_{k_3}}{\omega_{k_3}}\delta_{321}\right)(\tilde{\omega}_{k_1}+\tilde{\omega}_{k_2}+\tilde{\omega}_{k_3})[\omega^2-(\tilde{\omega}_{k_1}+\tilde{\omega}_{k_2}+\tilde{\omega}_{k_3})^2]^{-1}$$
$$+(1+\tilde{n}_{k_2}\tilde{n}_{k_3}-\tilde{n}_{k_1}\tilde{n}_{k_3}-\tilde{n}_{k_1}\tilde{n}_{k_2})\left(\frac{\tilde{\omega}_{k_1}}{\omega_{k_1}}\delta_{123}-\frac{\tilde{\omega}_{k_2}}{\omega_{k_2}}\delta_{213}-\frac{\tilde{\omega}_{k_3}}{\omega_{k_3}}\delta_{321}\right)$$

$$\times\ (\tilde{\omega}_{k_1}-\tilde{\omega}_{k_2}-\tilde{\omega}_{k_3})[\omega^2-(\tilde{\omega}_{k_1}-\tilde{\omega}_{k_2}-\tilde{\omega}_{k_3})^2]^{-1}+(\tilde{\omega}_{k_2}\leftrightarrow-\tilde{\omega}_{k_2})+(\tilde{\omega}_{k_3}\leftrightarrow-\tilde{\omega}_{k_3})\Big\}, \tag{A.9c}$$

where

$$\delta \equiv \delta_{123}+\delta_{213}+\delta_{321}$$

$$\delta_{123} \equiv \delta_{k_1'k_1}\delta_{k_2'k_2}\delta_{k_3'k_3}+\delta_{k_3'k_2}\delta_{k_1'k_1}\delta_{k_2'k_3}, \tag{A.9d}$$

δ_{213} and δ_{321} are obtained from δ_{123} by interchanging k_1 and k_2, k_1 and k_3 respectively. In (A.9) we have indicated by $(\tilde{\omega}_{k_2}\leftrightarrow-\tilde{\omega}_{k_2})$ and $(\tilde{\omega}_{k_3}\leftrightarrow-\tilde{\omega}_{k_3})$ that there are two further terms obtained by interchanging $\tilde{\omega}_{k_2}$ and $-\tilde{\omega}_{k_2}$, $\tilde{\omega}_{k_3}$ and $-\tilde{\omega}_{k_3}$ in the first term respectively.

If we substitute the two- and three-polariton Green's functions given by (A.5) and (A.9) into (27) we obtain the expression (32) for the polarization operator $P^{(c)}(k,\omega)$ where the coupling functions $\tilde{Q}_3^{(\pm)}(k_1,k_2,-k;\omega)$, $\hat{Q}_3^{(+)}(k_1,k_2,-k;\omega)$, $\tilde{Q}_4^{(\pm)}(k_1,k_2,k_3,-k;\omega)$ and $\hat{Q}_4(k_1,k_2,k_3,-k;\omega)$ have been defined as follows:

$$\tilde{Q}_3^{(\pm)}(k_1,k_2,-k;\omega) = \tfrac{1}{2}\sum_{k_1',k_2'}\Big[Q_3(k_1,k_2,-k;\omega)Q_3^{*}(k_1',k_2',-k;\omega)$$
$$\times\ (\delta_{k_1'k_1}\delta_{k_2'k_2}+\delta_{k_1'k_2}\delta_{k_2'k_1}) + 4\left(\frac{\tilde{\omega}_{k_1}}{\omega_{k_1}}\right)\tilde{\phi}_3(k_1,k_2,-k)\tilde{\phi}_3^{*}(k_1',k_2',-k)$$
$$\times\left(\frac{\tilde{\omega}_{k_1}}{\omega_{k_1}}\delta_{k_1'k_1}\delta_{k_2'k_2}\pm\frac{\tilde{\omega}_{k_2}}{\omega_{k_2}}\delta_{k_1'k_2}\delta_{k_2'k_1}\right)\Big]\left(\frac{\omega_{k_1}\omega_{k_2}}{\tilde{\omega}_{k_1}\tilde{\omega}_{k_2}}\right), \tag{A.10a}$$

$$\hat{Q}_3^{(\pm)}(k_1,k_2,-k;\omega) = 2\sum_{k_1',k_2'}Q_3(k_1,k_2,-k;\omega)\tilde{\phi}_3^{*}(k_1',k_2',-k)$$
$$\times\left(\frac{\tilde{\omega}_{k_1}}{\omega_{k_1}}\delta_{k_1'k_1}\delta_{k_2'k_2}\pm\frac{\tilde{\omega}_{k_2}}{\omega_{k_2}}\delta_{k_1'k_2}\delta_{k_2'k_1}\right)\left(\frac{\omega_{k_1}\omega_{k_2}}{\tilde{\omega}_{k_1}\tilde{\omega}_{k_2}}\right), \tag{A.10b}$$

$$\tilde{Q}_4^{(\pm)}(k_1,k_2,k_3,-k;\omega) = \tfrac{1}{4}\sum_{k_1',k_2',k_3'}\Big[Q_4(k_1,k_2,k_3,-k;\omega)$$
$$\times\ Q_4^{*}(k_1',k_2',k_3',-k;\omega)\delta + 9\left(\frac{\tilde{\omega}_{k_1}}{\omega_{k_1}}\right)\tilde{\phi}_4(k_1,k_2,k_3,-k)\tilde{\phi}_4^{*}(k_1',k_2',k_3',-k)$$
$$\times\left(\frac{\tilde{\omega}_{k_1}}{\omega_{k_1}}\delta_{123}\pm\frac{\tilde{\omega}_{k_2}}{\omega_{k_2}}\delta_{213}\pm\frac{\tilde{\omega}_{k_3}}{\omega_{k_3}}\delta_{321}\right)\Big]\left(\frac{\omega_{k_1}\omega_{k_2}\omega_{k_3}}{\tilde{\omega}_{k_1}\tilde{\omega}_{k_2}\tilde{\omega}_{k_3}}\right), \tag{A.10c}$$

$$\hat{Q}_4(k_1,k_2,k_3,-k;\omega) = \frac{3}{4} \sum_{k_1',k_2',k_3'} \tilde{\phi}_4(k_1,k_2,k_3,-k)$$
$$\times\, Q_4^*(k_1',k_2',k_3',-k;\omega) \left(\frac{\omega_{k_2}\omega_{k_3}}{\tilde{\omega}_{k_2}\tilde{\omega}_{k_3}}\right)\delta \,. \qquad \text{(A.10d)}$$

In general, using the Hamiltonian (31) or (35) one can derive the equation of motion

$$L(k_1,k_2,k_3,\ldots,k_n;\omega)\langle\langle\tilde{A}(k_1,k_2,k_3,\ldots,k_n);\tilde{A}_k^\dagger(k_1',k_2',k_3',\ldots,k_n')\rangle\rangle =$$
$$\frac{1}{2\pi}\left\langle[\tilde{A}(k_1,k_2,k_3,\ldots,k_n);\tilde{A}^\dagger(k_1',k_2',k_3',\ldots,k')]_-\right\rangle_{t=t'} , \qquad \text{(A.11)}$$

where the operator $\tilde{A}^\dagger(k_1,k_2,k_3,\ldots,k_n)$ is now a row vector having 2^n components of the form of

$$\tilde{A}_k^\dagger(k_1,k_2,k_3,\ldots;k_n) \equiv \Big(A_{k_1}^\dagger A_{k_2}^\dagger \ldots A_{k_n}^\dagger \quad B_{k_1}^\dagger A_{k_2}^\dagger \ldots A_{k_n}^\dagger \quad A_{k_1}^\dagger B_{k_2}^\dagger A_{k_3}^\dagger \ldots A_{k_n}^\dagger$$
$$\ldots\ldots \quad \ldots\ldots \quad \ldots\ldots \quad B_{k_1}^\dagger B_{k_2}^\dagger \ldots B_{k_{n-1}}^\dagger A_{k_n}^\dagger \quad B_{k_1}^\dagger B_{k_2}^\dagger \ldots B_{k_{n-1}}^\dagger B_{k_n}^\dagger \Big) ,$$
$$\text{(A.12)}$$

and $L(k_1,k_2,k_3,\ldots,k_n;\omega)$ is a 2^n by 2^n matrix with all diagonal elements equal to ω while the nonzero off-diagonal elements are easily established but they shall not be given here. Solving the system of equations (A.10) it is possible in principle to calculate polariton Green's functions for any desired value of n. In practice for n greater than three, though the calculation of the corresponding Green's functions is straightforward, the algebra is tedious.

REFERENCES

1. J.J. Hopfield, Phys. Rev. 112, 1552 (1958). See also U. Fano, Phys. Rev. 103, 1202 (1956); A.A. Lucas, Phys. Rev. 162, 801 (1967).

2. C.H. Henry and J.J. Hopfield, Phys. Rev. Letters 15, 964 (1965).

3. C. Mavroyannis and K.N. Pathak, Phys. Rev. 182, No. 3 (1969).

4. E. Burstein, S. Ushioda and A. Pinczuk, Sol. State Comm. 6, 407 (1968).

5. R. Wehner, Phys. Stat. Sol. 15, 725 (1966).

6. P.C.K. Kwok, Solid State Physics, edited by F. Sietz and D. Turnbull (Academic Press, New York, 1967), Vol. 20, p. 297.

7. R.F. Wallis, I.P. Ipatova, and A.A. Maradudin, Fiz. Tverd. Tela 8, 1064 (1966) [English transl.: Soviet Phys.-Solid State 8, 850 (1966)].

8. V.M. Agranovich, Zh. Eksperim. i Teor. Fiz. 37, 430 (1959) [English transl.: Soviet Phys. J.E.T.P. 37, 307 (1960)]. See also N.N. Bogoliubov, Lectures on Quantum Statistics, Vol. 1 (Gordon and Breach, Science Publishers Inc., New York, 1967), p. 213.

9. D.N. Zubarev, Usp. Fiz. Nauk 71, 71 (1960) [English transl.: Soviet Phys.-Uspekhi 3, 320 (1960)].

10. P.C. Martin and J. Schwinger, Phys. Rev. 115, 1342 (1959).

11. V.N. Kashcheev, Fiz. Tverd. Tela 5, 1358 (1963) [English transl.: Soviet Phys.-Solid State 5, 988 (1963)].

12. V.N. Kashcheev, Fiz. Tverd. Tela 5, 2339 (1963) [English transl.: Soviet Phys.-Solid State 5, 1700 (1964)].

13. I.P. Ipatova, A.A. Maradudin, and R.F. Wallis, Phys. Rev. 155, 882 (1967).

XVI

INFRARED ABSORPTION BY BOUND POLARONS

J. Devreese

Faculty of Science, University of Antwerp
and Solid State Physics Department
S.C.K.-C.E.N., MOL (Belgium)

R. Evrard* and E. Kartheuser
Institut de Physique, Physique Théorique,
Université de Liège (Belgium)

INTRODUCTION

Zero and multiphonon transitions due to color centers and excitons have been detected experimentally with relatively much detail. In general, and e.g. in alkali halides, the defect - and impurity centers couple to different types of phonons (e.g. transverse optical, transverse acoustical, longitudinal optical and longitudinal acoustical phonons are revealed in the multiphonon structure associated with the 5456 A and 5803 A zero phonon lines in NaF).

A general question which rises is: when can one expect to observe detailed structure in the phonon sidebands? Although it does not seem that a really satisfactory answer has been given to this question we indicate the following factors:

* - Chargé de Recherches du Fonds National Belge de la Recherche Scientifique

a) the resolution of the structure in the sidebands will increase with decreasing temperature. This is due i.e. to the increasing importance of anharmonic motion of the ions as the temperature increases.

b) the structure due to the phonons will be revealed especially if the zero-phonon transition is relatively sharp and strong. This means that the zero-phonon transition is towards a reasonably "stable" final state. If this condition is not fulfilled the n-phonon transitions ($n > o$) result in broad, overlapping absorption curves; the width of the zero-phonon line is reflected in the sidebands who lose their structure more and more as the zero phonon transition becomes less sharp. In general the number of phonons revealed in the sidebands is relatively small; $n \leq 5$. Indeed, if the zero phonon peak is relatively sharp and pronounced, the oscillator strength of the n-phonon curves decreases rapidly for large n. As far as the physical mechanism of the transitions is concerned, a sharp zero phonon line with a few phonons in the sideband, corresponds to the case where only an unimportant relaxation of the lattice occurs after the Franck-Condon transition.

Phenomenologically speaking one can say that a small Huang-Phys factor characterizes a favorable situation for the observation of a sideband with pronounced structure.

In fact the Huang-Phys factor is a measure for the amount of relaxation of the lattice "after" the "vertical" (Franck-Condon) transition. The Huang-Phys factor is nor-

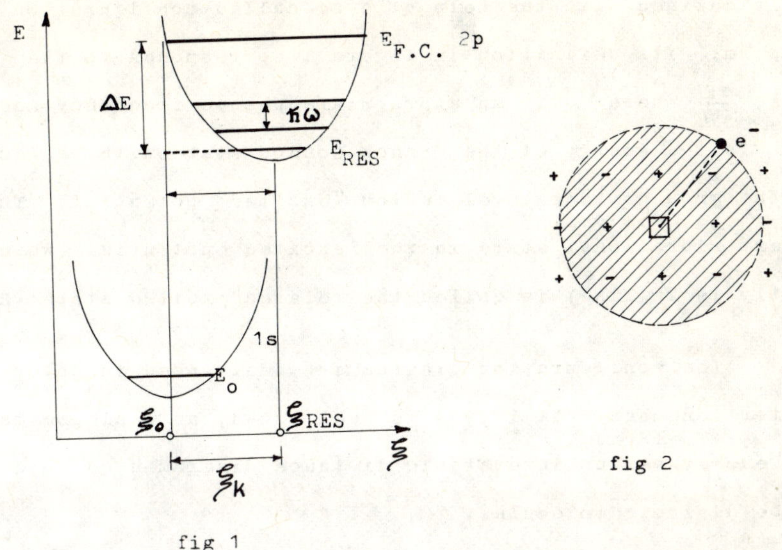

fig 1

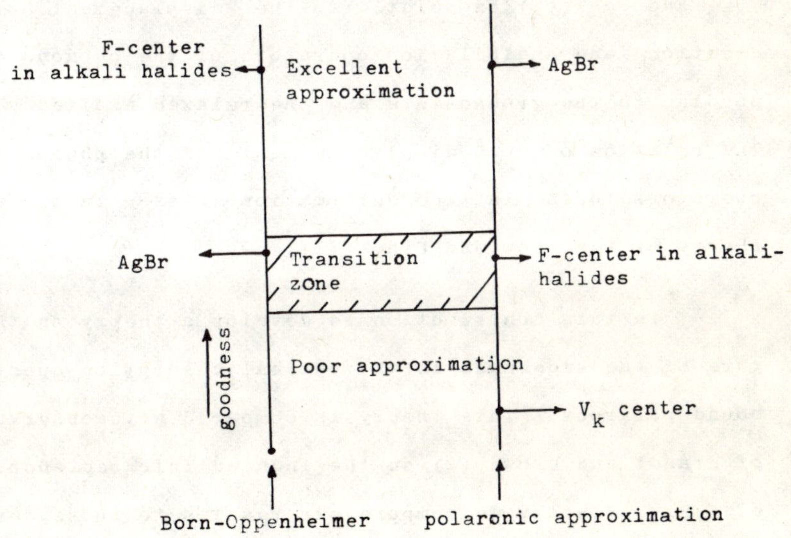

fig 3

Fig. 1 Configuration diagram

Fig. 2 Model for bound polaron

Fig. 3 Reliability of Born-Oppenheimer and Polaronic approximation

mally defined with the help of a so-called configuration diagram. Its definition in figure 1 corresponds to the ratio $\frac{\Delta E}{\hbar\omega}$ where $\hbar\omega$ is an appropriate phonon frequency and ΔE is the energy of the Franck-Condon state with respect to the ground-state level in the "excited" potential. The lowest vibrational state in the "excited" potential (which mostly is 2p-like) is called the relaxed excited state (R.E.S.).

The configuration diagram is widely used in color center language. It is, as is well known, an analogon to the energy-versus interatomic distance diagram used to describe diatomic molecules.

The difference between the configuration coordinates ξ_{RES} and ξ_o (ξ_k) is related to the "displacement" of the creation and annihilation operators of the phonons corresponding to the groundstate and the relaxed excited state. The relation of such displacements f_k of the phonon field operators to the lattice deformation is seen in the polaron theory of Lee, Low and Pines (1).

In this contribution we develop a theory on the structure of the sidebands in the optical absorption spectra of bound polarons. This theory is compared with observations of Brandt and Brown (2) on the induced infrared absorption of AgBr. Finally we compare our results to those deriving from the theory of Toyozawa and Hermanson (3). Brandt and Brown refer to the results of Toyozawa and Hermanson and to the results of the present authors as possible explanations of their data. We clarify the situation by discussing the approximations made in both theories.

1. A GENERAL THEORY ON THE STRUCTURE OF THE SIDEBANDS IN THE OPTICAL ABSORPTION SPECTRA OF BOUND POLARONS

1.a. The model and the approximations

The model we consider consists of a continuum (or Frohlich) polaron bound to a negative ion vacancy. The situation should be visualized as an electron cloud fixed at, or bound to the vacancy, but extending over enough space to justify the continuum approximation of Frohlich in which the field of induced polarization is treated as structure-independent i.e. as a continuum (figure 2). The model does not correspond to a system where the polaron "orbits" as an entity around the vacancy nor to a highly localized system in which the electron is "trapped" in a sphere with a radius of the order of one lattice parameter. The choice of this model is already a first approximation corresponding e.g. to the elimination of all phonons from the problem except for the long wavelength longitudinal optical phonons. More details about the continuum polaron model are found in our first contribution presented by E. Kartheuser (ref. 4).

A second approximation which is made is the Born-Oppenheimer approximation. This approximation, which is made in almost all theories in this field in order to conserve tractable mathematics, is justified as long as the frequency of the electron at the defect is large with respect to the longitudinal optical phonon frequency.

In figure 3 the "goodness" of the polaronic approximation (large polarons) and the Born-Oppenheimer approximation for different crystals and different types of defect centers is plotted. (Nothing is plotted along the abscis). The model used here is that of a diffuse F-center. The radius is large compared with a lattice parameter. From figure 3 one sees that this model is reasonable in the case of a bound polaron e.g. in AgBr. It might also give information about the polaronic contributions to the properties of F centers in alkali halides.

We will be interested here in the application of the results to the case of AgBr where the zero phonon frequency is about 1.4 larger than the frequency of a longitudinal optical phonon. Free polarons in AgBr have a radius of about 20 Å. The continuum approximation is very suitable therefore. The model treated here is a "pure" case in the sense that only longitudinal optical phonons are taken into account. From the measurements of Brandt and Brown it is seen that the L. O. phonons are easily identified in the induced infrared absorption. This means that these measurements provide also a test for the Fröhlich potential.

1.b. Solution of the model taking into account phonon dispersion and phonon-phonon interaction

To solve the continuum model for the bound polaron within the Born-Oppenheimer approximation some results about the internal structure which we obtained in previous work (5) (6) (7) and some of which was exposed by E. Kartheuser (4) are used. This means that one considers three types

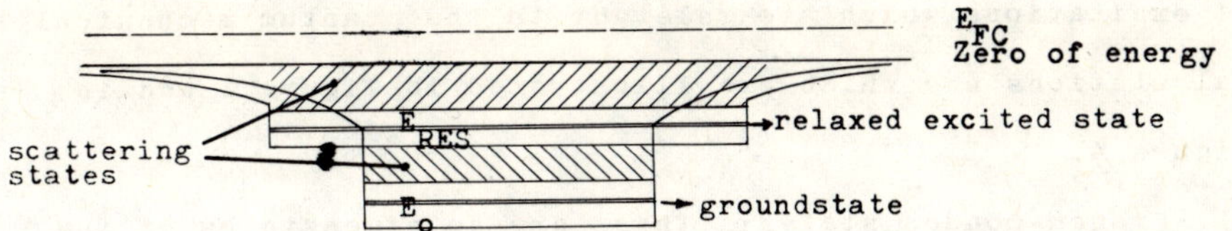

fig 4

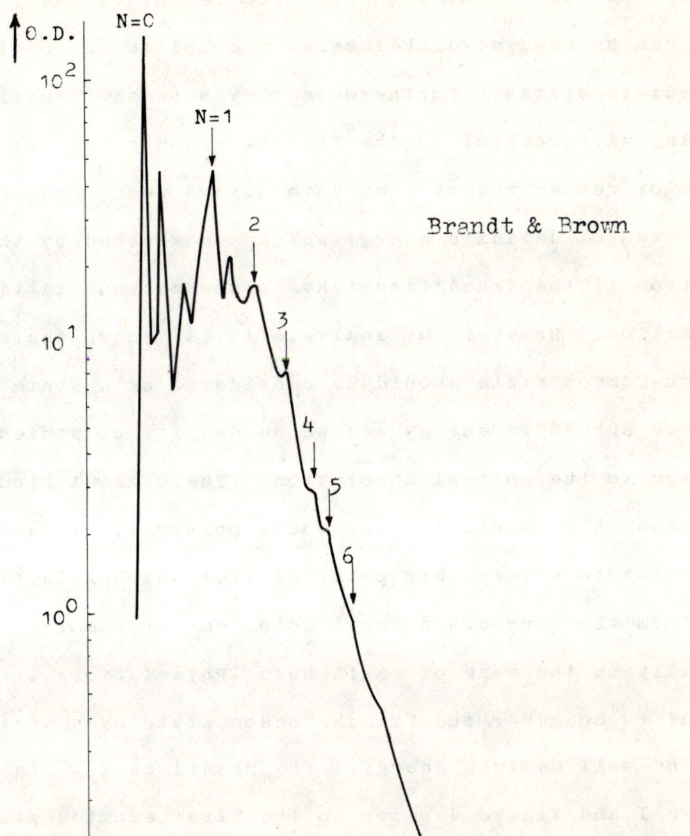

O.D.
N=C
N=1
2
3
4
5
6
Brandt & Brown
10^2
10^1
10^0
10^{-1}
10^2
10^3
10^4
Ω

fig 5

Fig. 4 Schematic representation of the low-lying states of the bound polaron

Fig. 5 Induced infrared absorption in AgBr (measurements of Brandt and Brown)

of excitations which are relevant to the quantum mechanical calculations and which are illustrated in figure 1 and in figure 4.

a) Franck-Condon states: these are no eigenstates of the system; they do not diagonalize the total hamiltonian because in these states the lattice is not relaxed. They can be used nevertheless as a complete set of intermediate states. Furthermore they have an intuitive meaning with respect to the Franck-Condon principle. In color center theory they have always been considered as states of definite energy which are reached by the electron if the transition takes place without lattice relaxation. However our analysis in (4) shows that the Franck-Condon state should be considered as a state with a large spread in energy and whose density of states is related to the optical absorption. The present study confirms this conclusion for bound polarons. It seems important to stress this point of view which is different from the approach of most color center theory. Especially in the case of small Huang-Phys factors it is useful to consider the Franck-Condon state as a state with no well defined energy. The notations E_{FC} in figure 1 and figure 4 refer to the first excited state in the original potential.

b) Relaxed-excited states: those are the approximate eigenstates of the hamiltonian in which the lattice and the electron are mutually adapted to the excitation. In our calculations it is assumed that those states are fairly stable: i.e. they have a long lifetime compared with the lattice period. The sharp peak in AgBr at 168 cm^{-1} is attributed by Brandt and Brown to a zero phonon transition. From figure 5 it is seen that the R.E.S. is very stable in this case. Those relaxed excited states can be considered as the lowest states in a potential modified by lattice relaxation (figure 1 and 4).

c) Scattering states: in general we define scattering states as approximate eigenstates which are fairly stable and which contain a whole number of free phonons $(n > 0)$ in the field. These states form a continuum as indicated in figure 3. It is seen that a transition from the groundstate to the first R.E.S. has to be considered as a resonance because of the continuum provided by the phonon-bath. The potentials drawn in figure 4 are of the Fowler type (8).

The essential features of this potential are quite similar for free polarons and for bound polarons. One of the main differences is that whereas the free polaron potential "moves" in real space because of recoil effects, the bound polaron potential is fixed at a point of the lattice. The Coulomb attraction between the vacancy and the electron makes it also reasonable to use the Born-Oppenheimer approximation for bound polarons in AgBr, e.g. where the coupling constant for polarons is only 1.7. This would not be possible for free polarons. In the case of bound polarons however the extra Coulomb force increases the electron frequency.

We now calculate the optical density for bound polarons. The transition probabilities from the groundstate to the final states, which we calculated here by perturbation theory, provide the optical density. Perturbation theory can be used if the final states are good approximations for the eigenstates of the problem. The dipolar interaction (analyzed in 4) is used. The optical absorption coefficient P will be calculated to within a constant factor:

$$P \div \Omega \sum_{\substack{\text{final}\\ \text{states}}} |\langle f \,|\, \vec{\varepsilon} \cdot \vec{D} \,|\, in \rangle|^2 \, \delta(E_{fin} - E_{in}) \quad (1)$$

$\vec{\varepsilon}$ is the electric field vector of the incident light. $\vec{D}$ is the dipole operator associated with the electron in the bound polaron. The δ-function conserves energy between the initial and final states.

More explicitly one has:

$$P \div \Omega \sum_n \int d\vec{k}_1 \ldots \int d\vec{k}_n \, |\langle 2p, \vec{k}_1, \ldots, \vec{k}_n \,|\, \vec{\varepsilon} . \vec{D} \,|\, 1s, 0 \rangle|^2 \, \delta(E_{fin} - E_{in}) \quad (2)$$

So the groundstate is assumed to be s-like and the excited state is assumed to be a 2p state. Polaron Zeeman effects obtained recently by Brandt et al. (9) have definitely established the 2p-type character of the relaxed excited state. Formula (2) corresponds to zero temperature because there are no phonons in the field. Summations over the final states result in summations over all possible k vectors in the final states. The final states which are considered consist of the first R.E.S. with one, two, three, ..., etc. phonons. The zero-phonon line is not studied here. First the different implications for free and bound polarons due to the conservation of energy are analyzed.

If one goes to a final state which is the R.E.S. with two phonons (wave vector $\vec{k}_1$, $\vec{k}_2$), the conservation of energy gives:

$$\hbar\Omega = E_{RES} - E_o + 2\hbar\omega_o + \frac{\hbar^2}{2m^x_{RES}} (\vec{k}_1 + \vec{k}_2)^2 \qquad (3)$$

here Ω is the frequency of the incident light, ω_o is the L.O. phonon frequency for $\vec{k} = o$ and m^x_{RES} is the mass of the polaron in the R.E.S. (6) (10). If one considers the case where one phonon is in the final state this means that the absorption curve sets in at $E + \hbar\omega_o$ (where $E = E_{RES} - E_o$) and tends to higher energies.

If the case of a bound polaron is considered, the recoil accompanying phonon emission is taken by the whole crystal. In that case m^x_{RES} can be replaced by ∞. In doing so however, one obtains a series of delta functions as sidebands. It is necessary in the case of bound polarons to introduce phonon dispersion. It is supposed here that the dispersion is "downward" at $k \approx o$ and writes $w_k = w_o - ak^2$ with $a > o$.

The cut-off in the electron phonon interaction eliminates the unrealistic behavior of this dispersion for large k.

Because of the term $- ak^2$ the one-phonon absorption has a critical point at $\hbar w = E + \hbar w_o$ but the absorption is only different from zero for $\Omega < \frac{E + \omega_o \hbar}{\hbar}$. This difference

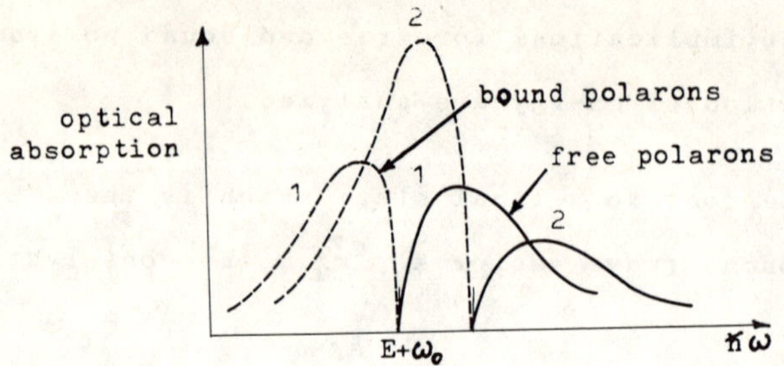

Fig. 6

Behavior of optical absorption for bound and free polarons

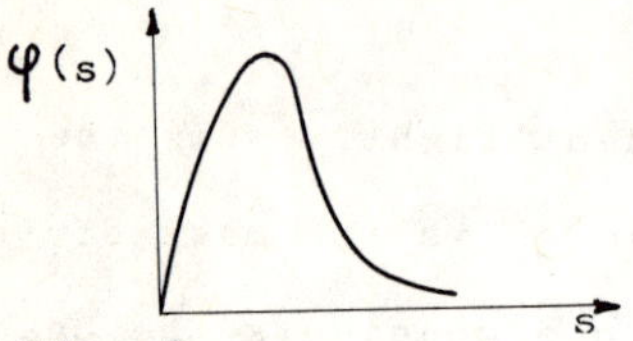

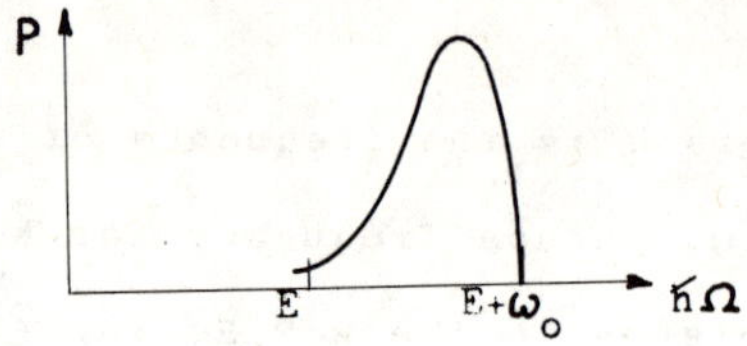

Fig. 7
and
Fig. 8 General behavior of the one-phonon contribution

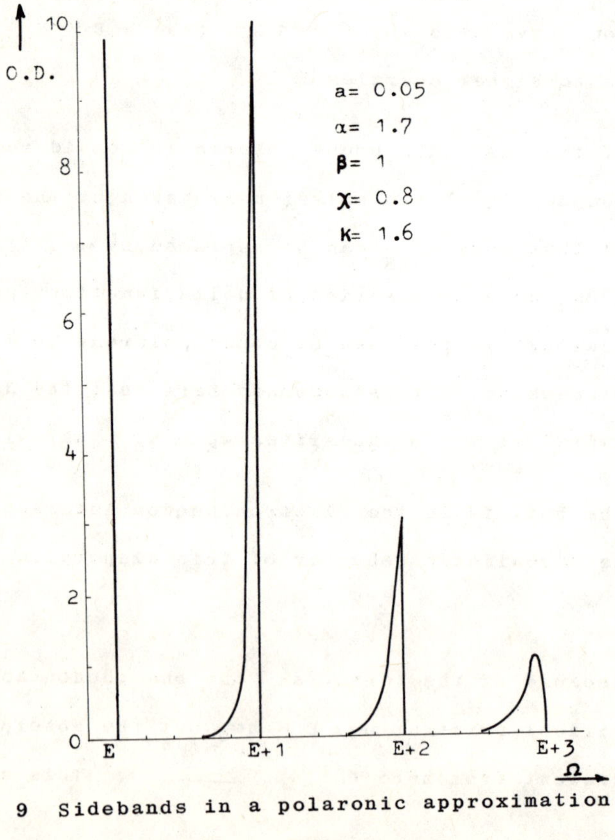

Fig. 9 Sidebands in a polaronic approximation

in behavior is illustrated in figure 6.

For free polarons a calculation of the absorption in the one phonon sideband does not interfere between $h\Omega = E + \omega_o$ and $h\Omega = E + 2\omega_o$ with the two phonon sideband. For bound polarons this is not the case. It will also turn out that the integrals to be performed are easier for bound polarons that for free polarons. This is mainly due to the directional dependence of $(\vec{k}_1 + \vec{k}_2)^2$ (see eq. 3) and similar terms involving more phonons.

The form of the wave functions will be discussed now. Because of the polaron-like and the Born-Oppenheimer approximation the following wave functions are used (see also 1):

$$| \Psi > = | \text{ electron } > | \text{ field } > \quad (4)$$

$$|\text{field}_{\text{initial}} > = e^{\sum_k \left(\frac{V_k^* \rho_k^*(in)}{e\hbar\omega} a_k^+ - \frac{V_k \rho_k(in)}{e\hbar\omega} a_k \right)} | o > \quad (5)$$

$$|\text{field}_{\text{final}} > = e^{\sum_k \left(\frac{V_k^* \rho_k^*(fin)}{e\hbar\omega} a_k^+ - \frac{V_k \rho_k(fin)}{e\hbar\omega} a_k \right)} a_{k_1}^+ a_{k_2}^+ \ldots a_{k_n}^+ | o > \quad (6)$$

$| o >$ is defined by $a_k | o > = o$ and $< j/o > = \delta_{jo}$.

The $\rho_k(in)$ and $\rho_k(fin)$ are related to the charge distribution of the electron in the groundstate and the excited states:

$$\begin{cases} \rho_k = e \int d\vec{r}\, \rho(\vec{r})\, e^{i\vec{k}.\vec{r}} \\ \rho(\vec{r}) = |\varphi(e)|^2 \end{cases} \quad (7)$$

For the electronic wave functions $\varphi(e)$, Coolomb type wave functions are taken:

$$\varphi_{1s}(e) = \varphi(in) = \frac{(\chi)^{5/2}}{\pi^{1/2}2^{5/2}} e^{-\frac{\chi r}{2}}$$
$$\varphi_{2p}(e) = \varphi(fin) = \frac{(x)^{3/2}}{\pi^{1/2}2^{3/2}} z\, e^{-\frac{xr}{2}} \quad (8)$$

For the electronic part of the wave functions the results derived by Buimistrov and Pekar (11) which have been analyzed in the case of AgBr by Brandt and Brown (2) can be used. Buimistrov and Pekar calculate the groundstate energy level and the R.E.S. energy level for a bound polaron.

As we are interested in the sidebands, which had not been considered before, we will use the results obtained from (11) to find χ and x.

Making use of the Born-Oppenheimer approximation eq. (2) becomes:

$$P \div \Omega \sum_n \int dk_1 \ldots \int dk_n \delta (\hbar\Omega - E - \hbar\omega_1 - \hbar\omega_2 \ldots - \hbar\omega_n)$$
$$|\langle e_{fin} |Z| e_{in} \rangle |2| \langle field_{final} | field_{in} \rangle |^2 \quad (9)$$

because $|\langle e_{fin} |Z| e_{in}\rangle|^2$ does not depend on $\vec{k}$ one can omit this constant factor as only the dependence of the absorption coefficient on Ω and not its absolute value is calculated.

The calculation of the matrix element for the overlap of the field-wave functions is straightforward and leads to:

$$P \div \Omega \sum_n \int d\vec{k}_1 \ldots \int d\vec{k}_n \delta(\hbar\Omega - E - \hbar\omega_1 - \hbar\omega_2 \ldots - \hbar\omega_n)$$

$$\times \; |P_{k_1}|^2 \, |P_{k_2}|^2 \ldots |P_{k_n}|^2 \qquad (10)$$

where the P_k are defined as:

$$P_k = \frac{V_k}{e\hbar\omega} (\rho_{fin} - \rho_{in}) \qquad (11)$$

Writing (10) in more detail gives:

$$P \div \Omega \sum_n \int_0^\infty k_1^2 dk_1 \ldots \int_0^\infty k_n^2 dk_n \delta\,(\hbar\Omega - E - \hbar\omega_1 - \hbar\omega_2 \ldots - \hbar\omega_n)$$

$$\times \int_0^{2\pi} d\varphi_1 \int_0^\pi \sin\theta_1 d\theta_1 \; |P_{k_1}|^2 \ldots \int_0^{2\pi} d\varphi_n \int_0^\pi \sin\theta_n d\theta_n \; |P_{k_n}|^2 \qquad (12)$$

with $\omega_i = \omega_o - ak_i^2$

and $x_i = \omega_i - \omega_o = -ak_i^2$

the δ-function is put in a suitable form:

$$P \div \sum_n \frac{(-)^n}{2^n a^{3n/2}} \int_{-\infty}^{o} dx_1 \ldots \int_{-\infty}^{o} dx_n \delta\,(\hbar\Omega - E - n\hbar\omega_o - \hbar x_1 - \hbar x_2 \ldots - \hbar x_n)$$

$$\phi(x_1)\,\phi(x_2) \ldots \phi(x_n) \qquad (13)$$

with:

$$\phi(x_j) = \int_0^{2\pi} d\varphi \int_0^\pi \sin\theta d\theta \; \frac{|P(\sqrt{-\frac{x_j}{a}})|^2}{\sqrt{-x_j}} \, x_j \qquad (14)$$

The following transformation:

$$\begin{cases} s_n = x_1 + x_2 + \ldots x_n \\ s_{n-1} = x_1 + x_2 + \ldots x_{n-1} = s_n - x_n \\ s_1 = x_1 = s_2 - x_2 \end{cases} \qquad (15)$$

allows one to find P with the aid of multiple convolution integrals:

$$P \div \sum_n \frac{(-)^n}{2^n a^{3n/2}} \int_o^\infty ds_n \delta (\hbar\Omega - E - n\hbar\omega_o + \hbar s_n) \int_o^{s_n} ds_{n-1} \varphi(s_n - s_{n-1})$$

$$\ldots \int_o^{s_3} ds_2 \varphi(s_3 - s_2) \int_o^{s_2} ds_1 \varphi(s_2 - s_1)\, \varphi(s_1) \qquad (16)$$

where:

$$\varphi(s_n - s_{n-1}) = \phi(s_{n-1} - s_n)$$

For (s) one finds, after some algebra, the following expression:

$$\varphi(s) = (-)\frac{2\alpha}{\pi} \frac{a^{1/2} (\frac{\hbar}{2m\omega_o})^{1/2}}{(\frac{s}{a})^{1/2}} \left[\frac{\chi^{12}}{(\chi^2 + \frac{s}{a})^6} + \frac{\frac{36}{5}\frac{s^2}{a^2}}{(\chi^2 + \frac{s}{a})^8} \right. \qquad (17)$$

$$\left. + \frac{x^8}{(x^2 + \frac{s}{a})^4} - \frac{4\chi^{12}\frac{s}{a}}{(\chi^2 + \frac{s}{a})^7} - \frac{2x^4\chi^6}{(\chi^2 + \frac{s}{a})^3 (x^2 + \frac{s}{a})^2} + \frac{4\chi^6 x^4 \frac{s}{a}}{(\chi^2 + \frac{s}{a})^5 (x^2 + \frac{s}{a})^2} \right]$$

$(s < o \rightarrow \varphi(s) = o)$.

α is the polaron coupling parameter ; $\sqrt{\frac{\hbar}{2m\omega_o}}$ is the polaron radius in the case of a free polaron.

The physical significance of $\varphi(s)$ is quite simple. To see this consider the first term of the sum (16):

$$\frac{(-)}{2a^{3/2}} \int^\infty ds_1\, \delta(\hbar\Omega - E - \hbar\omega_o + \hbar s_1)\, \varphi(s_1) \qquad (18)$$

or :

$$\frac{(-)}{2a^{3/2}} \varphi(E + \omega_o - \hbar\Omega) \qquad (19)$$

It is easy to see that $\varphi(s) = o$ if $s = o$, the divergent terms cancelling. It is obvious, furthermore, that $\varphi(s) \rightarrow o$

as s $\to \infty$. The function has one maximum. Its general appearance is given in fig. 7.

It follows that the one phonon peak given by (19) behaves as shown in fig. 8. This behavior is consistent with the one that was inferred from energy conservation considerations.

Until now phonon-phonon interaction or anharmonicity in the ionic motion was neglected. For some ionic crystals this anharmonicity is rather important in particular for the silver halides and the alkali halides. The anharmonicity (finite lifetime of the phonons) is revealed by a broadening of the phonon dispersion curves especially at k = o. At this stage of our work we have taken anharmonicity into account by a phenomenlocial device: a Gaussian distribution for the phonon frequency distribution was considered. One then has to integrate over the different L.O. frequencies for each phonon sideband. Consequently taking into account the dispersion of the L.O. phonons and anharmonic effects one derives the following expression for the absorption coefficient:

$$P\ (\text{anharmonic effects}) \div \int_o^\infty .. \int_o^\infty d\,\omega_o^1 d\omega_o^2 \ldots d\omega_o^n C_{n-1}(E + \hbar\omega_o^1 + \hbar\omega_o^2 + \hbar\omega_o^3 + \hbar\omega_o^4 + \ldots - \hbar\Omega)\, e^{-\lambda(\omega_o - \omega_o^1)^2} \ldots e^{-\lambda(\omega_o - \omega_o^n)^2} \qquad (20)$$

With an appropriate change in variables, similar to that given by (15) one obtains:

$$P\ (\text{anharmonic effects}) \div \int_0^{\infty} ds_n C_{n-1}(E+s_n-\Omega) \int_0^{s_n} ds_{n-1} e^{-\lambda(\omega_o - s_n + s_{n+1})^2}$$

$$\ldots \int_0^{s_3} ds_2 e^{-\lambda(\omega_o - s_3 + s_2)^2} \int_0^{s_2} ds_1 e^{-\lambda(\omega_o - s_1)^2} e^{-\lambda(\omega_o - s_1 + s_2)^2} \qquad (21)$$

One has in (20) and (21) : $C_o = \varphi(s) \frac{1}{2a^{3/2}}$ etc.

The effect of anharmonicity, of course, is a broadening of the different peaks in the sidebands.

Using properties of Laplace transforms of convolutions one can write (16) and (21), the general expressions proportional to the absorption coefficient, as condensed forms. These forms especially useful to treat the case of large Huang-Phys factor.

Although some analytical progress can be made starting from (16) and (21), these expressions are quite involved and for most of the applications computer work is needed.

2. APPLICATION OF THE THEORY TO THE INDUCED INFRARED ABSORPTION IN AgBr.

In applying our results to the case of AgBr one has to evaluate the parameters a, α, x, χ, $\sqrt{\frac{\hbar}{2m\omega_o}}$ and λ . The polaron coupling constant is equal to ~ 1.7. The analysis of Brandt and Brown (2) using Buimistrov's and Pekar's (11) theory gives a measure for the extension of the bound polaron in the groundstate and in the first R.E.S.; $x \simeq 1.6$, $\chi \simeq 0.8$. Units $\hbar = w_o = 2m = 1$ are used. No neutron data about the phonon dispersion in AgBr are available. We conjectured a large curvature for k = o. This conjecture leads

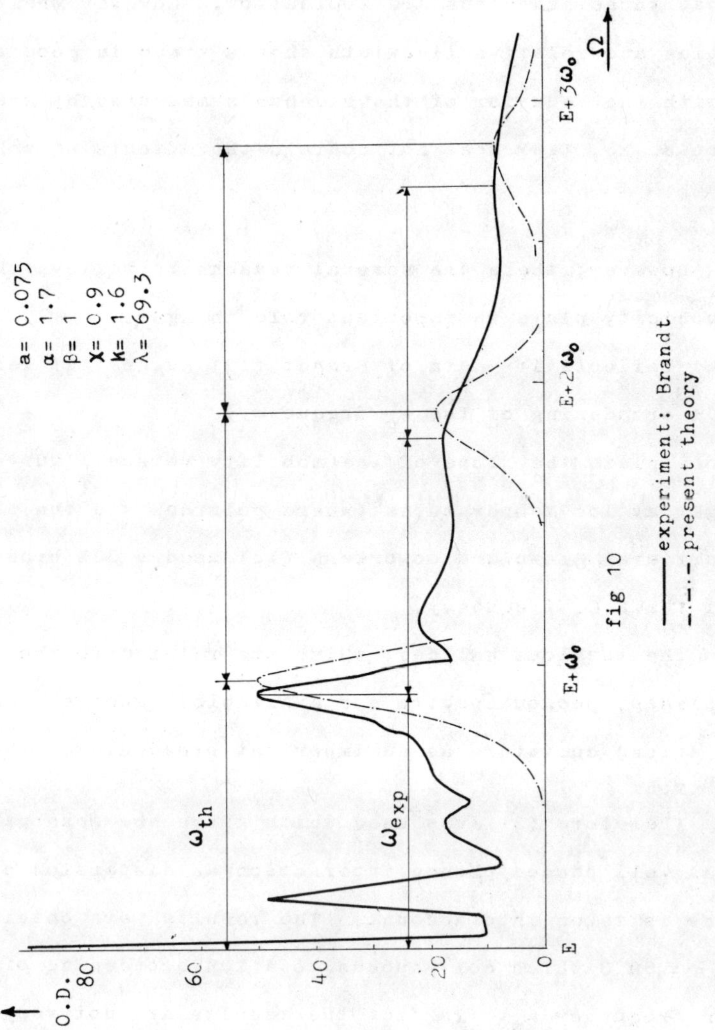

Fig. 10 Induced infrared absorption in AgBr, theory and experiment.

to reasonable results. Taking a = 0.05 and using eq. (16) one obtains the three sidebands given in fig. 9. It is obvious that those sidebands are too narrow. However their intensities and relative linewidth show a trend in good agreement with the behavior of the sidebands measured by Brandt and Brown. Eq. (16) does not contain the effects of anharmonicity.

However, there are several reasons to believe that anharmonicity plays an important role in AgBr.

a) The reflectivity data of Brandt (12) suggest at least a 10% broadening of the w_o frequency.

b) To explain the slope of the mobility versus $\frac{1}{T}$ curve in AgBr at low temperatures (where polarons are the charge carriers) Brown and coworkers (13) used a 30% broadening of the L.O. frequency.

c) In the thallous halides, which are related to the silver halides, phonon spectra are available. They show as well a strong curvature as an important broadening.

Therefore (21) was used which gives the absorption when as well phonon-phonon interaction as dispersion of the phonons is taken into account. The results were obtained with $\lambda = 69.3$ which corresponds to a 10% broadening of the phonon frequency w_o. In fact the results are not very sensitive to the choice of λ as long as one remains in the realistic range of about 10 a 20% broadening of w_o. In fig. 10 the results which have been obtained for the first three sidebands are shown and compared with the data of Brandt and Brown.

Consistent with our conjecture that the L.O. phonon dispersion in AgBr is large we choose a = 0.075. This is the only parameter in this application which is fitted.

From fig. 10 one sees that the theoretical widths of the first three peaks are in good agreement with experiment. The relative intensities of the peaks, the first one of which has been normalized, also agree with experiment.

An important effect revealed in fig. 10 is the 10% reduction or shift in the frequencies of the phonons which are emitted with the largest probability during the absorption process. All the experimental absorption peaks have a distance (w_{exp}) which is only 90% of the frequency w_o given by reflectivity or reststrahlen data (12). Our theoretical results also reveal a shift, which is about 5%. The shift occurring in the theoretical values for w_o is due "in se" to dispersion and enhanced by anharmonicity. It is seen that the distances between the theoretically predicted sidebands remain constant as is the case for the experimental peaks.

If the radius of the electron "orbit" in the 2p-state tends to that of the ls-state, ($\chi \rightarrow x$) the results are largely in error. This confirms Fowler's (8) considerations on the diffuse character of the excited state of the F center.

It is necessary to discuss the limitations of the present theory and to examine other factors determining the sideband structure. This leads to suggestions to improve the theory. The polaronic approximation a la Fröhlich

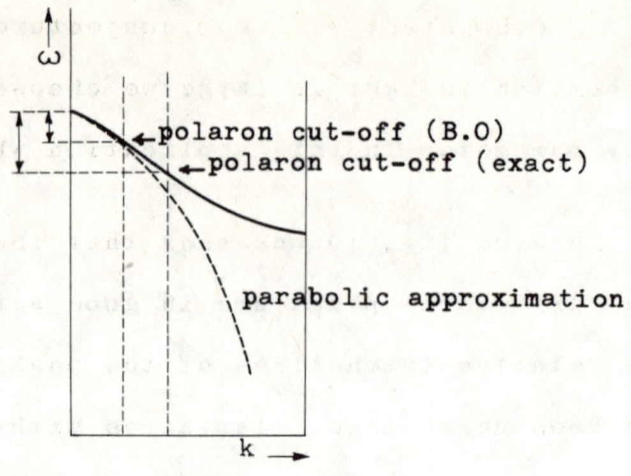

fig.11

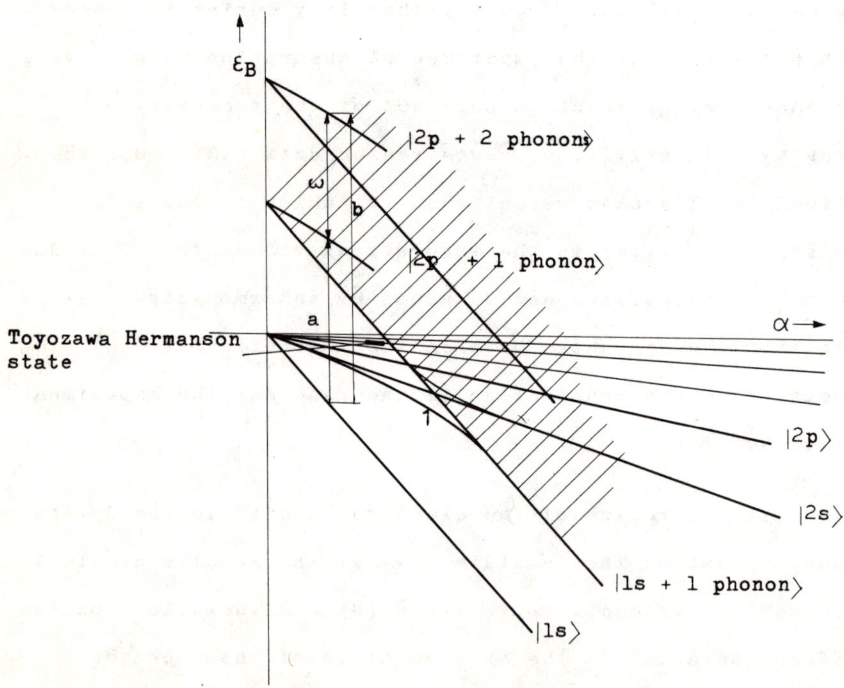

fig.12

Fig. 11 Conjecture about the phonon dispersion in AgBr

Fig. 12 Modified Bachrach scheme for the Toyozawa-Hermanson quasiparticle state.

seems unquestionable in the case of AgBr. The Born-Oppenheimer approximation on the contrary is only to be considered as a first step. Exact wave functions would have a tendency to delay the cut-off value of the wave vector to larger values. (This can be seen from ref. 14 by using a simplified exactly soluble polaron model.)

Such a delay of the cut-off wave vector would enhance the role of dispersion of the L.O. phonons. In this case smaller values of the parameter would still give an appreciable shift. (This point is illustrated in fig. 11.) Another restriction of the present theory is the symmetrical character we have assumed for the phonon density distribution. A many body treatment of tho phonon-phonon interaction is necessary to obtain the lineshape.

Other factors which might lead to a shift of the L.O. frequencies are the Doppler effect (the motion of the electron with respect to the L.O. models tends to decrease the frequency) and the possible binding between an exciton and one phonon states considered by Toyozawa and Hermanson (3) which we will discuss in the following paragraph.

3. A REMARK ON THE THEORY OF TOYOZAWA AND HERMANSON (T.H.) (3) AND ITS APPLICABILITY TO THE ANALYSIS OF THE INDUCED INFRARED ABSORPTION IN AgBr.

The main reason to bring up this theory in this context is the fact that Brandt and Brown (2) invoke it as a possible explanation of their infrared data on AgBr and especially of the shift of frequencies in the sidebands.

A brief resumé of the T. H. theory is given here; Toyozawa and Hermanson start from Haken's consideration that the exciton can be considered as consisting of two continuum polarons (with opposite charge) bound together. The T. H. description is made in the center of mass system of the exciton. The Hartree approximation is made; the wave function of the exciton is expressed as a product of a field function (containing the phonon variables) for which recoil is neglected (as in the well known Lee, Low, Pines wave functions), and a function containing the distance between the electron and the hole as the only variable. If the mass of the hole tends to infinity this model is the same as ours and also the same approximations are made (especially the Hartree Fock approximation) except for the fact that T. H. neglect phonon dispersion and anharmonicity which are important factors in our approach. Especially for a bound polaron these factors are essential because in that case there is no dispersion as there is no recoil of the exciton. Therefore the T. H. approximation cannot give, as such, the line shape of the sidebands in the case of bound polarons. T. H. then calculate, in this framework of the adiabatic approximation, the groundstate energy E_o of the exciton. Subsequently they postulate the first excited state of the system to be a superposition of

a) states with one real phonon (energy $\hbar\omega_o$), added to the groundstate;
b) states where the exciton is excited to different

Coulomb-like states in the potential corresponding to the groundstate.

If it is assumed that such an excited state exists it follows that the energy of this state is shifted downward with respect to the level with energy $E_o + \hbar\omega_o$. Toyozawa and Hermanson consider this state as a quasiparticlestate:

:a state consisting of an exciton bound to a phonon T. H. predict that this state will be revealed especially for crystals in which the electron hole binding energy is nearly equal to the L.O.-phonon energy. Finally they examine the intensity of the absorption to such a quasi particle state and compare it to the zero-phonon transition intensity.

Bachrach (15) has given a schematic representation of the situation. He considers the energy of the system consisting of an exciton and the field of phonons plotted versus a coupling parameter characterizing the crystal. He indicated curve 1 in figure 12 as resulting from the description of Toyozawa and Hermanson. His interpretation is that the 2p-state is "pinned" by the continuum (1S + 1 phonon >) in which case a final state with energy lower than $E_o + \omega_o$ arises. It does not seem clear why such a "pinning" (although an intuitively appealing effect) would occur for the 2s state.

It seems to us that the question of the penetration of an isolated state like the 2s state into the continuum should be examined. Anyhow the curve 1 given by Bachrach does not correspond to what T. H. calculate. The T. H. state, according to the authors, results from the interaction between the phonon continuum and the host of states penetrating the continuum if $\varepsilon_B \sim -\hbar\omega_o$. In fig. 11 this state is indicated and it might play a role in the optical transitions in the thallous halides as discussed by Bachrach.

On the other hand it is clear that the zero-phonon transition in AgBr at 168 cm^{-1} does not correspond to a transition towards a state such as the one described by T. H. For one reason, the 2p-state is in the continuum (168 $cm^{-1} > \omega_o$). Furthermore it does not correspond to a host of Coulomb like states interacting with a continuum by "penetrating it." In the description of fig. 11 the 168 cm^{-1} transition in AgBr corresponds to the transition indicated by a↑. The one phonon sideband is dominated by transitions like b↑. There might be a "binding" effect reducing the distance between the two final states from ω_o. However this is not given by the T. H. theory in its present form.

CONCLUSION

We have proposed a general theory for the calculation of the sideband structure in the optical spectra of bound polarons. This theory is based upon a continuum polaron picture and a Born-Oppenheimer approximation. Dispersion of the phonons and anharmonic effects are essential in this theory.

With this theory we can explain the general behavior of the sidebands in the phonon assisted transitions in AgBr observed by Brandt and Brown. An apparent shift of the phonon frequency emitted during the absorption is partly explained.

The theory of Toyowaza and Hermanson about the exciton-phonon interaction is briefly analyzed. Although this theory might account for the one phonon sideband in the thallous halides it is certainly not adapted as such to treat the transitions in AgBr which we discussed and which were observed by Brandt and Brown. Indeed these transitions are towards a continuum even for the zero-phonon transition.

Nevertheless the situation which respect to the case of AgBr is not completely resolved. It is possible that binding effects, such as described by Toyozawa and Hermanson, when calculated for this problem, contribute e.g. to the shift of the phonon frequencies.

The clarification of the situation would be possible if neutron spectra for the phonons in AgBr were available. It would be clear how important dispersion and anharmonicity are exactly from such measurements.

So although the general approach exposed here explains the essential features encountered in AgBr it also rises questions about the phonon spectra of this substance. Measurements of the phonon dispersion curves or "bands" are therefore suggested.

Finally we also draw some conclusions concerning the exciton spectra of CuCl (16) and ZnO (17). One might think a priori that the maxima in the phonon sidebands occur at $n = 3$ resp. $n = 5$ because of a large coupling parameter . (This is in fact the only possibility left in the framework of our approximations.) However both materials have a small α . We suggest that the broad background is due to interaction between the exciton and acoustical phonons and that the L.O. phonon structure is a superposition in this background. Our first calculations in which we used an appropriate potential confirm these ideas.

REFERENCES

1. T. Lee, F. Low, P. Pines, Phys. Rev. 90, 297 (1953)
2. R. Brandt and F. Brown in "Localized Excitations in Solids," edited by R. F. Wallis (Plenum Press 1968), p. 332.

   Ibid: to be published in Phys. Rev.
3. Y. Toyozawa and S. Hermanson, Phys. Rev. Letters 21 1637 (1968)
4. E. Kartheuser, R. Evrard, J. Devreese, "The Internal Structure of Frohlich Polarons and the Optical Absorption Process," Proceedings of the present conference.
5. J. Devreese and R. Evrard, Phys. Letters 11, 278 (1964)
6. R. Evrard, Phys. Letters 14, 295 (1965)
7. E. Kartheuser, R. Evrard and J. Devreese, "On the excitations of Polarons," Contribution to the meeting on "Electrical and Magnetic Ceramics," London (1966), Proc. Brit. Cer. Soc. 10, 165 (1968)
8. W. Beall Fowler, Phys. Rev. 135, A1725 (1965)
9. R. Brandt, D. Larsen, P. Crooker and G. B. Wright, to be published
10. E. Kartheuser, Thesis, University of Liege (1968)
11. V. Buimistrov andS. Pekar, Zh ETF. 32, 1193 (1957)
12. R. Brandt, Thesis, University of Illinois (1967)
13. T. Masumi, Ahrenkiel, F. Brown, Phys. stat. sol. 11, 163 (1965)
14. J. Devreese and R. Evrard, Phys. stat. sol. 9, 403 (1965)
15. R. Bachrach, Thesis, University of Illinois (1969)

16. T. Goto and M. Ueta, J. Phys. So. Japan 22, 112 (1967)

17. Chr. Solbrig, Zeitschrift fur Physik 211, 429 (1968)

XVII

DISPERSION RELATION APPROACH TO THE THEORY OF OPTICAL TRANSITIONS WITH AN APPLICATION TO THE POLARON MODEL

R. EVRARD, E. KARTHEUSER

Institut de Physique, Université de Liège
Sart-Tilman, Liège 1 (Belgium)

and

J. DEVREESE

Faculty of Science, University of Antwerp (Belgium) and Solid State Physics Department, Studiecentrum voor Kernenergie, Mol-Donk (Belgium)

INTRODUCTION

In ionic crystals, the interaction with the longitudinal optical phonons modifies the properties of free or bound charge carriers drastically. It is well known for instance that the observed effective mass can be quite different from the calculated band mass. The energy levels are shifted also, although in a less spectacular way. This is easily explained by the following considerations :

The charged particle polarizes the crystal, inducing a distortion in the lattice. This distortion reacts on the motion of the particle through the electric field due to the polarization. As the charge carrier lies in the potential well from which this electric field derives, its energy is lowered by a quantity that constitutes a kind of "self-energy". Moreover, the distortion must follow the displacements of the charge carrier in the crystal giving rise to an increase in its inertia.

In the present study, we restrict ourselves to the case where the wave packet representing the charge carrier is large compared with the dimensions of the lattice cell. In this case, the system constituted by the particle and its surrounding phonon cloud (the lattice distortion) is called a "large" or Fröhlich's polaron. The interaction is then restricted to long-wave longitudinal optical phonons. The effective mass, the energy and the mobility of Fröhlich's polarons have been studied by several authors. A survey of their works can be found in the proceedings of the Summer School held at St.Andrews (Scotland) in 1962 [(1)].

There is however no reason why the same interaction would have no effect on other properties such as the absorption of electromagnetic radiation. The study of this latter problem had not been undertaken up to now although it looks highly interesting. This study could indeed give informations on the polaron properties in a "differential" manner, we mean as functions of the frequency of the incident photon, rather than in an "integral" way as for the effective mass, the energy shift or the mobility. For this reason, the knowledge of the absorption spectrum of charge carriers in ionic crystals should in the near future lead to a more detailed check of polaron theory and a more complete understanding of the effects of the coupling with phonons. Especially it should give a definitive answer to the question of the existence of the internal excited states predicted by different authors [2-5]. A discussion of these internal excited states of free polarons and their role in the absorption of electromagnetic radiation can be found in Kartheuser's talk at the present conference.

Another case of perhaps more practical importance is that of polarons bound either in excitons or in color centers. Experimental data are available at the present moment for bound polarons on the contrary of free polarons.

These above considerations explain our interest in the problem of the evaluation of the absorption of electromagnetic radiation by free and bound polarons. We have succeeded in solving this problem [6,7] in the framework of the "adiabatic" approximation which is nothing else than the Born-Oppenheimer approximation transposed to the polaron theory by Landau and Pekar* : The correlations between the motions of the charged particle and the phonons are neglected. A report of our work with the adiabatic method is given in the talks by Devreese and Kartheuser.

The present paper is devoted to a new approach to the same question, but using now techniques which do not rest on the Born-Oppenheimer approximation. Therefore the results of free polarons will no longer be restricted to the region of strong coupling, but will probably be quantitatively valid also in the intermediate

---

* See for instance ref. 1.

range of coupling strengths where alkali-halides, silver-halides, etc... are found. In the same way, it will be possible to apply the method to the case of bound polarons with a transition frequency not much larger than the phonon frequency.

It should be pointed out that we cannot simply extend the calculations exposed in Devreese's and Kartheuser's talks to the case of intermediate coupling strengths. This would indeed require the knowledge of a polaron wave-function valid in this region of coupling. As far as we know such a wave-function is not available for the moment. Moreover, polaron wave functions obtained by the usual techniques as variational methods would probably not be of great use for the computation of the absorption spectrum. In effect, the wave functions obtained that way are adjusted to lead to the best value of the ground state energy compatible with their form. This does not mean in any way that they are suited for the computation of the absorption spectrum. For example the Lee, Low, Pines wave function [8] gives a better value of the ground state energy than the Landau-Pekar approximation for α (the coupling constant) smaller than 10. However it does not predict the existence of resonances in the absorption of electromagnetic radiation, although these resonances are known to exist and to be relatively stable for values of α as low as probably 3 (see Kartheuser's paper in these proceedings).

For our purpose, it seems therefore better to seek new methods in which the approximations are made directly on the elements of the matrix which gives the probability of transition rather than on the wave function. As the transition matrix is nothing else than the scattering matrix for the process of absorption of a photon, we are actually looking for a S-matrix theory of the polaron problem. This kind of theory has been extensively used under different forms in elementary particle physics. However, its extension to solid state physics has seldom been attempted.

In the present paper, we give a preliminary report of our attempt to develop such a method for the problem of the interaction of polarons with electromagnetic radiation.

It seems to us that the simplest approach to this question is to write and then solve equations of the kind of those established by Chew and Low [9] for the static model of nucleons in interaction with a pion field. We'll first develop such equations for the elastic scattering of a phonon by the polaron. We'll then show that the S matrix for this process is related in a simple way to the matrix which gives the rate of transition induced by an electromagnetic radiation field in the usual electric dipole approximation. Finally we'll solve the Chew-Low equations in the framework of a small oscillation approximation and say a few words about the prospects of the method.

LOW EQUATIONS FOR POLARONS

The starting point is the usual Fröhlich's hamiltonian written in notations of second quantization(*) :

$$H = H_0 + V \tag{1}$$

where H_0 is the hamiltonian of the electron-phonon system in the lack of interaction and V is the interaction term :

$$V = \sum_{\vec{k}} [V_k a_{\vec{k}} \exp(i\vec{k}\vec{r}) + V_k^* a_{\vec{k}}^+ \exp(-i\vec{k}\vec{r})] \tag{2}$$

The interaction coefficients V_k have been defined in Kartheuser's and Devreese's talks. They are :

$$V_k = \frac{i}{k} (4\pi/v)^{1/2} \hbar\omega (\hbar/2m\omega)^{1/4} (\omega_0/\omega)^{3/2} \alpha^{1/2} \tag{3}$$

where v denotes the volume of the crystal, $\omega = \omega(k)$ the frequency of the longitudinal optical phonon with wave vector $\vec{k}$ ω_0 the limit of this frequency for $\vec{k} \to 0$ and α the usual coupling constant between the charge carrier and the long wave longitudinal optical phonons. The free part H_0 of the hamiltonian is the sum of two terms : one for the particle and one

---

(*) see ref.(8). See also the papers by Kartheuser and Devreese in these proceedings.

for the phonon field :

$$H_0 = H_p + \sum_k \hbar\omega \, a_{\vec{k}}^{+} a_{\vec{k}} \quad (4)$$

with

$$H_p = \frac{p_e^2}{2m} + U(\vec{r}) \quad (5)$$

In this expression, $\vec{p}_e$ is the momentum of the particle, $\vec{r}$ its position and m its band mass. In the case of a free polaron $U(\vec{r}) = 0$. For a bound polaron, $U(\vec{r})$ is the interaction energy between the charge carrier and the vacancy to which it is bound. The extension to excitons is straightforward.

The interaction is linear in the phonon operators. This enables us to express causality in a very simple way. Indeed let us suppose that we know a solution of the Schrödinger equation, for instance a state $|p\rangle$ describing a polaron alone (no real phonon present) :

$$H|p\rangle = E(p)|p\rangle \quad (6)$$

where p denotes the polaron momentum for free polarons and the quantum numbers of the bound state for bound polarons. Then, it is easy to show* that :

$$|p;\vec{k}_{\mp}\rangle = \left[a_{k}^{+} - \frac{V_k}{H - E(p) - \hbar\omega \pm i\varepsilon} e^{i\vec{k}r} \right] |\vec{p}\rangle \quad (7)$$

---

* See for instance Ref. (11) p. 359 ff.

are also solutions of the Schrödinger equation with eigenvalue

$$E(p;\vec{k}_{\mp}) = E(p) + \hbar\omega$$

The solution $|p;\vec{k}_+\rangle$ is an incoming state : It describes a causal scattering of a real phonon by the polaron. The solution $|p;\vec{k}_-\rangle$ is an outgoing state, which can be obtained from the preceding solution by means of a time reversal.

The element of the S matrix for the scattering of a phonon $\vec{k}'$ by the polaron initially in the state $|p'\rangle$ leading to a final state with a phonon $\vec{k}$ and the polaron in the state $|p\rangle$ is given by

$$\langle p;\vec{k}|S|p';\vec{k}'\rangle = \langle p;\vec{k}_-|p';\vec{k}'_+\rangle \tag{8}$$

It is now possible to "contract" the phonon $\vec{k}'_+$, i.e. to replace in (8) the state $|p';\vec{k}'_+\rangle$ by its expression (7). This leads to

$$\langle p;\vec{k}|S|p';\vec{k}'\rangle =$$

$$\delta_{i,f} - 2i\pi\delta(E_f - E_i)\langle p;\vec{k}_-|V_{k'}e^{i\vec{k}'\vec{r}}|p'\rangle \tag{9}$$

where i, f refer to the initial, respectively final state. The first term of the right member, which is a Kronecker delta, expresses that, in the lack of interaction, the final state is identical to the initial state. The second term is the part of the phonon wave scattered by the polaron.

As the conservation of energy appears explicitly in the expression (9), the element of the T matrix defined by :

$$\langle p ; \vec{k} | T | p' ; \vec{k}' \rangle = \langle p ; \vec{k} | V_{k'} e^{i\vec{k}'\vec{r}} | p' \rangle \qquad (10)$$

has a meaning outside the energy shell, i.e. for values of the final energy which do not fulfill the conservation of energy.

Finally, contracting now the phonon $\vec{k}$ in the final state, one obtains after straightforward algebraic manipulations :

$$\langle p ; \vec{k}_- | e^{i\vec{k}'\vec{r}} | p' \rangle =$$

$$-V_k^* \langle p | e^{-i\vec{k}\vec{r}} \frac{1}{H - E(p) - \hbar\omega - i\varepsilon} e^{+i\vec{k}'\vec{r}} | p' \rangle - V_k^* \langle p | e^{i\vec{k}'\vec{r}} \frac{1}{H - E(p') + \hbar\omega} e^{-i\vec{k}\vec{r}} | p' \rangle \qquad (11)$$

which are the Low equations for the phonon-polaron collisions. To show that they really are equations, one must introduce a complete set $|n\rangle$ in the right-hand member :

$$\langle p ; \vec{k}_- | e^{i\vec{k}\vec{r}} | p' \rangle =$$

$$-V_k^* \sum_n \frac{\langle p | e^{-i\vec{k}\vec{r}} | n \rangle \langle n | e^{i\vec{k}'\vec{r}} | p' \rangle}{E_n - E(p) - \hbar\omega - i\varepsilon} - V_k^* \sum_n \frac{\langle p | e^{i\vec{k}'\vec{r}} | n \rangle \langle n | e^{-i\vec{k}\vec{r}} | p' \rangle}{E_n - E(p') + \hbar\omega} \qquad (12)$$

If we notice that it is possible to write similar relations for scattering involving more than one phonon, it is then clear that (12) are equations linking the matrix elements of the operators $e^{\pm i\vec{k}\vec{r}}$ together.

CONNECTION BETWEEN THE ABSORPTION OF ELECTROMAGNETIC RADIATION AND THE SCATTERING OF PHONONS

This connection is very similar to that existing between the photoproduction and the scattering of pions by nucleons at low energy (10). Let us consider the case where both incident and scattered phonons have a wave length large compared with the dimensions of the wave packet representing the particle. The exponential $e^{i\vec{k}\vec{r}}$ can thus be expanded in a power series around the mean position $\langle\vec{r}\rangle$ of the particle at the moment of the collision

$$e^{i\vec{k}\vec{r}} = e^{i\vec{k}\langle\vec{r}\rangle} \left(1 + i\vec{k}\cdot\Delta\vec{r} + \cdots\cdots \right) \tag{13}$$

with

$$\Delta\vec{r} = \vec{r} - \langle\vec{r}\rangle \tag{14,}$$

Moreover, if the direction $\vec{k}'$ of the incoming phonon is choosen as Z axis, one has :

$$\frac{-i}{k_z V_k^* e^{i\vec{k}<\vec{r}>}} < p ; \vec{k}_- | \Delta z | p' > =$$

$$< p | \Delta z \frac{1}{H - E(p) - \hbar\omega - i\varepsilon} \Delta z | p' > + < p | \Delta z \frac{1}{H - E(p') + \hbar\omega} \Delta z | p' > \quad (15)$$

where selection rules have been taken into account. Let us call :

$$f(\omega) = \frac{-i}{k_z V_k^* e^{-i\vec{k}<\vec{r}>}} < p ; \vec{k}_- | \Delta z | p' > \quad (16)$$

Taking the imaginary part of $f(\omega)$ leads to :

$$I_m f(\omega) = \pi \sum_n | < p | \Delta z | n > |^2 \delta [\omega - (E_n - E(p))] \quad (17)$$

which is nothing else than the probability of transition induced by an electromagnetic radiation of frequency ω in the dipole approximation, except for a constant factor. It is then evident that the knowledge of the cross section for the scattering of long wave longitudinal phonons allows the calculation of the absorption spectrum.

BEHAVIOR OF THE T-MATRIX AT HIGH ENERGY OF THE OUTGOING PHONON.

The usual ways of solving dispersion relations lead to an infinity of different solutions. It is therefore necessary to know the behavior of T at large $\hbar\omega$ (the energy of the outgoing phonon), in order that we can select the actual solution.

At this point most of the authors declare : from the Low equations (12), it is clear that the element of the T matrix under consideration goes as $1/\hbar\omega$ for large $\hbar\omega$. The situation is of course not so simple. To obtain the asymptotic behavior of the right hand side of (12), one cannot just neglect the other terms of the denominator with respect to $\hbar\omega$. These terms can indeed be as important as $\hbar\omega$, since one has to integrate over the whole energy range of the intermediate states. We think that it is worthwhile to show what is the correct way to proceed. Let us consider the expression :

$$I(z) = \sum_n \frac{\langle p|e^{-i\vec{k}.\vec{r}}|n\rangle\langle n|e^{i\vec{k}'.\vec{r}}|p'\rangle}{E_n - E(p) - z} \tag{18}$$

obtained from the first term of the right hand side of (12) by replacing $\hbar\omega - i\varepsilon$ by the complex variable z. One has rigorously:

$$I(z) = -\frac{1}{z}\sum_n \langle p|e^{-i\vec{k}.\vec{r}}|n\rangle\langle n|e^{i\vec{k}'.\vec{r}}|p'\rangle\left(1+\frac{1}{z}\,\frac{E_n - E(p)}{1-(E_n-E(p))/z}\right)$$

$$= -\frac{1}{z}\langle p|e^{i(\vec{k}'-\vec{k}).\vec{r}}|p'\rangle + J(z) \tag{19}$$

with

$$J(z) = -\frac{1}{z^2}\sum_n \frac{\langle p|e^{-i\vec{k}.\vec{r}}|n\rangle\langle n|e^{i\vec{k}'.\vec{r}}|p'\rangle\,(E_n - E(p))}{1-(E_n - E(p))/z} \tag{20}$$

But :

$$|J(z)| \leqslant \frac{1}{z^2}\sum_n \frac{\left|\langle p|e^{-i\vec{k}.\vec{r}}|n\rangle\langle n|e^{i\vec{k}'.\vec{r}}|p'\rangle(E_n - E(p))\right|}{\sqrt{1+\left|\frac{E_n-E(p)}{z}\right| - 2\left|\frac{E_n - E(p)}{z}\right|\cos\varphi}} \tag{21}$$

where φ is the argument of $\dfrac{E_n - E(p)}{z}$

As the denominator has $|\sin\varphi|$ for lower bound, one can write

$$|J(z)| \leqslant \frac{1}{|z|^2|\sin\varphi|}\sum_n \left|\langle p|e^{-i\vec{k}.\vec{r}}|n\rangle\langle n|e^{i\vec{k}'.\vec{r}}|p'\rangle\,(E_n - E(p))\right| \tag{22}$$

so that the assertion cited above is correct if

$$\langle p|e^{i(\vec{k}'-\vec{k}).\vec{r}}|p'\rangle \tag{23}$$

and

$$\sum_n \left| \langle p | e^{-i\vec{k}.\vec{r}} | n \rangle \langle n | e^{i\vec{k}'.\vec{r}} | p' \rangle (E_n - E(p)) \right| \qquad (24)$$

are finite quantities, what we will here assume without proof.

SMALL OSCILLATION APPROXIMATION

Let us now suppose that the amplitude of oscillation of the electron around its mean position is small :

$$e^{i\vec{k}.\vec{r}} = e^{i\vec{k}.\langle\vec{r}\rangle} \left(1 + i\vec{k}.\Delta\vec{r} + \dots\right) \qquad (25)$$

This is of course a rough approximation which should be understood as a first step towards the solution of the Low equations. Indeed the quantum-mechanical cut-off in the interaction occurs for $\vec{k}$-vector such that

$$k \sqrt{\langle \Delta r^2 \rangle} \simeq 1 \qquad (26)$$

However, as this approximation gives a simple example of the path to follow in solving the equations, we will describe it in details here. With the approximation (25), the equations reduce to :

$$\frac{-i}{k_z V_k^* e^{-i\vec{k}.\langle\vec{r}\rangle}} \langle p; \vec{k}_- | \Delta z | p' \rangle =$$

$$\langle p | \Delta z \frac{1}{H - E(p) - \omega - i\varepsilon} \Delta z | p' \rangle + \langle p | \Delta z \frac{1}{H - E(p') + \omega} \Delta z | p' \rangle \qquad (27)$$

It must be noticed that the operators $e^{i\vec{k}.\vec{r}}$ and Δz , which is the z – component of $\vec{r} - \langle \vec{r} \rangle$, modify the total momentum of the states under consideration in different ways. As a result, the conservation of total momentum for free polarons has been lost in performing the small oscillation approximation. This means that this approximation is valid when the recoil of the polaron during a collision with phonons is negligeable. This is the case at strong or intermediate coupling strengths where the polaron effective mass has a relatively high value*, but not at weak coupling.

From now on, our study will be restricted to energies below the threshold of production of a second phonon and to the case where p = p'.

Let us then write :

$$\frac{-i}{k_z V_k^* e^{-i\vec{k}\langle\vec{r}\rangle}} \langle p ; \vec{k}_- | \Delta z | p \rangle = f(\omega) \tag{28}$$

As we neglect the recoil of the polaron, we can suppose that it ever remains in the same state $|p\rangle$, if it is a free polaron. The case of bound polarons will be discussed later. With this approximation the optical theorem is simply obtained in taking the imaginary part of the Low equation (27). This leads to

$$\operatorname{Im} f(\omega) = \frac{v}{6\pi\hbar} k^4 |V_k|^2 \frac{dk}{d\omega} |f(\omega)|^2 \tag{29}$$

---

* see ref. (2) p. 100.

Let us now proceed further in the approximations. We assume that the contribution of states with more than one phonon to the first term of the right hand side of (27) is negligeable. In the denominator of the second term, we replace the hamiltonian H by a mean energy $\bar{E}$. This parameter can be determined by means of a variational procedure similar to the one developed by Kirkwood [12] in atomic physics. This leads to the following equation :

$$f(\omega) = \frac{v}{6\pi^2} \int_{\omega_0}^{\infty} d\omega' k'^4 |V_{k'}|^2 \frac{dk'}{d\omega'} \frac{|f(\omega')|^2}{\hbar\omega' - \hbar\omega - i\varepsilon} + \frac{\mu_0}{\hbar\omega + \bar{E}} \tag{30}$$

with :

$$\bar{E} = \frac{\mu_1}{\mu_0} \tag{31}$$

$$\mu_0 = \langle p | \Delta z^2 | p \rangle \tag{32}$$

$$\mu_1 = \langle p | \Delta z (H - E(p)) \Delta z | p \rangle \tag{33}$$

These parameters have a simple physical meaning : μ_0 is the square of the extension of the wave packet representing the electron. Its value varies from $\hbar/2m\omega_0$ at weak coupling to $\hbar/2m\Omega$ at strong coupling, where Ω is the Landau-Pekar frequency of internal polaron oscillation. The quantity μ_1 can be evaluated as follows :

$$\mu_1 = \langle p | \Delta z [H, \Delta z] | p \rangle = \langle p | [\Delta z, H] \Delta z | p \rangle$$

$$= \frac{1}{2} \langle p | [\Delta z, [H, \Delta z]] | p \rangle \tag{34}$$

But :

$$[H, \Delta z] = [H, z] = -i\hbar \frac{p_z}{m} \qquad (35)$$

$$[\Delta z, [H, \Delta z]] = \frac{\hbar^2}{m} \qquad (36)$$

So that

$$\mu_1 = \frac{\hbar^2}{2m} \qquad (37)$$

and $\overline{E}$ lies between $\hbar\omega_0$ and $\hbar\Omega$, depending on the coupling strength. At strong coupling, it is equal to the quantum of internal polaron oscillation $\hbar\Omega$, as expected.

In (30) $\hbar\omega$ now denotes the total energy of the state containing a polaron and a phonon. Let us suppose that the total momentum is zero. Then, one has :

$$\omega = \omega_0 + \frac{\hbar^2 k^2}{2m^*} \qquad (38)$$

if the dispersion of the phonon frequency is neglected. In (38), m^* is the effective polaron mass.

Let us then consider the analytic continuation of (30) obtained in replacing $\omega + i\varepsilon$ by the complex variable Z. After use of the optical theorem expressed by (29), this leads to

$$f(z) = \frac{1}{\pi}\int_{\omega_0}^{\infty} d\omega \, \frac{\operatorname{Im} f(\omega)}{\omega - z} + \frac{\mu_0}{\hbar z + \overline{E}} \qquad (39)$$

As $f(\omega)$ is a real function, its imaginary part is equal to :

$$\mathrm{Im}\, f(\omega) = \frac{1}{2i}\left\{ f(\omega + i\varepsilon) - f(\omega - i\varepsilon) \right\} \tag{40}$$

Since we assumed that $f(z)$ goes as Z^{-1} when $Z \to \infty$, we can rewrite (39) as :

$$f(z) = \frac{1}{2i\pi}\int_C dz' \frac{f(z')}{z' - z} + \frac{\mu_o}{\hbar z + \bar{E}} \tag{40}$$

where the contour C is the following :

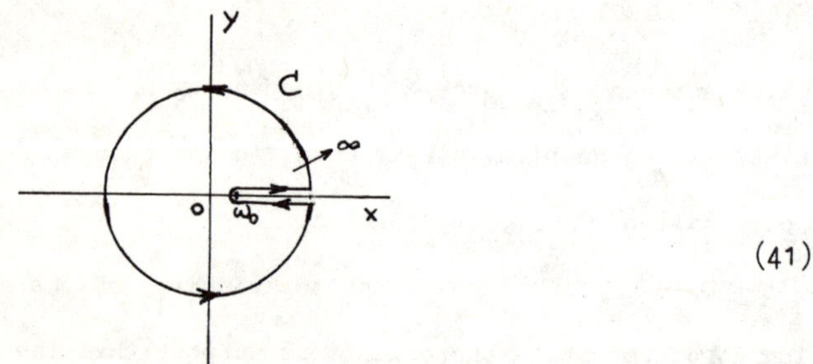

(41)

It is clear that this relation expresses the analytic properties of $f(z)$: $f(z)$ has a cut on the real axis from ω_o to ∞, with a discontinuity whose value is given by the optical theorem, and a pole at $z = -\bar{E}/\hbar$ whose residue is $\mu_o/\hbar$. With the knowledge of these properties and of the asymptotic behavior, the solution is readily built up. It is :

$$f(z) = \frac{1}{(\hbar z + \bar{E})\left(a - \frac{v}{6\pi^2}\int_{\omega_o}^{\infty} d\omega\, k^4 |V_k|^2 \frac{dk}{d\omega} \frac{1}{(\hbar\omega + \bar{E})(\hbar\omega - \hbar z)}\right)} \tag{42}$$

with

$$a = \frac{1}{\mu_0} + \frac{v}{6\pi^2} \int_{\omega_0}^{\infty} d\omega \, \frac{k^4 |V_{kl}|^2 \frac{dk}{d\omega}}{(\hbar\omega + \bar{E})^2} \tag{43}$$

The numerical study of $f(z)$ as a function of the different parameters is in progress at the present time. It is however already clear from (42) that the cross section (or the probability of transition induced by electromagnetic radiation) can present a resonance for some values of the parameters.

BOUND POLARONS

Similar methods can be used to solve the Low equations in the case of bound polarons. As there may be several bound states, one is led to a system of coupled equations (one for each bound state). The calculations are therefore more complicate, even if the basic principles are similar, and they fall outside the scope of the present paper. As a matter of fact, our purpose was here to give a general idea of how the dispersion relations can be used in the problem of the absorption of electromagnetic radiation by solids and not to go deeply in the details of the calculations.

CONCLUSIONS

The main interest of the dispersion relation approach described in this paper is that it does not rest on the preliminary calculation of wave-functions. However, it is too early to conclude whether it leads to a satisfactory solution of the problem of the absorption of electromagnetic radiation by free or bound polarons.

To be able to conclude definitively we have to obtain better solutions of the Low equations. For this purpose we can or solve them by successive iterations, starting with the rough solution described here, or use variational principles, as the Lippmann-Schwinger ones. We are now studying the different possibilities.

REFERENCES

1. C.G. Kuper, G.D. Whitfield : Polarons and Excitons, Oliver and Boyd (1963).

2. T.D. Schultz : Electron-lattice interactions in polar crystals. M.I.T. Solid State and Molecular Theory Group Technical Report n° 9 (1958).

3. R. Evrard, Ph.D. Thesis, Liège (1964)
Phys.Letters, 14, 295 (1965)

4. J. Devreese, R. Evrard, Phys. Letters 11, 278 (1964).

5. E. Kartheuser, Ph.D. Thesis, Liège (1968).

6. E. Kartheuser, R. Evrard, J. Devreese, Phys. Rev. Letters 22, 94 (1969).

7. J. Devreese, R. Evrard, E. Kartheuser, Solid State Communications 7, 767 (1969).

8. T.D. Lee, F.E. Low, D. Pines, Phys. Rev. 90, 297 (1953).

9. G.F. Chew, F.E. Low, Phys. Rev. 101, 1570 (1956).

10. G.F. Chew, F.E. Low, Phys. Rev. 101, 1579 (1956).

11. P. Roman : Advanced Quantum Theory, Addison-Wesley (1965).

12. J.G. Kirkwood, Z. Phys. 33, 57 (1932).

XVIII

OPTICAL ABSORPTION LINE SHAPE AND INTERNAL STRUCTURE OF FRÖHLICH CONTINUUM POLARONS

E. KARTHEUSER et R. EVRARD*

Institut de Physique, Physique Théorique,
Université de Liège, Sart-Tilman, Liège,
Belgium

J. DEVREESE

Faculty of Science, University of Antwerp, Belgium,
and Solid State Physics Department, Studiecentrum
voor Kernenergie, Mol-Donk, Belgium

The ionic polarization plays an important role in the optical properties of ionic crystals. We will analyze here the effect of this polarization in the case of absorption of light by free charge carriers in ionic crystals. This study is interesting for two reasons :

1°) From the experimental point of view it enables one to obtain direct information about the internal structure of free charge carrier in ionic crystals.

2°) From the theoretical point of view this study might confirm the predicted internal excitations of polarons.

As is well known the quanta of the quasi-harmonic vibrations of the lattice are called "phonons". In most polar crystals, the interaction between a slow conduction electron (or hole) and the longitudinal optical phonons is very important. The strength of this interaction can be characterized by a dimensionless coupling constant:

*Chercheur qualifié du Fonds National Belge de la Recherche Scientifique.

$$\alpha = \frac{1}{2} \frac{\frac{e^2}{\left(\frac{\hbar}{2m\omega}\right)^{1/2}}}{\hbar\omega} \left(\frac{1}{\varepsilon_\infty} - \frac{1}{\varepsilon_s} \right) \tag{1a}$$

where ε_s is the static dielectric constant, ε_∞ the electronic dielectric constant and ω the frequency of the longitudinal optical phonons at wave vector $k=0$.

α can be considered as the energy (expressed in unit $\hbar\omega$) of a particle with a dimension given by $\left(\frac{\hbar}{2m\omega}\right)^{1/2}$, moving in a dielectric. This constant takes values between 10^{-2} and 6 in polar crystals. The small and large α values correspond respectively to semi-conductors of the III-V components and to the alkali-halides. In the latter case α is two orders of magnitude larger than the electron-photon interaction constant

$$\alpha = \frac{e^2}{\hbar c} \tag{1b}$$

Consequently, the electron-phonon interaction modifies considerably the motion of the conduction electron:

By Coulomb interaction, the electron induces a polarization in the lattice which can extend over a certain number of lattice parameters. This polarization produces in turn an electric field which reacts on the electron, so that the electron moves through the crystal, taking along a cloud of virtual phonons. This complex system consisting of the electron and the polarization can be thought of as a quasi-particle which is called a "polaron".

Optical Absorption Line Shape Frohlich Continuum Polarons

If the extension (radius) of the polaron is much larger than the lattice constant, the atomicity of the lattice does not have to be taken into account and one may consider the crystal as a vibrating continuum. This is the case of the large polaron which we consider in the present investigation.

The hamiltonian describing the continuum polaron has been derived by Fröhlich [1]. With the notations of Lee, Low, Pines [2], this hamiltonian takes the form :

$$H = \frac{p^2}{2m} + \int d\vec{k}\, \hbar\omega\, a^{+}(\vec{k})\, a(\vec{k}) + \int d\vec{k} \{ V(k)\, a(\vec{k})\, e^{i\vec{k}\vec{r}} + c.c. \} \quad , \qquad (2)$$

where $\vec{r}$ and $\vec{p}$ are the position and the momentum of the electron and m is its bare mass (also called band mass). The operators $a(\vec{k})$ and $a^{+}(\vec{k})$ are the annihilation and creation operators of L.O. phonons with wave vector $\vec{k}$ and frequency ω. The Fourier coefficients $V(\vec{k})$ of the hamiltonian are given by

$$|V(k)|^2 = \frac{\alpha}{2\pi^2} \left(\frac{\hbar}{2m\omega} \right)^{1/2} \left(\frac{\hbar\omega}{k} \right)^2 \qquad (3)$$

For the present study, the most interesting range of α values is that for which $\alpha > 3$, i.e. the intermediate or strong coupling region :

If the coupling constant takes large values, the polarization induced by the conduction electron becomes very important. This gives rise to an appreciable "restoring force" acting on the electron, which results in internal oscillations of the electron. In the limit of strong coupling the translational motion of the polaron can be neglected in many situations. A good approximation is to suppose that the electron undergoes only oscillations in the static potential well it induces itself in the lattice. In this strong coupling region, the frequency of oscillation of the electron is much larger than the optical phonon frequency, so that the polarization field cannot follow the instantaneous motion of the electron and can only adapt to the mean charge distribution of the electron. This enables us to use the Landau-Pekar's [3] adiabatic harmonic approximation. The wave function of the polaron in this approximation can be written as a product of electronic $\varphi(\vec{r})$ and field contributions Φ :

$$\Psi = \varphi(\vec{r}) \cdot \Phi \qquad , \tag{4}$$

where the electronic part is a harmonic oscillator wave function.

The properties of the groundstate have been studied by Landau-Pekar [3,4]. From a variational method they obtain for the energy E_0 the oscillation frequency Ω_0 of the electron and the polaron mass m_0^* the results listed in the first line of table I :

Fundamental state			
	$E_0/\hbar\omega$	Ω_0/ω	m_0^*/m
Pekar (3)	$-\frac{\alpha^2}{3\pi}$	$\frac{4}{9}\frac{\alpha^2}{\pi}$	$16\frac{\alpha^4}{81\pi^2}$
FRANCK-CONDON like state			
	$E_{FC}/\hbar\omega$	Ω_{FC}/ω	m_{FC}^*/m
Pekar (3)	$\frac{E_0}{\hbar\omega}+\frac{\Omega_0}{\omega}$	$\frac{4}{9}\frac{\alpha^2}{\pi}$	$16\frac{\alpha^4}{81\pi^2}$
First relaxed excited state "RES"			
	$E_1/\hbar\omega$	Ω_1/ω	m_1^*/m
Pekar (3)-Schultz (5)	$-0{,}0422\ \alpha^2$	$10{,}67\ 10^{-2}\ \frac{\alpha^2}{\pi}$	–
Evrard (6)	$-0{,}0358\ \alpha^2$	$9{,}3\ 10^{-2}\ \frac{\alpha^2}{\pi}$	$0{,}625\ \frac{\alpha^4}{81\pi^2}$
Kartheuser (7)	$-0{,}0425\ \alpha^2$	$9{,}14\ 10^{-2}\ \frac{\alpha^2}{\pi}$	$1{,}205\ \frac{\alpha^4}{81\pi^2}$

Table I : Properties of the internal structure of the polaron.

The polaron wave function in the groundstate is given by :

$$\Psi_0 = \varphi_0(\vec{r}, \Omega_0) \cdot \Phi_0(\rho_0) \tag{5a}$$

where the electronic part is :

$$\varphi_0(\vec{r}, \Omega_0) = \left(\frac{m\Omega_0}{\hbar}\right)^{3/4} \pi^{-3/4} \exp\left[-\frac{m\Omega_0}{\hbar}\frac{r^2}{2}\right] \tag{5b}$$

and the field part :

$$\Phi_0(\rho_0) = S_0 |0\rangle \tag{5c}$$

Here $|0\rangle$ is the vacuum state and the operator S_0 can be expressed in terms of annihilation and creation operators $a(\vec{k})$, $a^{+}(\vec{k})$:

$$S_0 = \exp\left[\int d\vec{k} \left\{ \frac{V^*(k)\rho_0^*(k)}{e\hbar\omega} a^{+}(\vec{k}) - \frac{V(k)\rho_0(k)}{e\hbar\omega} a(\vec{k}) \right\}\right] \tag{5d}$$

The expression :

$$\rho_0(k) = e \cdot \exp\left[-\frac{\hbar}{2m\Omega_0} \cdot \frac{k^2}{2}\right] \tag{5e}$$

is the Fourier transform of the electronic charge distribution in the grounastate.

More recently the possibility of internal excitations has been treated by different authors (5-7). The main conclusion of these theoretical investigations is that there exist certainly two types of internal excitations at strong coupling strength.

1°- The states where the electron is excited in the groundstate potential which we call "FRANK-CONDON" states (because of the analogy of these excitations with those occurring in molecules). During the excitation of the electron towards those states, the ionic polarization does not relax and remains adapted to the initial electronic charge configuration. The properties of this state are listed in the second line at table I.

The wave function of this state is given by the following expressions :

$$\Psi_{FC} = \varphi_{FC}(\vec{r}, \Omega_0) \cdot \Phi_0(\rho_0) \tag{6a}$$

where the electronic part is now a p state :

$$\varphi_{FC}(\vec{r}, \Omega_0) = \left(\frac{m\Omega_0}{\hbar}\right)^{3/4} \pi^{-3/4} \frac{z}{\left(\frac{\hbar}{2m\Omega_0}\right)^{1/2}} \exp\left[-\frac{m\Omega_0}{\hbar}\frac{r^2}{2}\right] \tag{6b}$$

2°- A self-consistent excited state which we denote as "R.E.S." (relaxed excited state). This is a state where the potential well of the lattice polarization is adapted to the new electronic charge configuration. The properties E_1, Ω_1 and m^*_1 of this state have been evaluated by different authors [5-7]. The results of this various investigations are listed on the three last lines of table I. The wave function of the "RES" state has the following form :

$$\Psi_1 = \varphi_1(\vec{r}, \Omega_1) \cdot \Phi_1(\rho_1) \tag{7a}$$

with :

$$\varphi_1(\vec{r}, \Omega_1) = \left(\frac{m\Omega_1}{\hbar}\right)^{3/4} \pi^{-3/4} \frac{z}{\left(\frac{\hbar}{2m\Omega_1}\right)^{1/2}} \exp\left[-\frac{m\Omega_1}{\hbar} \cdot \frac{r^2}{2}\right] \tag{7b}$$

$$\Phi_1(\rho_1) = S_1 |0\rangle \tag{7c}$$

and

$$S_1 = \exp\left[\int d\vec{k} \left\{ \frac{V^*(k)\, \rho_1^*(\vec{k})}{e\hbar\omega} a^+(\vec{k}) - \frac{V(k)\, \rho_1(\vec{k})}{e\hbar\omega} a(\vec{k}) \right\}\right] \tag{7d}$$

In this case, the Fourier transform of the electronic charge density depends on the direction of the wave vector $\vec{k}$ in the following way :

$$\rho_1(\vec{k}) = e\left\{1 - \frac{\cos^2\theta}{\frac{2m\Omega_1}{\hbar}} k^2\right\} \cdot \exp\left[-\frac{\hbar}{2m\Omega_1} \frac{k^2}{2}\right] \tag{7e}$$

where θ is the angle between the z axis and the direction of $\vec{k}$.

Until now only qualitative results about these internal excitations at intermediate coupling are available. Therefore we restrict ourselves to the strong coupling region in the present study. Nevertheless it can be expected that some specific properties obtained at strong coupling strength, especially those related to internal excitations, "survive" at intermediate coupling. One of us (J. Devreese) has also illustrated this point by means of calculations on an exactly soluble polaron model [8].

The main purpose of this work is to find a mechanism which reveals the existence of the internal polaron structure discussed above. We shall show in the following that this can be done by absorption of light by free polarons.

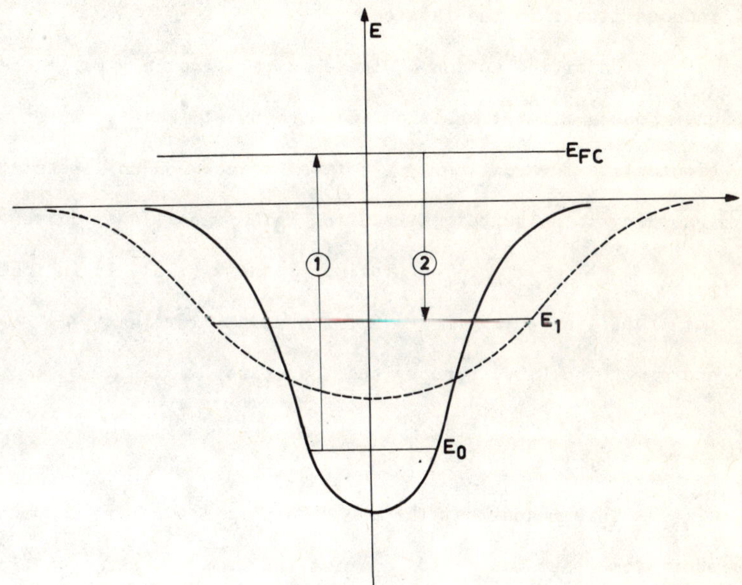

Fig. I : Internal excited states of the polaron at strong coupling

E_{FC} : FRANCK-CONDON like state

E_1 : RELAXED EXCITED STATE

From the physical point of view this absorption process can be easily understood by a transition in two steps :

Initially the electron stays at the groundstate level E_o; oscillating in a potential well due to the static polarization it induces itself in the lattice.

Under the action of the electro-magnetic wave, the electron undergoes a first transition (denoted by (1) in fig. I) from the groundstate towards the Ψ_{FC} state (Franck-Condon like state). According to Heisenberg's relation this transition requires a time $\tau_{FC} \sim \hbar/(E_{FC}-E_o)$. The previous results (table I) indicate that the transition energy $E_{FC} - E_o$ is much smaller than $\hbar\omega$ (the energy of the L.O. phonons) and therefore

$$\tau_{FC} \ll \frac{1}{\omega}$$

This means that the photon absorption occurs in such a short time that the ionic polarization remains unchanged because of the inertia of the ions. This transition is quite similar to the optical transition arising in molecules which is usually interpreted by the "Franck-Condon" rule. This "FC" state of the polaron is highly unstable, it decays in a time of the order of $\frac{1}{\omega}$ (the relaxation time of the lattice). Nevertheless it is always possible to expand the "FC" state in terms of more stable states [(o)] the "R.E.S." :

$$\Psi_{FC} = \Psi_1 + \sum_k c_k \Psi_{1;1_k} + \sum_{k,k'} c_{kk'} \Psi_{1;1_k;1_{k'}} + \qquad (8a)$$

[(o)] We will demonstrate this point in the following.

where
$$\Psi_{1;1_k} = \varphi_1(\vec{r}, \Omega_1) \cdot S_1 |\vec{k}\rangle \tag{8b}$$

$$|\vec{k}\rangle = a^{\dagger}(\vec{k}) |0\rangle \tag{8c}$$

is the "R.E.S." state with in addition one phonon of wave vector $\vec{k}$.

Another way of looking at this is to say that the incident light creates a wave packet Ψ_{FC} having no well defined energy. This wave packet deforms itself and after a time $\frac{1}{\omega}$ tends towards the "R.E.S." state accompanied by several phonons at large distance from the polaron. This decay constitutes the second step (denoted by (2) in fig. I) of the absorption mechanism. In a time of the order of $\frac{1}{\omega}$ the lattice polarization relaxes and adapts itself to the new $(\rho_1(\vec{k}))$ charge distribution of the electron. This leads to a decrease of energy and we finally find the "R.E.S." state.

From the previous discussion we can expect an absorption line shape with a strong and sharp peak at the energy level E_1 (if this state is stable) corresponding to a direct transition from the Ψ_0 state towards the Ψ_1 state. On the high photon energy side, this peak can be followed by one side band (in the case of a decay of the Ψ_{FC} state with emission of one phonon) or by more side bands. The whole structure presumably is a large asymetrical band starting at $\hbar\omega$ above the energy level E_1.

Before going over on to the quantitative evaluation of the transition probabilities due to the absorption of light, it is very important to realise that the Ψ_1 state is really a more stable one than the "FE" like state (as it has been supposed in previous discussions).

This can be easily seen with an intuitive argumentation based on the conservation of energy and momentum required during the decay of the "R.E.S." state.

For instance, we suppose that the "R.E.S." state decays towards the ground state with emission of one phonon.

From table I it follows that for we have the following relation :

$$E_1 \leq E_0 + 2\hbar\omega \tag{9}$$

Therefore at lower α values $\alpha \leq 6$ the transition between the R.E.S. state and the groundstate is accompanied by emission of one phonon at most.

If we suppose that the particle is initially at rest, it will get a momentum $-\hbar\vec{k}$ after the transition. Moreover the energy conservation leads to the relation

$$E_1 = E_0 + \frac{\hbar^2 k^2}{2m_0^*} + \hbar\omega \tag{10}$$

It follows that

$$k \sim \left(\frac{2m_0^* \omega}{\hbar} \right)^{1/2} \tag{11}$$

Substituting the polaron mass, we finally find that the vector defined in this way is greater than $\left(\frac{2m\Omega_0}{\hbar} \right)^{1/2}$, the quantum k cutt-off value at strong coupling. Therefore we can conclude that the transition between the Ψ_1 state towards the Ψ_0

state becomes highly improbable because after the cutt-off value the electron-phonon interaction falls rapidly of. This result can be confirmed by a quantitative calculation starting from Wigner-Weisskopf line width theory which gives the following expression for the half line width :

$$\frac{\gamma}{2} = \frac{2\pi}{\hbar\omega} \sum_i |<\Psi_1 | V | \Psi_i >|^2 \delta(E_1 - E_i) \tag{12}$$

where Ψ_1 and E_1 design respectively the wave function and the energy corresponding to the initial state and Ψ_i , E_i represent the same quantities for the final states. A complete set of these states may be described by the ground state Ψ_0 and this state with in addition one or more phonons $\Psi_{0,1_k}$, ...

Here the interaction V means the perturbation term inducing the decay of the "R.E.S.". The existence of this perturbation can be easily understood by the following considerations :

The function Ψ_0 , Ψ_1 , ... are correctly speaking not exact eigenstates of Fröhlich's polaron hamiltonian (2). In fact, these functions are rather eigenstates of the hamiltonian $\mathcal{H}$ corresponding to Pekar's harmonic adiabatic approximation.

Consequently, the interaction V can be defined by :

$$H = \mathcal{H} + V \tag{13}$$

The adiabatic hamiltonian $\mathcal{H}$ can easily be obtained from Pekar's results [3], using the parity properties of the Ψ_0 and Ψ_1 states [7]. We find immediately that the half line-width is given

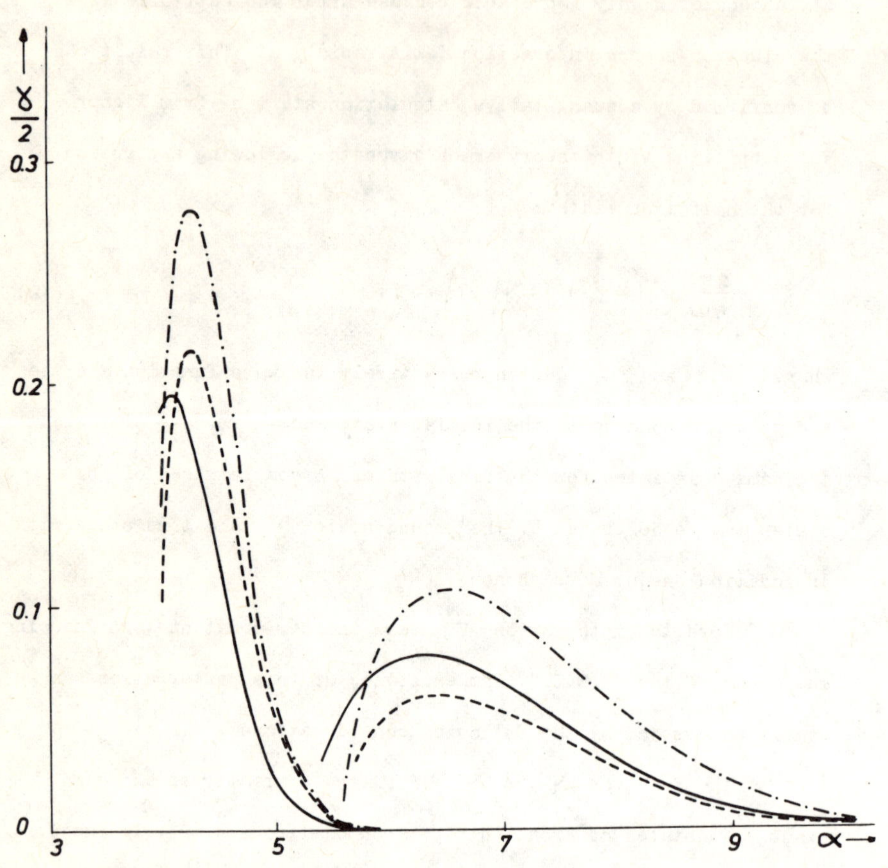

Figure II : Line width of the "R.E.S." as function of the coupling constant

——— : Evrards results on table I

-·-·- : Schultz " " "

----- : Kartheusers " "

by

$$\frac{\gamma}{2} = \frac{2\pi}{\hbar\omega} \sum_i |\langle \Psi_1 | \{ \int d\vec{k}\, V(k)\, a(\vec{k})\, e^{i\vec{k}\vec{r}} + c.c. \} | \Psi_i \rangle|^2 \, \delta(E_1 - E_i) \qquad (14)$$

Owing to the conservation relation of energy and momentum (presehce of the delta function), the contribution due to a transition without emission of phonons cancels.

Again, the part resulting from a decay of "R.E.S." towards the groundstate accompanied by emission of one phonon can be written :

$$\frac{\gamma(1)}{2} = \frac{2\pi}{\hbar\omega} \int d\vec{k}_1 |\langle \Psi_{0,1_{k_1}} | \{ \int d\vec{k} \{ V(k)\, e^{i\vec{k}\vec{r}}\, a(\vec{k}) + c.c. \} | \Psi_1 \rangle|^2 \, \delta[E_1 - (E_0 + \hbar\omega + \frac{\hbar^2 k^2}{2m_0^*})] \qquad (15a)$$

Using the commutation relations of phonon operators $a^\dagger(\vec{k})$ and $a(\vec{k})$ we obtain after integration over k space :

$$\frac{\gamma(1)}{2} = \frac{64}{3} \frac{(\Omega_0 \Omega_1)^{3/2}}{(\Omega_0 + \Omega_1)^5}\, \omega . \Omega_1 \left(\frac{m_0^*}{m} \right)^{3/2} \left(\frac{\Delta\omega_1}{\omega} \right)^{1/2} \exp\left[- \frac{2m_0^*}{m} \frac{\Delta\omega_1}{\Omega_0 + \Omega_1} \right] \qquad (15b)$$

This gives rise to the peak in fig. II. which starts at about $\alpha = 4$ (due to the presence of the delta function) and becomes very small at $\alpha > 6$

For this high coupling range it is necessary to evaluate the contribution to $\frac{\gamma}{2}$ due to a decay of "R.E.S." towards the groundstate with emission of two phonons. This leads to the following expression :

$$\frac{\gamma(2)}{2} = \frac{2\pi}{\hbar\omega} \iint d\vec{k}_1 d\vec{k}_2 \left| \langle \Psi_{0,1_{k_1},1_{k_2}} \right| \int d\vec{k} \{ V(k) e^{i\vec{k}\vec{r}} a(\vec{k}) + c.c. \} \left| \Psi_1 \rangle \right|^2 \delta\left[E_1 - \left(E_0 + 2\hbar\omega + \frac{\hbar^2 (\vec{k}_1 + \vec{k}_2)^2}{2m_0^*} \right) \right] \quad (16a)$$

Here we have performed an analytic calculation only just beyong the threshold at which two phonons are emitted. We obtain the expression :

$$\frac{\gamma(2)}{2} = 2^6 \frac{\alpha^2}{\sqrt{\pi}} \left(\frac{m_0^*}{m} \right)^{3/2} \left(\frac{\Delta\omega_2}{\omega} \right)^{1/2} e^{-\varkappa\alpha^2} \left\{ \frac{3}{7} \left(\frac{\Omega_0 + \Omega_1}{\Omega_0 + 2\Omega_1} \right)^{5/2} + \frac{8}{5} \left(\frac{\Omega_0^2 + \Omega_0\Omega_1}{8\Omega_0\Omega_1 + 2\Omega_0^2 + 2\Omega_1^2} \right)^{3/2} - \frac{1}{5} \left(\frac{\Omega_0 + \Omega_1}{\Omega_0 + 2\Omega_1} \right)^{3/2} \right\} \quad (16b)$$

which holds only for small values of

$$\frac{\Delta\omega_2}{\omega} = \left[\frac{E_1}{\hbar\omega} - \frac{E_0}{\hbar\omega} - 2 \right] \quad (16c)$$

The parameter $\varkappa$ is given on table II.

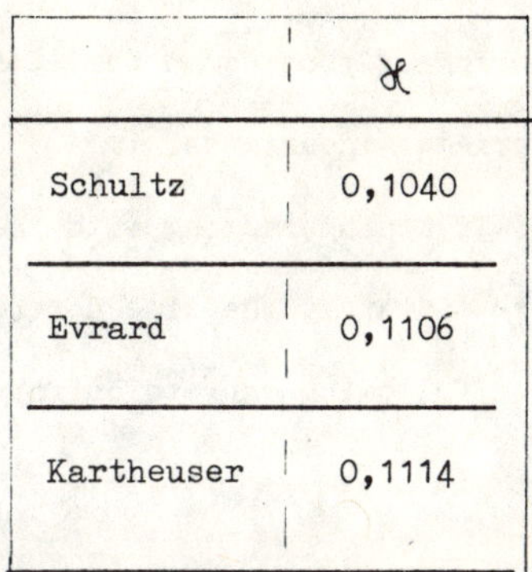

	χ
Schultz	0,1040
Evrard	0,1106
Kartheuser	0,1114

Table II :

values of the parameter χ obtained from Schultz, Evrard's or Kartheuser's results given in table I.

As is seen on fig. II., this contribution starts at about $\alpha = 6$

We can conclude from these calculations that

1°) the "R.E.S." is much more stable than the "F.C." state;

2°) the life time has an oscillating behaviour as a function of the coupling constant α .

The stability of the "R.E.S." is of great importance in order to be able to apply perturbation theory to the absorption of light. In fact from the formal point of view this theory postulates that the interaction of light can only produce transitions between eigenstates of the unperturbed hamiltonian. Therefore it may be a good approximation to assume that the Ψ_0, Ψ_1 states and the states describing scattering of phonons on the latter, would form a complete set of appropriate eigenstates.

The intensity of light interacting with the system is usually small, so that we can apply the time dependant perturbation theory which leads to the following expression for the transition probability per unit time :

$$W = \frac{2\pi}{\hbar} \sum_f |\langle \Psi_f | H_1 | \Psi_0 \rangle|^2 \, \delta[E_f - E_0 - \hbar\Omega] \tag{17a}$$

where H_1 is the external perturbation and Ω the frequency of the incident light.

The electromagnetic wave may be described as a transverse monochromatic plane wave. Their interaction with the longitudinal optical lattice modes can therefore be neglected. Moreover in the case of slow conduction electrons, we are in position to omit the influence of the magnetic component. Finally, since the incident wave length is much greater than the dimension of the scattered system (the polaron), H_1 is the electric dipolar interaction between the electron and the external electric field which, if we assume propagation in the direction, becomes :

$$H_1 = e\, z \,.\, \mathcal{E} \tag{17b}$$

with

$$\mathcal{E} = \mathcal{E}_m (e^{i\Omega t} + e^{-i\Omega t}) \tag{17c}$$

At first examination, it appears not clearly how the expression (17a) has some connection with our physical description of absorption process. In fact this connection really exists because it is always possible to introduce in this expression a complete set of intermediate states (the "F.C." state and this state with in addition one or more phonons). Here we recall that in our adiabatic approximation the electronic part is given by harmonic functions. As a result of this, the selection rules and orthonormality properties lead to cancellation of all terms in the summation over intermediate states with the exception of one, the "F.C." state contribution.

Another way of looking at this is simply to consider the state as a superposition of the final relaxed excited states as assumed in previous discussions.

We are now in position to see how the indirect transition is clearly revealed in the expression (17a). Indeed we can now write this expression as follows :

$$W = 2\pi^2 \cdot \frac{e^2}{\hbar c} I_0(\Omega) \sum_f | \langle \Psi_f | \Psi_{FC} \rangle \langle \Psi_{FC} | z | \Psi_0 \rangle |^2 \; \delta[E_f - E_0 - \hbar\Omega] \quad (17d)$$

in terms of the incident light intensity $I_0(\Omega)$.

From this it can be seen that, first the optical transition excites the electron to a "FC" state (matrix element of z), second the "FC" state decays towards the "R.E.S." state or towards this state with in addition one or more phonons.

The calculations are straightforward, we do not give here the mathematical details (which will be published elsewhere [9]). We only outline the main results :

The first contribution of expression (17^d) corresponds to a direct transition towards the Ψ_1 state. Because of the high stability of this state this leads to a sharp peak (a Lorentz wave characterized by the line width γ (14). This "zero phonon" contribution can be written

$$W(0) = 2\pi^2 \frac{e^2}{\hbar c} I_0(\Omega) |\langle \Psi_1 | z | \Psi_0 \rangle|^2 \frac{\gamma}{\pi \hbar\omega \left[\left(\frac{E_1}{\hbar\omega} - \frac{E_0}{\hbar\omega} - \frac{\Omega}{\omega} \right)^2 + \frac{\gamma^2}{4} \right]} \quad (18a)$$

After integration this can be written in arbitrary units of

$$P_0 = 2\pi^2 \frac{e^2}{\hbar c} I_0(\Omega) \frac{\hbar}{2m\omega} \quad (18b)$$

as :

$$\Gamma(0) = \frac{P(0)}{P_0}$$

$$= \frac{9}{4} \left(2 \frac{\sqrt{\Omega_0 \Omega_1}}{\Omega_0 + \Omega_1} \right)^5 \frac{e^{-\ell\alpha^2}}{\alpha^2} \left(\frac{\Omega}{\omega} \right) \frac{\gamma}{\left[\left(\frac{E_1}{\hbar\omega} - \frac{E_0}{\hbar\omega} - \frac{\Omega}{\omega} \right)^2 + \frac{\gamma^2}{4} \right]} \quad (18c)$$

where $P(0) = \hbar\Omega \, . \, W(0)$

is the power absorbed during this transition.

The second contribution of the summation (17d) can be called the "one phonon" transition. This means that this contribution is due to a transition from the groundstate to the "R.E.S." accompanied by emission of one phonon.

Hence we have :

$$W(1) = 2\pi^2 \frac{e^2}{\hbar c} I_0(\Omega) \int d\vec{K} \, | \langle \Psi_{1,1_K} | \Psi_{FC} \rangle \langle \Psi_{FC} | z | \Psi_0 \rangle |^2 \, \delta\left[E_1 + \hbar\omega + \frac{\hbar^2 k^2}{2m_1^*} - E_0 - \hbar\Omega \right] \qquad (19a)$$

From the conservation relations of energy and momentum it results that this part starts at   above the energy level of the "R.E.S.". We finally obtain (7) with the notations of expressions (18) :

$$\Gamma(1) = \frac{9}{4} \left(2 \frac{\sqrt{\Omega_0 \Omega_1}}{\Omega_0 + \Omega_1} \right)^5 \left(\frac{m_1^*}{m} \right)^{1/2} \left(\frac{\Delta\Omega_1}{\omega} \right)^{-1/2} \frac{e^{-\ell\alpha^2}}{\alpha} \frac{\Omega}{\omega} A(1) \qquad (19b)$$

where

$$A(1) = \Bigg[\left\{ \exp\left(-\frac{1}{2}\frac{m_1^*}{m}\frac{\Delta\Omega_1}{\Omega_0}\right) - \exp\left(-\frac{1}{2}\frac{m_1^*}{m}\frac{\Delta\Omega_1}{\Omega_1}\right) \right\}^2 + \frac{2}{3}\frac{m_1^*}{m}\frac{\Delta\Omega_1}{\Omega_1} \left\{ \exp\left(-\frac{1}{2}\frac{m_1^*}{m}\frac{\Omega_0+\Omega_1}{\Omega_0}\frac{\Delta\Omega_1}{\Omega_1}\right) - \exp\left(-\frac{m_1^*}{m}\frac{\Delta\Omega_1}{\Omega_1}\right) \right\}$$

$$+ \frac{1}{5}\left(\frac{m_1^*}{m}\right)^2 \left(\frac{\Delta\Omega_1}{\Omega_1}\right)^2 \exp\left(-\frac{m_1^*}{m}\frac{\Delta\Omega_1}{\Omega_1}\right) \Bigg] \qquad (19c)$$

and

$$\frac{\Delta\Omega_1}{\omega} = \left[\frac{\Omega}{\omega} - \left(\frac{E_1}{\hbar\omega} - \frac{E_0}{\hbar\omega} + 1 \right) \right] \qquad (19d)$$

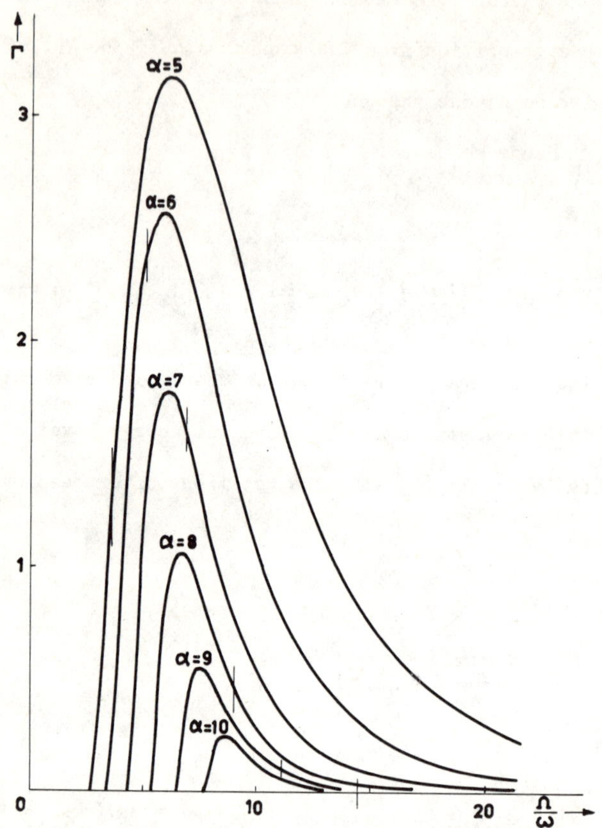

Figure III: Absorption (arbitrary units) with emission of a single phonon as a function of the photon frequency Ω and at different coupling strengths.

Figure III shows the "one phonon" contribution as a function of the frequency Ω of the incident light.

It turns out that the "one phonon" part starts at the threshold of emission of one phonon. At this point, the curves show a horizontal tangent. The maximum of the curves appears at an energy of about a few quantum $\hbar\omega$ above the one phonon emission threshold. The exponential decrease after this maximum is due to the cutt-off in the electron-phonon interaction. This leads to an asymetrical behaviour of the curves. The vertical line indicates the position of the E_{FC} energy level as calculated from table I. We remark that this level undergoes a shift to high energy values with increasing coupling constant and finally appears inside the region where the quantum cutt-off cancels the absorption rate. This explains clearly the fact that the transition probability associated with emission of more than one phonon becomes important at strohg coupling. Therefore in the strong coupling region the third contribution to the summation over final states in expression (17d) becomes important. In the same way, this "two phonon" contribution is given by

$$W(2) = 2\pi^2 \frac{e^2}{\hbar c} I_0(\Omega) \int d\vec{k}_1 d\vec{k}_2 |\langle \Psi_{1;1_{k_1};1_{k_2}} | \Psi_{FC} \rangle \langle \Psi_{FC} | z | \Psi_0 \rangle|^2 \delta\left[\frac{\hbar(\vec{k}_1 + \vec{k}_2)^2}{2m_1^* \omega} - \frac{\Delta\Omega_2}{\omega} \right] \qquad (20a)$$

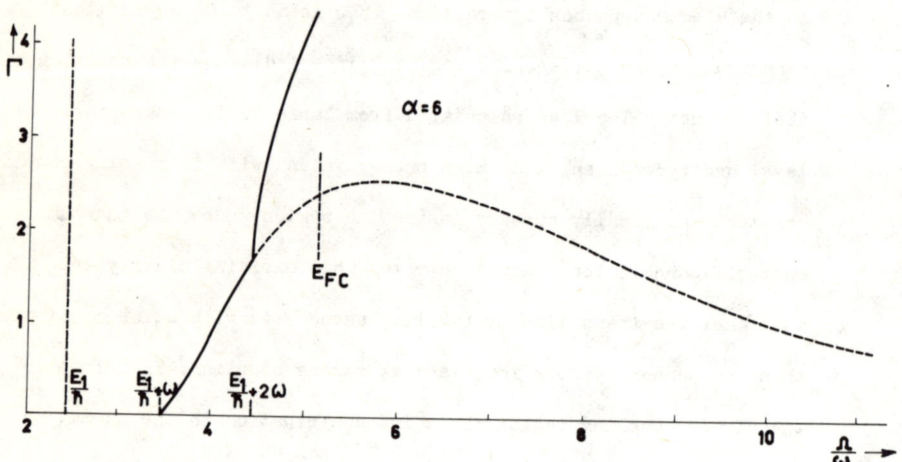

Figure IV Low energy part of the absorption (arbitrary units) by free polarons at = 6.

Here an analytical calculation is only possible near the emission threshold of two phonons, i.e. at small values of :

$$\frac{\Delta\Omega_2}{\omega} = \left[\frac{\Omega}{\omega} - \left(\frac{E_1}{\hbar\omega} - \frac{E_2}{\hbar\omega} + 2 \right) \right] \qquad \text{(20b)}$$

This leads to the expression :

$$\Gamma(2) \simeq \frac{9}{4} \left(2 \frac{\sqrt{\Omega_0 \Omega_1}}{\Omega_0 + \Omega_1} \right)^5 \left(\frac{m_1^*}{m} \right)^{3/2} \frac{e^{-\varkappa\alpha^2}}{\alpha} \left(\frac{\Delta\Omega_2}{\omega} \right)^{-1/2} \qquad \text{(20c)}$$

An exact value of this last expression can be obtained by numerical calculation. Fig. IV shows the part of the absorption curve at $\alpha = 6$.

In conclusion, the most important contributions of this work are the following :

1°) We have determined the general behaviour of the optical absorption line shape in the framework of Landau-Pekar's adiabatic approximation. These results are not directly applicable to usual continuum polaron crystals. However the general form of the structure obtained here presumably remains if the coupling constant decreases. Our results are in contradiction with the only known theoretical calculations due to Feynman and his coworkers [10]. In fact, it seems difficult to understand how this highly involved work might take into account the relaxed internal states.

2°) We have shown that the usual concept of the "FRANCK-CONDON" transition cannot be applied to the absorption of light by free polarons : Indeed, because of the instability of the "FC" state, the latter cannot be represented by a line-width contrary to the whole broad band which describes the actual "FC" state. We can look at this considering the absorption as due to a transition between the groundstate towards the "FC" state which is immediately followed by a decay of the "FC" state.

3°) The absorption curves obtained in the present work are important to the experimentalists; their observations would permit to confirm directly the existence of the internal structure predicted by different theoretical works and also confirm the whole polaron theory.

Finally, the basic ideas developped in these investigations can be extended to the interpretation of the absorption and emission spectrum in the case of bound polarons [11] (colour centers or excitons). Here [12] only the polarization effect has been analyzed, whereas in the case of excitons or colour centers, the binding effect should be taken into account.

References

1. Fröhlich H., Pelzer H. and Zienau S., Phil. Mag. 41, 221 (1950).
2. Lee T.D., Low F.E. and Pines D., Phys. Rev. 90, 297 (1953).
3. Pekar S.I., "Untersuchungen über die Elektronentheorie der Kristalle", Akademie-Verlag Berlin (1954). Summarizing the Russian work up to 1956. A book with much physical intuition. No second quantization (strong coupling).
4. R. Evrard, E. Kartheuser, J. Devreese, "A New Approach to Continuum Polaron Theory", contributed to the second conference on Solid State Physics, University of Bristol (January 1965), paper n° S114.
5. Schultz T.D., Electron-lattice interactions in polar crystals, M.I.T., Solid-State and Molecular Theory Group, Technical Report n° 9 (1958).
6. Evrard R., Thèse, Liège (1964) - Phys. Letters, 14, 295 (1965).
7. Kartheuser E., Thèse, Liège (1968).
8. J. Devreese, Thesis, Leuven (1964).
   J. Devreese, R. Evrard, Physics Letters 11 (1964), 278.
   Studying excitations of the polaron(model).
9. Kartheuser E., R. Evrard, J. Devreese (to be published in Physical Review 1970).
10. Feynman R.P., Hellwarth R.W., Iddings C.K. and Platzman P.M., Phys. Rev. 127, 1004 (1962).
11. J. Devreese, R. Evrard, E. Kartheuser, Solid State Communications 7, 767 (1969).
12. E. Kartheuser, R. Evrard, J. Devreese, Phys.Rev.Letters 22, 94(1969)

XIX

INFRA-RED VIBRATION FREQUENCIES OF THE H-BOND IN KH_2PO_4 FERROELECTRIC

J. SCHMIT
Department of Experimental Physics,
University of Liège, Belgium

and

A. LUCAS*
Department of Theoretical Physics,
University of Liège, Belgium

1. INTRODUCTION

In this paper, we plan to describe calculations we have made in the problem of ferroelectric transition of KH_2PO_4 and to show how our results can be relevant to the interpretation of the infra-red spectrum of this material and related ferroelectrics.

Today there are essentially two kinds of philosophy for the interpretation of ferroelectric transition.

In the first one, the transition is considered to be a disorder-order mechanism. For the Hydrogen-bonded class of ferroelectrics, in particular for KDP, this point of view has been elaborated by Slater[1]. The interactions involved in this mechanism are of short-range character and are adequately chosen to favor an ordered state of the protons along the hydrogen bond below the transition point. In Slater theory however, the origin of these short-range interactions

*Chargé de Recherches au Fonds National de la Recherche Scientifique.

is not explicitly elucidated and moreover, long-range Coulomb forces are completely ignored.

In the second kind of theory of ferroelectricity, the central idea has been given by Anderson and Cochran [2]. Here ferroelectricity is treated as a problem in Lattice Dynamics, the transition being thought of as resulting from some kind of singular behaviour in the phonon dispersion relation. More precisely, long-range dipolar forces between induced ionic dipoles are believed to be responsible for a drastic lowering of some optical phonon mode frequency around the transition temperature. The most representative ferroelectrics to which this theory seems to apply belong to the perovskite class ($CaTiO_3$ and related compounds). Qualitatively, the connection between vibrational modes and dielectric properties is most clearly exhibited in the so-called Lyddane-Sachs-Teller relation

$$\frac{\epsilon_s}{\epsilon_\infty} = \frac{\omega_L^2}{\omega_T^2} \qquad (1)$$

which is written here for a cubic diatomic crystal of NaCl type. ϵ_s and ϵ_∞ are the static and high-frequency dielectric constants respectively; ω_L and ω_T are the optical longitudinal and transverse frequencies for long wavelength.

Now if ω_T is temperature dependent according to the formula

$$\omega_T^2 \propto T - T_c \qquad (2)$$

one finds immediately the typical Curie-Weiss law for the para-electric phase

$$\epsilon_s = A + B/(T - T_c) \qquad (3)$$

This temperature softening (2) of a transverse optical phonon branch has been actually observed for perovskite ferroelectrics with at least two experimental techniques, namely with infra-red optical reflectivity measurements [3] and with neutron diffraction techniques [4].

Now let us come back to KDP ferroelectrics. In spite of the success of Slater theory in accounting for many features of the transition in this material, it is to be suspected that long-range dipolar interactions should play an important role in the transition mechanism. As long ago as 1949, another interpretation based entirely on long-range forces had been proposed by Pirenne [5]. Also, Cochran in his original paper on displacive ferroelectrics suggested that the notion of soft phonon modes could be of general application in ferroelectricity and some optical [3] and neutron diffraction [6] experiments to verify this in KDP have been performed.

In the next sections we study the lattice dynamics of a model crystal for KH_2PO_4 and we show that with reasonable values for the ionic and electronic polarizabilities of the various ions in this structure, softening of a phonon mode and corresponding $\frac{4\pi}{3}$-like catastrophe are likely to occur in this model.

2. MODEL CALCULATIONS

Due to the relativity intricate structure of KDP and the large number of atoms per unit cell, the lattice dynamics of this materials is fairly complicated. In KDP, there are four formulae per unit cell or 32 nuclei. For the general lattice dynamical problem, this makes a 96 X 96 matrix to be diagonalized. This however can be somewhat reduced with the following model.

1) We treat the PO_4 ions as a single spherical unit with an isotropic electronic polarizability α_0, centered at the P ion. In other words, we neglect the actual electronic structure of the group and disregard the possible ionic motion of the Posphor us and oxygen ions;

2) We disregard completely the Potassium ions;

3) We assign to each proton an isotropic ionic polarizability α_H.

Of course this simplified model cannot provide the full dispersion relation of the actual crystal, but it should give an adequate indication whether or not the KDP structure is favourable to a vibrational instability similar to the one which occurs in displacive type ferroelectrics.

Thus we have to solve the following set of equations of motion :

$$m_H \ddot{\underset{\sim}{X}}_{H_i} = -m_H \omega_H^2 \underset{\sim}{X}_{H_i} - e_H^2 \sum_j{}' T_{H_i H_j} \cdot \underset{\sim}{X}_{H_j} - e_H e_O \sum_j T_{H_i O_j} \cdot \underset{\sim}{X}_{O_j}$$

$$m_O \ddot{\underset{\sim}{X}}_{O_i} = -m_O \omega_O^2 \underset{\sim}{X}_{O_i} - e_O^2 \sum_j{}' T_{O_i O_j} \cdot \underset{\sim}{X}_{O_j} - e_O e_H \sum_j T_{O_i H_j} \cdot \underset{\sim}{X}_{H_j} \qquad (4)$$

where the various T matrices are the dipolar tensors

$$T_{ij} = \left(E - 3 \hat{\underset{\sim}{R}}_{ij} \hat{\underset{\sim}{R}}_{ij} \right) / R_{ij}^3$$

connecting the induced ionic (H) or electronic (O) dipole moments of amplitude $\underset{\sim}{X}$. It will be shown that the only parameters on which depend the characteristic frequencies of this system are ω_H and two polarizabilities :

$$\alpha_O = \frac{e_O^2}{m_O \omega_O^2} \quad , \quad \alpha_H = \frac{e_H^2}{m_H \omega_H^2} \qquad (5)$$

Then we introduce the usual canonical transformation towards the normal modes coordihates. Here, we will restrict ourselves to the $\underset{\sim}{k} = 0$ modes which are most likely to contain the ferroelectric soft mode :

$$\underset{\sim}{X}_{H_i} = \underset{\sim}{Q}_H \exp(i\omega t) \; ; \; \underset{\sim}{X}_{O_i} = \underset{\sim}{Q}_O \exp(i\omega t) \qquad (6)$$

The largest amount of work is required by the calculation of the various lattice dipole sums

$$T_{HH} = \sum_j T_{H_i H_j} \qquad \text{etc...} \qquad (7)$$

for which the well-known Ewald summation procedure has been used [7].

Of course the electronic degrees of freedom associated with the electronic motion of the PO_4 groups can be eliminated from the eigenvalue problem of the dynamical matrix because we have supposed that the polarization of this unit follows instantaneously the proton motion. The eigenfrequencies ω of the proton motions themselves are related to the eigenvalues of the following matrix

$$M = T_{HH} - T_{HO} \cdot \left(\frac{1}{\alpha_o} E + T_{OO} \right)^{-1} \cdot T_{HO} \tag{8}$$

according to

$$\underset{\cdot}{M} \cdot \underset{\sim}{Q} = \lambda \underset{\sim}{Q} \tag{9}$$

where

$$\lambda^{-1} = \alpha_H / \left(\frac{\omega^2}{\omega_H^2} - 1 \right) \tag{10}$$

or

$$\omega^2 = \omega_H^2 \left(1 + \lambda \alpha_H \right) \tag{11}$$

Thus the final proton frequencies in this model depend only on three parameters : the frequency associated with the "free" hydrogen bond ω_H , the ionic polarizability α_H and the electronic polarizability α_0 (which, besides the lattice structure, completely determines λ) of the PO_4 groups.

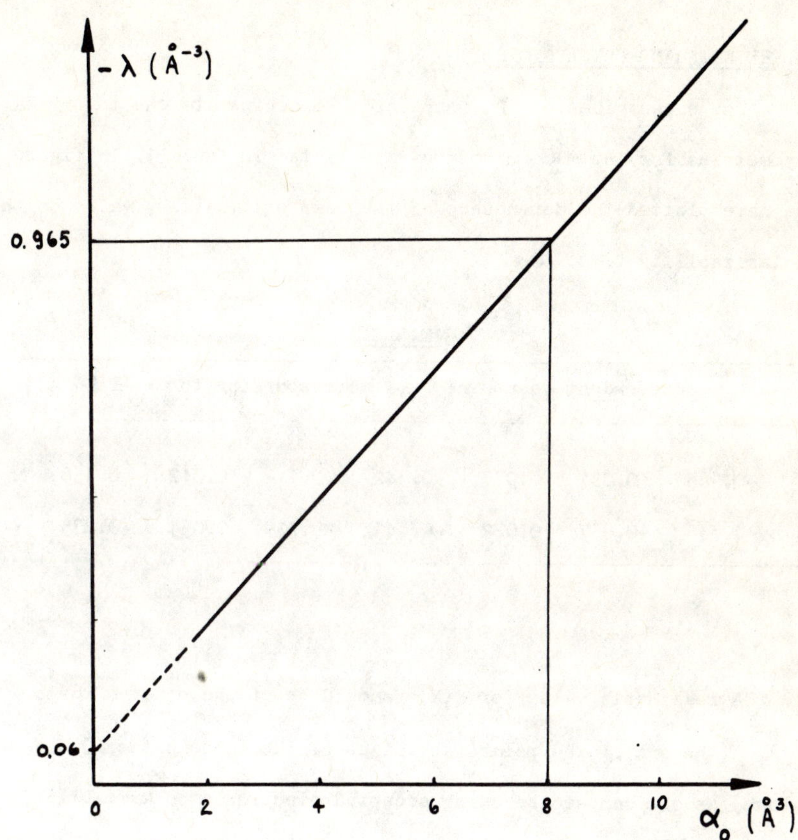

Figure 1 : Linear dependence of the λ parameter on α_0 for the soft ferroelectric mode at $\underset{\sim}{k} = 0$.

3. RESULTS AND CONCLUSIONS

The eigenvalues of $\mathcal{M}$ (some of them degenerate due to crystal symmetries for the $\underset{\sim}{k} = 0$ modes) are listed in table 1. In figure 1, we have plotted the dependence of the most critical λ value on the polarizability α_0.

TABLE 1

Independent values of λ corresponding to $\alpha_0 = 8\ (\mathring{A}^3)$

-0.965	0.081	0.077	-0.542	0.112	0.042	-0.156
-0.338	-0.057	0.022	-0.041	-0.319	0.054	-0.275

A reasonable value for α_0 should be of the order of 6 to 8 $\mathring{A}^3$ (the sum of the polarizabilities of the ions in the PO_4 group) as one can see be using accepted electronic polarizability of the O^{--} ion.

If we introduce for α_H the estimation

$$\alpha_H = \frac{e_H^2}{m_H \omega_H^2} \sim 1 \text{ to } 2\ \mathring{A}^3 \tag{12}$$

with $\hbar\omega_H \sim 10^3\ cm^{-1}$, we see that, already for this simple model, there exists a strong softening of the mode for which $\lambda = -0.965\ \mathring{A}^{-3}$. We find for this mode, according to (11)

$$\omega^2 \simeq 0.035\ \omega_H^2 \quad \text{or} \quad \omega_H \simeq 0.18\ \omega_H \tag{13}$$

We have used a second model where we do take into account the internal structure of the PO_4 groups, assigning an ionic polarizability to the phosphorus ion and including an ionic polarizability of the Potassium ion as well. For this more elaborated model and using reasonable values for the various parameters, most of the optical phonon frequencies turn out to be purely imaginary. This means that this model is completely unstable as a result of the strong Lorentz local fields coupling the various ions. One sees in the calculation that the main dipolar interaction responsible for these instabilities is the one between the oxygen and the hydrogen ions of a given H-bond. The Lorentz local field between these two ions is of the order of twenty times $\frac{4\pi}{3} \underset{\sim}{P}$, the usual value for cubic crystals.

We will conclude in the following way

1) The purely ionic model of KH_2PO_4 is certainly unstable with respect to optical proton motions.
2) The model in which the only source of dipole fields is the hydrogen bond is capable of exhibiting soft mode behaviour of the Cochran type.
3) The calculated softening is consistent with the theoretical value postulated in Pirenne's theory[5] and also with experimental evidences[8] in the infra-red absorption spectrum of KDP which shows a strong band located around 200 cm^{-1}.

Numerical computations have been performed on the IBM 7040 of the "Centre de Calcul et Traitement de l'Information", University of Liège, Belgium.

REFERENCES

(1) J.C. Slater

J. Chem. Phys. 9, 16 (1941).

(2) W. Cochran

Advanc. Phys. 9, 387 (1959)

(3) A.S. Barker, Jr.

Ferroelectricity. Proceedings of the Symposium on Ferroelectricity, General Motors Research Labo., Warren, Michigan, 1966. Edited by E.F. Weller, Elsevier (1967)

(4) R.A. Cowley

Phys. Rev. 134, 981 (1964)

(5) J. Pirenne

Physica 15, 1019 (1949)

(6) V.H. Schmidt, E.A. Vehling

Phys. Rev. 126, 447 (1962)

(7) All the dipolar tensors connecting the lattice sites of the actual structure of KDP have been computed and will be published elsewhere.

(8) F. Jona, G. Shirane

Ferroelectric crystals, Pergamon (1962)

XX

OPTICAL EXCITATION OF THE SPIN SYSTEMS IN MAGNETIC CRYSTALS   I LIGHT ABSORPTION

Toru MORIYA

Institute for Solid State Physics

University of Tokyo, Roppongi, Tokyo

1. INTRODUCTION

In a recent half decade, remarkable progress has been made in the optical studies of magnetic crystals, particularly the studies of excitations in spin systems or magnons by optical methods. This was started with the observation of far infrared absorption by two magnon excitations in an antiferromagnet (FeF_2)[1] and that of the optical absorption by simultaneous excitations of exciton-magnon pairs in an antiferromagnet (MnF_2)[2]. Subsequently, light scattering by one and two magnons in antiferromagnetic FeF_2 was reported.[3] We should also mention an earlier observation of infrared absorption by simultaneous excitation of a phonon and magnons.[4] Although many spectroscopic studies on ordered magnetic crystals have been reported earlier than these,[5] the possibility of studying elementary excitations and their interactions by means of optical methods has been realized impressively by the above mentioned observations.

An interesting point in all of these observations is that the coupling between the electric component of the radiation field and the spin system is of primary importance. This coupling arises from the spin-orbit interaction[5] and the inter-ionic Coulomb interaction of mainly exchange character[7] as will be discussed later.

It is well known that the direct magnetic interaction between radiation field and the spin system gives rise to the magnetic resonances. These are limited, however, to the excitation of $\underset{\sim}{k} = 0$ components of the spin waves in bulk crystals and several standing spin waves in films. When we look into the possible couplings of the radiation field and the spin system in more detail, we find many possible types of interactions which is related with a single or pleural number of elementary excitations: magnons, excitons, and phonons. When a pleural number of elementary excitations are excited simultaneously, only the total momentum is to be conserved. Therefore, The absorption or scattering line shapes in these cases give information on the spectra of these elementary excitations. In addition, we may expect to study the interaction between elementary excitations from the analysis of the spectral line shape, which is influenced significantly by this interaction. In some cases, it can even give rise to a bound state.

Studies in this field so far have been limited mainly to insulators. The magnetic properties of insulator compounds are considered to be understood very well owing to the powerful spin Hamiltonian approach. We can extend this type of approach to include the radiation field and treat conveniently the problem of far-infrared excitations of two magnons in the spin systems (Section 2). Local symmetry arguments are particularly useful in this case. When excitons are involved, the problem becomes more complicated, since we have additional degrees of freedom. Symmetry arguments should take account of the symmetries of the orbital states involved. The treatment should therefore be specific to each substance, though group theoretical considerations simplify the problem a great deal. In the case of magnon sidebands associated with the spin-allowed transitions, where the spins and excitons have separable degrees of freedom, we can give a fairly general discussion (Section 3. 3) as in the case of two-magnon problems.

In the case of metals, it seems that no study has been reported so far. Although there is a disadvantage due to the skin effect, it is expected that the present day experimental techniques will make it possible soon.

In this paper, we shall give a brief survey on the mainly theoretical accounts of two-magnon absorptions and magnon-sidebands associated with the spin-allowed and the spin forbidden transitions. We put particular emphasis on the coupling of various elementary excitations to the radiation field, interaction among them, and the consequences as observed by optical absorptions. A few remarks are made on some future possibility in this field of research. No attempts are made here to cover experimental results exhaustively. Discussions based on the magnetic space group are also omitted. Readers are referred to the literature cited.

In the following paper (II) we will survey on the light scattering by spin excitations in insulators and then extend discussions to the case of magnetic metals.

2. FAR-INFRARED ABSORPTION BY TWO-MAGNON EXCITATIONS (8, 9, 10

2.1. Spin dependent electric dipole moment

2.1.1. Let us first consider transition metal compounds. When the orbital ground state of a magnetic ion is non-degenerate, as is usually the case, and the excited orbital states are well separated from it, we can make use of a usual spin Hamiltonian approach; i.e., the low energy levels corresponding to the spin degrees of freedom can be represented effectively by polynomials of the components of spin operators. The spin Hamiltonian under the external electric field $\mathbf{E}$ may be expanded in powers of E. We generally have a term linear in $\mathbf{E}$, and therefore, the spin-dependent electric dipole moment $\mathbf{P}$. We have

$$H = H_0 - \mathbf{E}\cdot\mathbf{P}, \qquad \mathbf{P} = \sum_j \mathbf{P}_j + \sum_{j>\ell} \mathbf{P}_{j\ell} + \ldots \tag{1}$$

where $\mathbf{P}_j$ is the moment associated with the j-th ionic spin and $\mathbf{P}_{j\ell}$ the one associated with the j,ℓ pair of ionic spins. The most general expressions for these moments are

$$P_j^{\alpha} = \sum_{\beta,\gamma} K_j^{\alpha,\beta\gamma} S_{j\beta} S_{j\gamma},$$

$$P_{j\ell}^{\alpha} = \pi_{j\ell}^{\alpha}(\mathbf{S}_j\cdot\mathbf{S}_\ell) + \sum_{\beta,\gamma} \Gamma_{j\ell}^{\alpha,\beta\gamma}[(S_{j\beta}S_{\ell\gamma} + S_{j\gamma}S_{\ell\beta})/2]$$

$$+ \sum_{\beta} d_{j\ell}^{\alpha,\beta}[\mathbf{S}_j \times \mathbf{S}_\ell]_\beta, \tag{2}$$

retaining only quadratic terms in the spin components. Here α, β, and γ denote the coordinates x, y, and z. P_j is related with the linear Stark effect and vanishes when the j-th ion is located at the center of inversion. Possible forms for the coefficient tensors in Eq. (2) are determined from considerations of the local symmetry of the crystal around the j-th ion (for K_j) and around the j, ℓ pair of ions (for $\pi_{j\ell}$, $\Gamma_{j\ell}$, and $d_{j\ell}$). For example, when a center of inversion is located at the midpoint on the line connecting the two ions j and ℓ, $\pi_{j\ell}$ and $\Gamma_{j\ell}$ vanish, while $d_{j\ell}$ does not. This is the case, for example, in rock salt-type (MnO, etc.) and perovskite type ($RbMnF_3$, etc.) structures. In the case of rutile-type iron group fluorides, all the terms in Eq. (2) are non-vanishing for a pair of ions, one at a corner and the other at a body-center sites.

The same symmetry arguments apply to the cases of rare earth compounds if we replace S_j by the total angular momentum J_j, etc.

In the cases of iron group salts with the orbital angular momentum not well quenched, i.e. the crystalline filed splittings of the orbital levels are of the same order of magnitude as the spin-orbit coupling, we have to extend the above arguments to include the orbital degeneracy. This is generally complicated though feasible in principle.*

---

* For example, a Co^{2+} (Fe^{2+}) ion in an octahedral site has three-fold degenerate ground orbital states (t_{2g}). The orbital angular momentum within the t_{2g} manifold is effectively represented by an angular momentum operator ℓ with ℓ=1. When we take account of the crystalline field of lower symmetry and the interionic exchange interactions, the effective Hamiltonian may be expressed in terms of ℓ_j and S_j. We can thus apply symmetry arguments similar to the above.

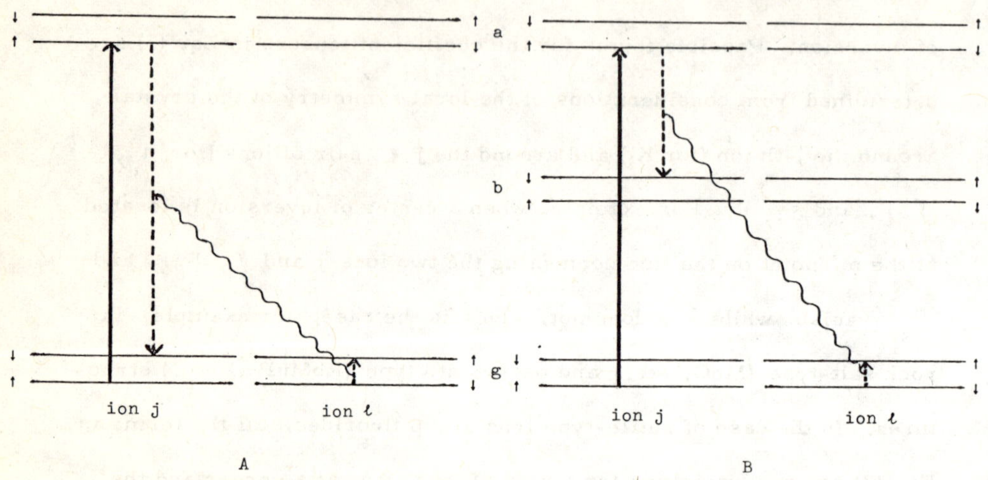

Fig. 1

Figure Captions

Fig. 1. Predominant mechanisms for the magnon-induced electric dipole transitions; A. two-magnon excitation, B. exciton-magnon excitation. g, a, and b represent the ground state, excited states of odd and even parities, respectively.

Straight arrows indicate the virtual excitations due to the radiation field and dashed lines connected by wavy lines represent the transitions (virtual) due to the non-diagonal exchange interaction.

2.1.2. Quantum mechanical calculation of the spin-dependent electric dipole moments reduces to that of the spin Hamiltonian under the external electric field. Particularly, the term $-\underset{\sim}{E}.\underset{\sim}{P}_{j\ell}$ is regarded as a change, linear in the field, of the inter-ionic exchange energy of isotropic and anisotropic natures. It is easy to extend the theory of isotropic and anisotropic exchange interactions in insulators to include the effect of orbital polarization due to the external electric field or electron-radiation interaction. An important point in this perturbation procedure is to take account of the non-diagonal part of the exchange energy[7] as in the mechanism of antisymmetric exchange interaction.[11] This most important mechanism which gives rise to the first term of $P_{j\ell}$ in Eq. (2) is shown schematically in Fig. 1a.

The order of magnitude of $\underset{\sim}{\pi}_{j\ell}$ may be

$$|\underset{\sim}{\pi}_{j\ell}| \sim (a \mid er \mid g)\, J_{n.d.}/\Delta E, \tag{4}$$

where ΔE is the odd parity excitation energy, $J_{n.d.}$ the non-diagonal exchange, and $(a \mid er \mid g)$ is the matrix element of the electric dipole moment operator. For MnF_2 and FeF_2 we estimate $\pi \sim 10^{-5} - 10^{-4}$ au., in agreement with the observed intensity for two-magnon absorptions.

When the orbital angular momentum is well quenched, relative importance of the three terms in Eq. (2) is the same as that of the usual exchange interactions of isotropic, anisotropic (symmetric), and antisymmetric characters:

$$|d| \sim (\lambda/\Delta E_e)|\pi| \sim (\Delta g/g)|\pi|, \qquad |\Gamma| \sim (\lambda/\Delta E_e)^2|\pi| \sim (\Delta g/g)^2|\pi|, \tag{3}$$

where λ is the spin-orbit coupling constant, ΔE_e the relevant excitation energy of even parity, and g the g-factor of the ion and $\Delta g = g-2$.

One important point here may be the expected short range character of $P_{j\ell}$, similar to that of the usual exchange interactions in insulators. This originates in the localized character of the electronic states in insulators, [12] and simplifies the problem a great deal.

2.1.3. In the case of rare earth compounds, all terms in Eq. (2) can be of equal importance. Moreover, the terms of higher order in the components of J_j need not necessarily be small. The arguments again are similar to those of interionic exchange interactions. [13] $P_{j\ell}$ is expected to be of short range.

2.2. Two-magnon absorptions

2.2.1. Absorption of the electro-magnetic radiation due to the spin-dependent electric dipole moment is obtained from the imaginary part of the complex polarizability:

$$\chi(\omega) = (i\hbar)^{-1}\int_0^\infty dt\ e^{i\omega t}\langle [P, P(t)] \rangle,$$

with

$$P(t) = e^{itH/\hbar}\ P\ e^{-itH/\hbar},$$
$$\langle A \rangle = Tr(e^{-\beta H}A)/Tr\ e^{-\beta H}, \qquad (5)$$

where H is the spin Hamiltonian ($E=0$), and $\beta = 1/k_B T$. This expression is quite general and by using Eq. (1) for P one can calculate the absorption spectrum at any temperatures. Various approximation methods in the statistical mechanical theory of the spin systems (Heisenberg model) may be employed.

2.2.2. At low temperatures, P can be expanded in terms of the spin wave annihilation and creation operators. All the terms should conserve the wave vector. The terms bilinear in the creation operators, for spin waves of opposite wave vectors, give rise to simultaneous excitations of two magnons. In the case of a simple two-sublattice antiferromagnet the absorption coefficient may be calculated as follows, taking only the first term of $P_{j\ell}$ in Eq. (2) and neglecting the interaction between magnons:

$$A(\omega) = -(4\pi^2 \omega S^2/\hbar)\rho(\omega/2)\langle \pi(k)\pi(k)\rangle_{\omega/2},$$

$$\pi(k) = \sum_j \pi_{j\ell} \exp[-ik.(r_j - R_\ell)], \qquad (6)$$

where $\rho(\omega)$ is the density of states for antiferro-magnons per branch and $\langle\ \rangle_{\omega/2}$ means an average over the energy surface $\omega = \omega_k$. In this case two magnons in different branches (running mainly in the different sublattices) are excited simultaneously. We expect that

(1) the magnon density of states reflects on the absorption line shape; the absorption line shape has strong peaks and sharp breaks coming from the large density of states near the Brillouin zone boundary and the Van Hove critical points on the boundary surface.

(2) The selection rule in the k space is given by $\pi(k)$ which generally has anisotropic k-dependence. When the magnetic ions have centers of inversion we have $\pi(k) = -\pi(k)$, and therefore $\pi(0)=0$.

(3) The absorption coefficient tensor, Eq. (6), is generally anisotropic and the intensity of absorption depends on the direction of polarization of radiation.

(4) The external magnetic field has no influence in its first order, since the frequencies of two magnons in different branches are shifted just oppositely. These features are consistent with the experimental observations.

2.2.3. Detailed calculation and measurement of the absorption line shapes have been carried out for MnF_2,[14] and agreement between them was satisfactory only in qualitative sense. Discrepancies in detailed line shapes were attributed first to the extended range of $\underset{\sim}{P}_{j\ell}$ [14] and subsequently to the effect of spinwave interactions.[15]

Since Eq. (5) is a retarded Green function for $\underset{\sim}{P}$ we need to study the Green functions for the quadratic forms in the spin components. Simplest approximation may be to decouple properly the higher order Green functions in the equations of motion. The result of a recent calculation [15] [15.a.] suggests a satisfactory improvement on the line shape, without taking account of the extended range of $\underset{\sim}{P}_{j\ell}$ Note that there is no ambiguous parameter for the magnon-magnon interactions once spin Hamiltonian is built up.

Physically, a local picture helps qualitative understanding of the effect.[15] For an Ising interaction with the nearest neighbor exchange J, two spin intervals cost an energy 2zJS, where z is the number of nearest neighbors, unless they are on neighboring sites, when it costs only (2zS-1) J. Since magnon pairs are excited by the electric dipole moments associated with neighboring pairs of ionic spins, the absorption peak will be observed at the frequency lower than the maximum by about $J/\hbar$. This is qualitatively consistent with the observation.

It is interesting to point out that the two-magnon absorption may be a good method to study the spin wave interactions in antiferromagnets. Since the neutron scattering experiments give information on single magnon dispersions, one can extract the effect of magnon interactions from detailed analysis of the two-magnon absorption line shape and the neutron results.

2.2.4. Temperature dependences of the line shape and the absorption intensity have not yet been studies satisfactorily, although the Green function method seems to be appropriate for their study.

Also, experimental studies have been limited to low temperatures. When $P_{j\ell}$ is large, we may expect to see the absorption even above the Curie temperature. The study near the Curie temperature, if possible, would be of particular interest.

2.2.5. Two-magnon absorption is generally allowed in antiferro- and ferrimagnets. In antiferromagnets, absorption intensity is usually expected to be small when there is an inversion center between a neighboring pair of ions. In ferromagnets, it will be observed only when the second or third terms in Eq. (2) is sufficiently large, i.e., when the effect of orbital angular momentum is significant and the anisotropy is strong.

3. MAGNON-EXCITON ABSORPTIONS[16][10][9]

3.1. Excitons

Let us consider the cases of transition metal compounds. For divalent ions, excited states of odd parity are separated from the ground state more than about 10^5 cm^{-1}. Therefore, all the transitions of lower energies are parity-forbidden. Hereafter we shall restrict ourselves to the configurations consisting of d electrons only. Magnetic dipole transitions are of course possible.

When the excited and ground states have the same spin multiplicity, the intensity of absorption is significant (spin-allowed transitions). Even when the spin multiplicities are different for the two states, transition of week intensity is expected owing to the spin-orbit coupling (spin-forbidden transitions).

Since there are inter-ionic Coulomb interactions among the ions in the crystal, the excited states should naturally be described in terms of excitons (Frenkel excitons in the present case). The exciton dispersion in the present cases is generally not very large. The most effective contribution to the excitation transfer may be the inter-ionic exchange interactions. Next comes the quadrupole-quadrupole interactions; when the spin is changed in the excitation, one should take additional account of the spin-orbit interaction. The dispersion depends on the types of excitations. In MnF_2, for example, the overall dispersion from the center to the zone boundary (along the c axis) in the k space is about a few cm^{-1} for ${}^6A_{1g} \rightarrow {}^4T_{1g}$ excitons and about -70 cm^{-1} for one of the ${}^6A_{1g} \rightarrow {}^4E_g$, ${}^4A_{1g}$ transitions.

The above mentioned optical transitions should be regarded as the excitations of $\underset{\sim}{k}=0$ excitons. When there are more than two equivalent magnetic ions in a unit cell, we have degenerate branches of excitons or nearly degenerate branches with the Davydov splitting at the zone center. Observation of the Davydov splitting gives information on the excitation transfer between the ions [17][18] in a unit cell.

3.2. Magnon sidebands

3.2.1. There are two types of mechanisms for the simultaneous optical excitations of an exciton and a magnon (magnons).

(1) The parity of a pair of an exciton and a magnon with opposite wave vectors can be odd. In this case the electric dipole transition by simultaneous excitations of an exciton and a magnon is allowed. This is similar to the electric dipole transitions induced by odd parity phonons; odd parity phonons are here replaced by magnons. These types of transitions are described by using the spin-dependent electric dipole transition moments. Almost all magnon-sidebands observed so far are considered to be due to this mechanism. They are associated with spin-forbidden magnetic dipole lines.

(2) When an ion occupies an excited orbital state in a crystal, the spin states of the entire crystal are generally different from those in the ground orbital state. In other words, an n-magnon states in the ground orbital state is generally non-orthogonal to (n+m) - magnon states in an orbitally excited state. Therefore, the magnetic dipole transition for a zero-magnon line will accompany magnon sidebands of generally weaker intensities. This type of magnon sidebands have been observed as the spin-allowed magnetic dipole transitions in FeF_2 [19] and CoF_2 . [20] The same effect will also be important in explaining possible many-magnon sidebands of electric dipole character. [21]

Also, this final state interaction between an exciton and magnons has strong influence on the line shape of exciton-magnon absorption, as will be discussed later.

3.2.2. Let us consider the case where the sites of magnetic ions have centers of inversion, as is usually the case. The magnetic and electric dipole transition moments may be written as follows:

$$\underset{\sim}{M} = \underset{\sim}{M}^{(0)} + \underset{\sim}{M}^{(1)} + \underset{\sim}{M}^{(2)} + \dots$$

$$\underset{\sim}{P} = \underset{\sim}{P}^{(2)} + \dots,$$

where $\underset{\sim}{M}^{(n)}$ and $\underset{\sim}{P}^{(n)}$ are of the n-th order in the spin components. $\underset{\sim}{M}^{(0)}$ is of the order of 1 μ_B per ion for spin-allowed transitions and is very much reduced from this for spin-forbidden transitions. $\underset{\sim}{M}^{(1)}$, $\underset{\sim}{M}^{(2)}$, etc., are usually smaller than $\underset{\sim}{M}^{(0)}$, particularly for the spin-allowed transitions.

The simplest and most important contribution to the spin dependent electric dipole transition moment, $\underset{\sim}{P}^{(2)}$, may be given by

$$\underset{\sim}{P}_{j\ell} = \underset{\sim}{\pi}_{j\ell}(b \leftarrow g) \sum_{m,m'} c_{jbm}{}^{+} c_{jgm'} (m|\underset{\sim}{S}|m') . \underset{\sim}{S}_{\ell}, \qquad (8)$$

in a simple case of one-electron ions.[7] Here b is the even parity excited state concerned, g the ground state, m, m' specify the spin states, and c_{jbm} is the annihilation operator for the b, m state of the j th ion, etc. This type of transition moment is derived from the second order perturbation theory bilinear in the electron-radiation interaction and the non-diagonal exchange energy.[7] This process is shown in Fig. 1b. Of course we have many other terms in $\underset{\sim}{P}_{j\ell}$ allowed from symmetry, which can be derived from perturbation theory taking account of the spin-orbit interaction and the interionic Coulomb interaction of of exchange as well as multipole-multipole characters.[9][22] Full expressions are generally quite complicated. Usually the term in Eq. (8) predominates.

3.3. Case of spin-allowed transitions.[23]

3.3.1. We first consider the excitons arising from spin-allowed transitions in a two-sublattice antiferromagnet. The ground state g as well as the relevant excited state e of the ion are assumed to be non-degenerate and have the same spin multiplicity. Thoy are also assumed to be well saparated from the other excited states.

Under these circumstances we can easily extend the spin Hamiltonian approach to include the excited orbital state. For this purpose we introduce the Pauli spin matrices $\underset{\sim}{\sigma}_j$, etc., to describe the orbital degrees of freedom. The orbital energy for the entire crystal is given by

$$-(E_0/2)\{\sum_j \sigma_{jz} + \sum_\ell \sigma_{\ell z}\}, \tag{9}$$

where $\sigma_{jz} = 1$ and -1 correspond to the ground and the excited orbital states of the j th ion, respectively. j and ℓ specify the sites in the 1 st and 2 nd sublattices, respectively.

The spin Hamiltonian now depends on the occupation of the orbital states. When all the ions are in their ground orbital states, we have a usual spin Hamiltonian H_S. When the j th ion is excited (to the state e), the spin Hamiltonian is changed by $\delta H^{(j)}$. The spin Hamiltonian diagonal in the orbital states may be given by

$$\dot{H}_S + \sum_j \sigma_{j-}\sigma_{j+}\delta H^{(j)} + \sum_\ell \sigma_{\ell-}\sigma_{\ell+}\delta H^{(\ell)},$$

where

$$\sigma_{j\pm} = (\sigma_{jx} \pm i\sigma_{jy})/2. \tag{10}$$

Here we take account of only the states with no or one excited ion. This is sufficient for the optical problem of the present interest.

Then we have excitation transfer terms which conserve the orbital energy:

$$H_{tr.} = \sum_{j,j'} b_{jj'}\sigma_{j-}\sigma_{j'+} + \sum_{\ell\ell'} b_{\ell\ell'}\sigma_{\ell-}\sigma_{\ell'+}$$
$$+ \sum_j \sum_\ell (b_{j\ell}\,\sigma_{j-}\sigma_{\ell+} + b_{j\ell}{}^*\sigma_{j+}\sigma_{\ell-}). \tag{11}$$

We also have spin dependent transfer terms which are written as

$$H_{tr.}{}^{(S)} = \sum_j \sum_\ell (\sigma_{j-}\sigma_{\ell+}c_{j\ell} + \sigma_{j+}\sigma_{\ell-}c_{j\ell}{}^+)$$
$$+ \sum_{jj'} \sigma_{j-}\sigma_{j'+}c_{jj'} + \sum_{\ell\ell'} \sigma_{\ell-}\sigma_{\ell'+}c_{\ell\ell'} \tag{12}$$

where $c_{j\ell}$, etc., are functions of spins. Thus we get the starting Hamiltonian: Eqs. (7) - (10). Although there are many other terms which do not conserve the orbital energy, they can be shown to be unimportant or effectively included in the above Hamiltonian as the contribution of higher order perturbations.

Eqs. (9) and (11) are diagonalized by Fourier transformations:

$$\sigma_{j-} = (2/N)^{1/2} \sum_k A_{1k}^+ \exp(-i\underset{\sim}{k}\cdot\underset{\sim}{R}_j).$$

$$\sigma_{\ell-} = (2/N)^{1/2} \sum_k A_{2k}^+ \exp(-i\underset{\sim}{k}\cdot\underset{\sim}{R}_\ell)$$

where A_{nk}, $A_{n'k'}{}^+$ are the annihilation and creation operators for the excitons in the n, and n' th branches, respectively. They satisfy Boson commutation relations to a good approximation.

The interaction arises between excitons and the spin system (or magnons) arises from the terms with $\delta H^{(j)}$, and $\delta H^{(\ell)}$ in Eq. (8) and from $H_{tr.}{}^{(S)}$. General forms for these terms are obtained from local symmetry considerations taking account of the symmetries of the orbitals g and e. The most general expression (up to the quadratic terms in the spin components) for $\delta H^{(j)}$ may be written as

$$\delta H^{(j)} = \delta V_j + \sum_j \{ \delta J_{j\ell}(\underset{\sim}{S}_j\cdot\underset{\sim}{S}_\ell) + \delta d_{j\ell}\cdot[\underset{\sim}{S}_j \times \underset{\sim}{S}_\ell] + \underset{\sim}{S}_j\cdot\delta K_{j\ell}\cdot\underset{\sim}{S}_\ell \}, \tag{13}$$

where δV_j is the change in the one-ion spin Hamiltonian due to the orbital excitation and the other terms are the changes in the isotropic and anisotropic (antisymmetric and symmetric) exchange energies; $\delta K_{j\ell}$ is a symmetric tensor. The usual quantum mechanical perturbation theory for providing the spin Hamiltonian applies to these terms.

The spin-dependent transfer terms $H_{tr.}{}^{(S)}$ can include terms of odd power in the spin components. The most general expression may be

$$c_{j\ell} = \Gamma_{j\ell}\cdot\underset{\sim}{S}_j + \Gamma_{\ell j}{}^*\cdot\underset{\sim}{S}_\ell + \tilde{J}_{j\ell}(\underset{\sim}{S}_j\cdot\underset{\sim}{S}_\ell) + \tilde{\underset{\sim}{d}}_{j\ell}\cdot[\underset{\sim}{S}_j \times \underset{\sim}{S}_\ell] + \underset{\sim}{S}_j\cdot\tilde{K}_{j\ell}\cdot\underset{\sim}{S}_\ell,$$

where

$$\tilde{J}_{j\ell} = \tilde{J}_{\ell j}, \quad \tilde{\mathbf{d}}_{j\ell} = -\tilde{\mathbf{d}}_{\ell j}, \quad \tilde{K}_{j\ell} = \tilde{K}_{\ell j}. \tag{14}$$

These terms can also be calculated quantum mechanically in a similar way to the one constructing the spin Hamiltonian. Eqs. (13) and (14) are usually reduced to simpler forms by symmetry considerations for a particular problem.

At low temperatures, we describe the spin system in terms of spin waves or magnons. The exciton-magnon interaction arises from Eqs. (10) and (12). We have five types of processes as shown in Fig. 2.

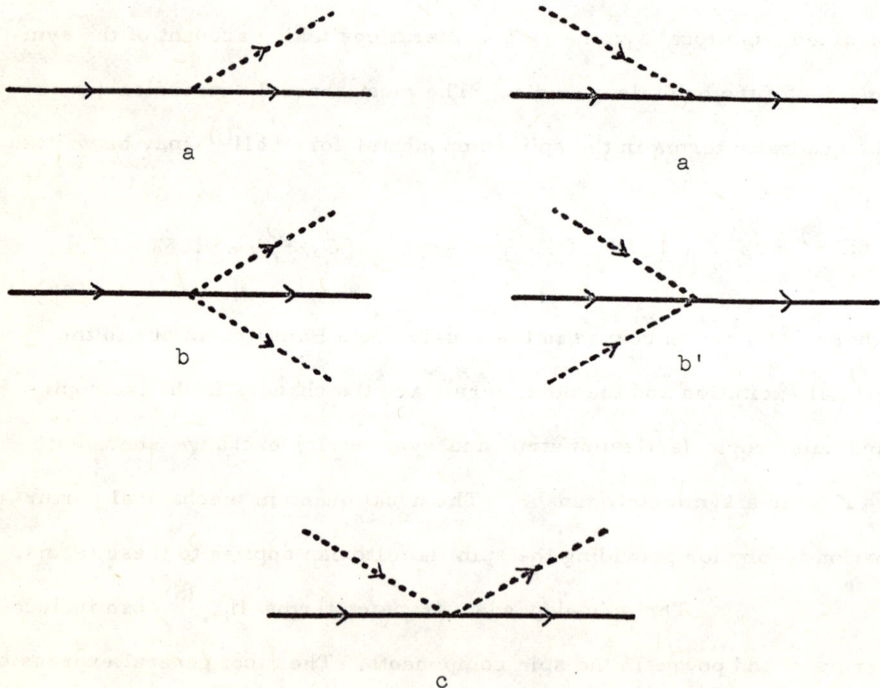

Fig. 2. Possible types of exciton-magnon interactions. Solid and dashed lines represent excitons and magnons, respectively.

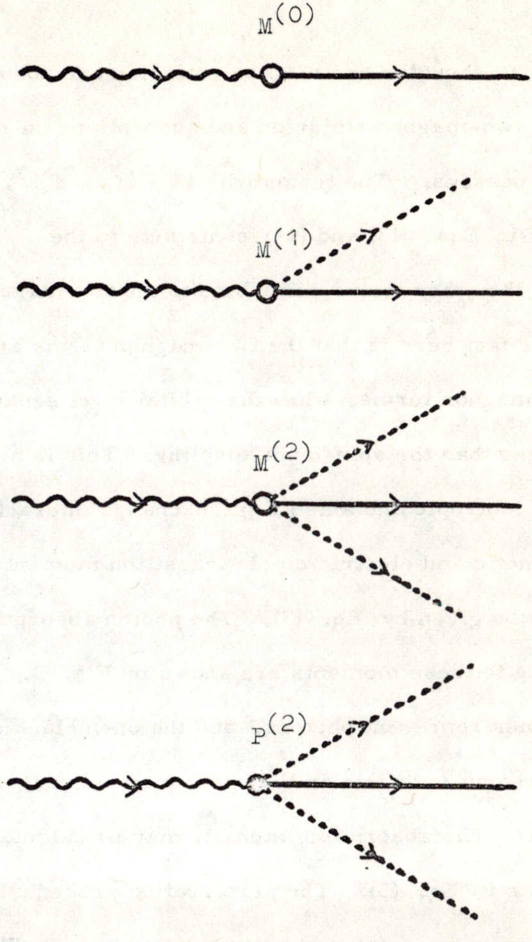

Fig. 3. Possible single processes for light absorption by an exciton and magnons. The wavy lines represent photons, open and black circles at the vertices indicate magnetic and electric interactions, respectively.

Fig. 2a and a' show the one-magnon emission and absorption processes, b and b' the two-magnon emission and absorption and c the magnon scattering processes. The terms with $\delta d^{x,y}$, $\tilde{d}^{x,y}$, $\Gamma^{x,y}$, $\delta K^{xz,yz}$, and $\tilde{K}^{xz,yz}$ in Eqs. (13) and (14) contribute to the magnon process and all the other terms contribute to the two magnon processes. A remarkable fact here is that the two-magnon terms are usually larger than the one magnon terms, when the orbital level separations of the ions are larger than the spin-orbit coupling. This is a consequence of the theory of isotropic and anisotropic exchange interactions.

3.3.2. Magnetic and electric dipole transition moments for the present problem may be given by Eq. (7). The photon absorption processes corresponding to these moments are shown in Fig. 3.
Here wavy lines represent photons, and the open black circles at the vertices represent magnetic and electric interactions, respectively.

The absorption intensity may be calculated by using these moments in Eq. (5). The perturbation procedure may be represented by diagrams constructed by connecting repeatedly those in Fig. 2 to the right of one of those in Fig. 3. We should also take account of the magnon-magnon interactions which are not shown in the figures.

In this way, we can, in principle, calculate the absorption line shapes for various processes or n-magnon sidebands with $n = 0, 1, 2, \ldots$ In other words, we need to calculate the renormalized Green functions for excitons and magnon and the vertex parts for absorption of a photon and emission of an exciton and n-magnons. To carry this program through is generally a complicated task and in actuality the second order perturbation calculation for the cummulants has been made so far. Further developments made recently will be discussed in a later paragraph in connection with spin-forbidden transitions.

3.3.3. Observed magnon sidebands in near infrared spectrum of FeF_2 [19] have been analyzed along the line as discussed above.[23] The excitons concerned are assigned to correspond to $A_{1g} \rightarrow B_{3g}$ transitions in Fe^{2+} ions. We have two branches of excitons corresponding to two sublattices. Quantum mechanical estimation shows that the effect of spin-dependent transition moments (both magnetic and electric) is small as compared with the combined effect of direct magnetic dipole transitions and the exciton magnon interactions shown in Fig. 2. This explains the observed magnetic dipole character of the sidebands. The estimated intensity of two magnon sideband seems to be reasonable as compared with the experiment (sideband at 155 cm^{-1} from the main line). The sideband at 77.3 cm^{-1} from the main line might be assigned to the one-magnon line. Theoretically this interpretation is possible only if the separation between the excited B_{3g} and A_{1g} levels is of order 100 cm^{-1}. Also, there is a possibility of assigning this line to a phonon sideband together with the other observed sidebands, though the problem has not been settled as yet. The shift and broadening of the zero-magnon line has been studied. Here the magnon scattering processes as shown in Fig. 2c play a predominant role.[19][23] The estimated line width compares well with the experiment. Finally theoretical estimates indicate that the reduction of the excitation transfer (or exciton dispersion) due to exciton-photon coupling can be significant in this case.

3.4. Case of spin-forbidden transitions

3.4.1. The magnon sidebands associated with the spin-forbidden transitions have been observed in many antiferromagnets.(16) They all are electric dipole transitions and considered to be induced by the spin-dependent electric dipole transition moments, $\underset{\sim}{P}^{(2)}$ in Eq. (7) or Eq. (8). At low temperatures, this moment may be expanded in terms of excitons and magnons. For a two sublattice antiferromagnet, we have:

$$\underset{\sim}{P} = \sum_{n}\sum_{k} \{ \underset{\sim}{\pi}_{nk}^{(1)} A_{nk}^{+} b_{-k}^{+} + \underset{\sim}{\pi}_{nk}^{(2)} B_{nk}^{+} a_{-k}^{+} + \text{H. c.} \}, \quad (15)$$

where A_{nk}^{+} and B_{nk}^{+} (a_k^{+} and b_k^{+}) are the creation operators for excitons in the n th band (magnons) running in the first and second sublattices, respectively. These excitons are associated with excitations with the change of spin multiplicity in an ion, so are of different type from those associated with spin-allowed transitions. $\underset{\sim}{\pi}_{nk}$ are given by a proper linear combination of the Fourier transforms of $\underset{\sim}{\pi}_{j\ell}$ (b g), etc. The selection rules in the $\underset{\sim}{k}$ space is given by the $\underset{\sim}{k}$ dependence of $\underset{\sim}{\pi}_{nk}$, which is determined by the local symmetry of the crystal and the symmetries of the ground and excited orbital states.

3.4.2. The line shaped for magnon sidebands associated with $^6A_{1g} \rightarrow {}^4T_{1g}$ transitions in MnF_2 have been calculated first by using the above discussed transition moment without taking account of the exciton-magnon interactions.[24][25][10] The results were reasonable qualitatively; particularly the observed qualitative difference in the line shapes of two sidebands which appear for σ - and π - polarizations, respectively, were explained. However, the peak positions and the detailed line shapes were not satisfactorily reproduced.

Then the effect of exciton magnon interaction was invoked with and without the assumption of vanishing exciton band width. The exciton-magnon interaction in the present case is more complicated than in the case of spin-allowed transitions, though the interaction Hamiltonian is obtained in a similar way. Only the principal terms have been taken into account in actual theories.

When the exciton dispersion is neglected, the problem reduces to that of the impurity effect in the spin-wave theory.[26][27][28][29] Thus as far as the magnon-magnon interaction is neglected, the problem essentially is a one-body problem and can be solved by the Koster-Slater method,[30] or its equivalents. Actual calculation, though rather complicated, has been carried out for the sidebands associated with ${}^6A_{1g}$ ${}^4T_{1g}$ transitions in MnF_2 and $RbMnF_3$. These theories have improved the line shape reasonably well, particularly in the case of $RbMnF_3$.

A theoretical treatment to take account of the exciton dispersion has also been developed[31] for the ${}^6A_{1g}$ ${}^4A_{1g}$, ${}^4E_g$ transitions in MnF_2. What has essentially been done is to solve the exciton-magnon scattering problem by taking into account only the states with a pair of exciton and magnon. This corresponds to summing the exciton-magnon ladder-type diagrams. The self-energy correction due to magnon creation and annihilation processes are shown to be small in this case. The effect of exciton-phonon interaction is also shown to be small in this case.

This theory particularly deals with the exciton-magnon bound state. The sharp peak observed in the sideband of one of the $^6A_{1g} \rightarrow {}^4A_{1g}, {}^4E_g$ transitions [32] were assigned to be due to a bound state. Negative dispersions of the excitons and resulting high density of states and sharp edge at the low energy bound for the free exciton-magnon pairs seem to be responsible for this bound state formation, together with the attractive nature of the exciton-magnon interaction.

3.4.3. For further details, readers are referred to the original papers. We shall conclude here with some remarks on future possibility:

(1) Magnon sidebands will be observed in all antiferromagnets and ferrimagnets. For ferromagnets, observation may be possible when non-uniaxial anisotropy is sufficiently large.

(2) Many-magnon sidebands may be understood from combination of two mechanisms mentioned in 3.2.1. The effect of magnon-exciton and magnon-magnon interactions will reduce their intensity with increasing number of magnons participating.

Detailed account of this problem has not been worked out so far.

(3) Temperature dependence of the magnon sidebands has not been worked out. For this purpose we should retain the spin operators

in the transition moments and interaction Hamiltonians, without reducing them to magnon operators. Various statistical mechanical methods for spin systems may be employed; for example, the Green function approach seems to be convenient. When the spin-dependent electric dipole moment is large, one may expect to observe the spin (paramagnon) - induced electric dipole transitions even above the Curie temperature.[33]

REFERENCES:

1) J. W. Halley and I. Silvera, Phys. Rev. Letters 15, 654 (1965).

2) R. L. Green, D. D. Sell, W. M. Yen, A. L. Schawlow, and R. M. White, Phys. Rev. Letters 15, 565 (1965).

3) P. A. Fleury, S. P. S. Porto, L. E. Cheesman, and H. J. Guggenheim, Phys. Rev. Letters 17, 84 (1966).

4) R. Newman and R. M. Chrenko, Phys. Rev. 114, 1507 (1959); see for interpretation: Y. Mizuno and S. Koide, Phys. Kondens. Materie 2, 166 (1964).

5) Optical studies of ordered magnetic materials have been initiated by: I. Tsujikawa and E. Kanda, J. Phys. Radium 20, 352 (1959). For subsequent studies to 1963 see: S. Sugano and Y. Tanabe, "Magnetism" Vol. I (edited by G. T. Rado and H. Suhl, Academeic Press, Inc., New York, 1963) Chap. 6, p. 243.

6) R. J. Elliott and R. Loudon, Phys. Letters 3, 189 (1963).

7) Y. Tanabe, T. Moriya, and S. Sugano, Phys. Rev. Letters 15, 1023 (1965).

8) For surveys on experimental results see: R. G. Wheeler, F. M. Reames, and E. J. Wachtel, J. Appl. Phys. 39, 915 (1968); P. L. Richards, J. Appl. Phys. 38, 1500 (1967).
For theoretical survey see the following two references:

9) T. Moriya, J. Appl. Phys. 39, 1042 (1968).

10) R. Loudon, Advances in Phys. 17, 243 (1968.)

11) T. Moriya, Phys. Rev. 120, 91 (1960).

12) P. W. Anderson, Phys. Rev. 115, 2 (1959).

13) See for example: R. J. Elliott and M. F. Thorpe, J. Appl. Phys. 39, 802 (1968).

14) S. J. Allen, R. Loudon, and P. L. Richards, Phys. Rev. Letters 16, 463 (1966).

15) R. J. Elliott, M. F. Thorpe, G. F. Imbusch, R. Loudon, and J. B. Parkinson, Phys. Rev. Letters 21, 147 (1968).

15a) M. F. Thorpe and R. J. Elliott, Proc. International Conference on Light Scattering Spectra in Solids, New York, 1968, to be published.

16) For a survey on experimental data see: D. D. Sell, J. Appl. Phys. 39, 1030 (1968), experimental data to 1967 are tabulated here. See also: "Optical Properties of Ions in Crystals" (edited by H. M. Crosswhite and H. W. Moos, Interscience Publishers, 1967) Part III.

17) J. P. Van der Ziel, Phys. Rev. 161, 483 (1967).

18) K. Aoyagi, K. Tsushima, and S. Sugano, Solid State Communications 7, 229 (1969); S. Sugano, K. Aoyagi, and K. Tsushima, ibid. 7, 233 (1969).

19) J. Tylicki and W. M. Yen, Phys. Rev. 166, 488 (1968).

20) J. P. Van der Ziel and H. J. Guggenheim, Phys. Rev. 166, 479 (1968).

21) A. I. Belyaeva, V. V. Eremenko, N. N. Mikhailov, V. N. Pavlov and S. V. Petrov, Zh. Eksp. i Teor. Fiz. 50, 1472 (1966). translation: Soviet Phys. -JETP 23, 979 (1966) ; V. V. Eremenko, Yu. A. Popkov, V. P. Nivokov, and A. I. Belyaeva, Zh. Ekop. i Teor. Fiz 52, 454 (1967) translation: Soviet Phys. -JETP 25, 297 (1967) ; R. Stevenson, Phys. Rev. 152, 531 (1966).

22) J. W. Halley, Phys. Rev. 149, 423 (1966); 154, 458 (1967); order estimations in these articles do not seem to be sound: Y. Tanabe, Bussei 8, 651 (1967), and private communication.

23) T. Moriya and M. Inoue, J. Phys. Soc. Japan 24, 1251 (1968).

24) Y. Tanabe and K. I. Gondaira, J. Phys. Soc. Japan 22, 573 (1967).

25) D. D. Sell, R. L. Green, and R. M. White, Phys. Rev. 158, 489 (1967).

26) Possible importance of exciton-magnon interaction was pointed out previously by: T. Moriya, J. Phys. Soc. Japan 23, 490 (1966).

27) T. Tonegawa, Prog. Theor. Phys. 41, 1 (1969).

28) Y. Tanabe, K. I. Gondaira, and H. Murata, J. Phys. Soc. Japan 25, 1562 (1968).

29) J. B. Parkinson and R. Loudon, Proc. Phys. Soc. (London) C1, 1568 (1968).

30) G. F. Koster and J. C. Slater, Phys. Rev. 95, 1167 (1954); G. F. Koster, Phys. Rev. 95, 1436 (1954); G. F. Koster and J. C. Slater, Phys. Rev. 96, 1208 (1954).

31) S. Freeman and J. J. Hopfield, Phys. Rev. Letters 21, 910 (1968); S. Freeman, Technical Report, RCA, Laboratories, Princeton.

32) R. Meltzer, M. Y. Chen, D. S. McClure, and M. L. -Pariseau, Phys. Rev. Letters 21, 913 (1968).

33) This seems to be the case in MnO and -MnS: I. Harada and K. Motizuki, private communication.

OPTICAL EXCITATION OF THE SPIN SYSTEMS IN MAGNETIC CRYSTALS II LIGHT SCATTERING

Toru MORIYA

Institute for Solid State Physics

University of Tokyo, Roppongi, Tokyo

1. INTRODUCTION

Light scattering by spin systems in magnetic crystals has been studied theoretically since 1959. The mechanism first studied was the one due to direct magnetic interaction between the radiation field and the ionic magnetic moments.[1)]
This mechanism, however, turns out to be very small usually as compared with the one due to electric interaction by way of the spin-orbit coupling.[2)] The first observation of the spin Raman scattering was due to Pr^{3+} ions in LaF_3.[3)] Subsequently, light scattering by spin waves (one-magnon scattering) has been discussed theoretically,[4)] and the first observation of the light scatterings by one and two magnons in antiferromagnets (FeF_2) has been reported.[5)] In order to explain the observed strong intensity of two-magnon scattering, it was proposed[5,6)] as the mechanism to take account of the non-diagonal exchange interaction in place of the spin-orbit coupling in the one-magnon scattering, similarly to the case of two-magnon and exciton-magnon absorptions.[7)]

Subsequent developments in theoretical[8,9,10,11,12,13)] and experimental studies[6,11,14,15)] seem to have attained fairly satisfactory understanding of the problem as far as insulators are concerned.

As for magnetic metals, however, there has been reported neither experimental nor theoretical study so far. Although the skin effect stands in the way, observation of light scattering by spin fluctuations in magnetic metals does not seem to be impossible, considering the advanced experimental techniques in the present time.

We shall here give a brief summary of the theory of light scattering in magnetic insulators and then extend discussions to the case of metals.

2. INSULATORS

2.1 Spin-dependent electric polarizability and light scattering.

2.1.1. Light scattering by spin systems may be treated theoretically in two ways. One is the direct quantum mechanical calculation of the transition probability for light scattering, and the other is an approach to make use of the spin-dependent electric polarizability. The latter seems to be convenient in treating the problem from a general point of view: (1) One can develop a phenomenological theory taking account of only the crystal symmetry; the quantum mechanical theory is reduced to that for the spin-dependent electric polarizability. (2) Calculation of the scattering intensity can be made, even without knowing the eigenstates for the system, by making use of various approximation methods in the statistical mechanics for the spin systems. This method may be regarded as a natural extension of the spin Hamiltonian approach to the light scattering problems.

2.1.2. An electric polarizability associated with the j th ionic spin may be expanded in terms of the components of $\underset{\sim}{S}_j$ as follows:

$$\alpha_j^{(\omega)} = \sum_{\mu} \alpha_{j\mu}(\omega) S_{j\mu} + \sum_{\mu,\nu} \alpha_{j,\mu\nu}(\omega) S_{j\mu} S_{j\nu} + \cdots \qquad (1)$$

Similarly the polarizability associated with the j,ℓ pair of ionic spins may be written as:

$$\alpha_{j\ell}^{(\omega)} = \sum_{\mu,\nu} \alpha_{j\ell,\mu\nu}(\omega) S_{j\mu} S_{\ell\nu} + \cdots \qquad (2)$$

From the microscopic mechanisms for these polarizabilities we advance that the terms of higher order than the above are less important when the ionic orbital angular momentum is well quenched. We also expect that $\alpha_{j\ell}$ is of short range. Therefore the local polarizability per ion may be given by $[\alpha_j^{(\omega)} + (1/2)\sum_{\ell} \alpha_{j\ell}^{(\omega)}]$.

The electrical component of the radiation field; $\underset{\sim}{E}_0(\underset{\sim}{r}, t) = E_0 \exp i(\underset{\sim}{k}_0 \cdot \underset{\sim}{r} - \omega_0 t)$, induces the electric dipole moment

$$\underset{\sim}{P}_j(t) = [\alpha_j^{(\omega_0)}(t) + (1/2)\sum_{\ell} \alpha_{j\ell}^{(\omega_0)}(t)] \underset{\sim}{E}_0 \exp i(\underset{\sim}{k}_0 \cdot \underset{\sim}{R}_j - \omega_0 t),$$

with

$$\alpha_j^{(\omega)}(t) = e^{itH/\hbar} \alpha_j^{(\omega)} e^{-itH/\hbar}, \qquad (3)$$

where H is the spin Hamiltonian for the system under study. Thus the relatively slow motion of the spin system is taken into account adiabatically.

The scattered light may be described as the electric dipole radiation from these induced moments modulated by the spin motion. Since the induced moment for an entire crystal has the Fourier components other than $\mathbf{k}_0$ owing to the spin fluctuation, we have light scattering by spin systems. The Fourier $\mathbf{k}$ component of the induced moment $\mathbf{P}_k(t)$ is given by the polarizability α_q for the wave vector change $\mathbf{q} = \mathbf{k} - \mathbf{k}_0$:

$$\mathbf{P}_k(t) = \alpha_{k-k_0}^{(\omega_0)}(t)\, \mathbf{E}_0 \exp(-i\omega_0 t),$$

$$\alpha_q^{(\omega)} = V^{-1} \sum_j [\alpha_j^{(\omega)} + (1/2)\sum_\ell \alpha_{j\ell}^{(\omega)}] \exp(-i\mathbf{q}.\mathbf{R}_j). \tag{4}$$

Thus the energy flux for the radiation of frequency $\omega \sim \omega + d\omega$ scattered into the solid angle $d\Omega$ around the direction $\mathbf{r}$ is given by

$$d\Omega d\omega [\omega^4/4(2\pi)^2 c^3] \int_{-\infty}^{\infty} dt\, e^{i\omega t} \langle \mathbf{P}_k^*(0).(1-\mathbf{rr}).\mathbf{P}_k(t)\rangle, \tag{5}$$

where $\mathbf{r} = \mathbf{R}/R = \mathbf{k}/k$ and $\langle\ \rangle$ means the statistical average. The differential extinction coefficient is given from Eqs. (4) and (5) as follows:

$$d^2h/d\Omega d\omega = (\omega_0 \omega^3/c^4)(2\pi)^{-1} \int_{-\infty}^{\infty} dt \exp[i(\omega - \omega_0)t]$$
$$\times \langle \mathbf{e}_0^* . \tilde{\alpha}_{k-k_0}^{*(\omega_0)}(0).(1-\mathbf{rr}).\alpha_{k-k_0}^{(\omega_0)}(t).\mathbf{e}_0\rangle, \tag{6}$$

where $\mathbf{e}_0 = \mathbf{E}_0/E_0$, $\tilde{\alpha}$ is the transposed tensor of α, and * means to take a complex conjugate. Usually it is a good approximation to use a uniform polarizability for α_q, though $\mathbf{q}$-dependence is important in the critical and the Brillouin scatterings.

2.2 Symmetry considerations

Possible forms for $\alpha_j^{(\omega)}$, $\alpha_{j\ell}^{(\omega)}$, etc., are determined from the local symmetry of the crystal as in the case of spin Hamiltonians and the spin dependent electric dipole moments discussed in the preceding article I. In the present case it is important to ask further symmetry requirements for general admittance tensors or kinetic coefficients, since α_j, $\alpha_{j\ell}$, etc., are polarizabilities. From the latter requirements [16] it immediately follows that $\alpha_{j\mu}^{(\omega)}$ is an antisymmetric tensor while $\alpha_{j,\mu\nu}^{(\omega)}$ and $\alpha_{j\ell,\mu\nu}^{(\omega)}$ are symmetric tensors when there is no external magnetic field. One important consequence of this argument is the gyrotropic nature of the polarizability tensor linear in the spin components, $\alpha_j^{(\omega)}$. For transparent crystals, where polarizability tensors are Hermitic, $\alpha_{j\mu}^{(\omega)}$ is purely imaginary, and therefore is gyrotropic, while $\alpha_{j,\mu\nu}^{(\omega)}$ and $\alpha_{j\ell,\mu\nu}^{(\omega)}$ are real and the imaginary components of the kinetic coefficients should vanish in the limit of $\omega \to 0$. These arguments lead to an important relation between the light scattering (one-magnon scattering) and the Faraday rotation.[17]

Eq. (2) may generally be rewritten as

$$\alpha_{j\ell} = P_{j\ell}(\underset{\sim}{S}_j \underset{\sim}{S}_\ell) + \sum_{\rho,\tau} Q_{j\ell,\rho\tau} S_{j\rho} S_{\ell\tau} + \sum_{\rho\tau} R_{j\ell,\rho\tau} S_{j\rho} S_{\ell\tau}, \quad (6)$$

where P, Q, R are symmetric tensors. Q and R are symmetric and antisymmetric with respect to the interchange of two ionic spins j and ℓ or ρ and τ, respectively. For further details and the forms of these tensors for rutile type iron group fluorides, as an example, see the references 9 and 10.

2. 3. Origin of the spin-dependent electric polarizability.

Quantum mechanical calculation for the spin-dependent polarizabilities are quite similar to those for the spin-dependent dipole moments; we need to pick up the terms of the second order in the electron-radiation interaction in this case. Therefore the most important terms in the perturbation series are the third order ones which are linear in the spin-orbit coupling and the non-diagonal exchange energy, respectively. These terms contribute to α_j and $P_{j\ell}$ in Eq. (6), respectively, and their orders of magnitude are given by

$$a_m \sim \theta |(g|er|a)|^2 \lambda/\Delta E_0 \Delta E, \text{ and } p_m \sim (J/\lambda)\, \theta^{-1}\, a_m, \tag{7}$$

where a_m, p_m represent this third order contribution to the matrix elements of $\alpha_j^{(\omega)}$ and $P_{j\ell}^{(\omega)}$ respectively, g and a the ground and the odd-parity excited states, λ and J the spin-orbit and the non-diagonal exchange energies, $\Delta E_0 = E_a - E_g - \hbar\omega$, ΔE the smaller one of the ΔE_0 and the even-parity excitation energy, and θ is a factor which vanishes in the limit of $\omega_0 \to 0$ and is given for small ($\hbar\omega_0/\Delta E_0$) by

$$\theta = \hbar\omega_0/(\Delta E_0 + 2\hbar\omega_0).$$

From the higher order perturbation terms we get many more terms of all possible forms, $R_{j\ell\rho\tau}$, $Q_{j\ell\rho\tau}$, etc. They are expected to be less important when $(\lambda/\Delta E)$ and $(J/\Delta E)$ are small.

2.4. Light scattering by spinwaves.

2.4.1. Light scattering by linear spin fluctuations, including one-magnon scattering at low termperatures, is described by the first term in Eq. (1), or the electric polarizability linear in the spin components. From the above symmetry arguments we see that in this case the polarization vector of the scattered radiation is rotated by 90 degrees from that of the incident one. The scattering intensity is expressed in terms of the space time correlation functions for the spin components as may be seen from Eq. (6), and therefore is re-expressed in terms of wave-vector and frequency-dependent magnetic susceptibilities $\chi(q,\omega)$ etc., by using a fluctuation-dissipation theorem. Thus the problem reduces to the calculation $\chi(q,\omega)$ and various statistical mechanical approximation methods may be employed for this purpose. For example, a simple molecular field approximation explains qualitatively the temperature dependence of the one-magnon scattering intensity in antiferromagnetic FeF_2.[5, 9] The order of magnitude of the scattering intensity of the extinction coefficient in this case is estimated as in 2.3, to be $\sim 10^{-12} - 10^{-10}$ for the photon energy of $2 \times 10^4\ cm^{-1}$, in agreement with the observation. The selection rules for one-magnon scattering may also be derived by a group theoretical consideration of magnon modes.[2, 17a] The present method applies to all kinds of magnetic crystals and at any temperatures.

2.4.2. Main contribution to the light scattering by two-magnon excitations comes from $\alpha_{j\ell}$ in Eq. (2), particularly the first term in Eq. (6). Expanding the spin operators in Eq. (2) or (6) in terms of spin wave annihilation and creation operators, we can pick up the relevant part of the polarizability. Thus the selection rules for various polarizations for the incident and scattered radiation are obtained from the coefficients of two-magnon creation terms. The latters are expressed in terms of the Fourier transforms of $P_{j\ell}$, $Q_{j\ell,\rho\tau}$ $R_{j\ell}$, $\rho\tau$, etc. The relative intensity of various components of the scattering can be discussed by using the arguments in 2.3. Detailed analyses along this line of the two magnon scatterings in antiferromagnetic MnF_2 and FeF_2 have given results consistent with the experiments including the relative intensity of one- and two-magnon scatterings. It is interesting to mention that the selection rules for two-magnon scatterings are generally different from either those for two-magnon absorptions or those for magnon sidebands.

Detailed calculation of the intensity spectra for the two-magnon scattering in MnF_2 was made first with the assumption of non-interacting magnons.[6, 11] The agreement was only qualitative. Discrepancies in the detailed line shapes were first attributed to the extended range of $P_{j\ell}$ and recently to the magnon interactions. According to a recent calculation[12] of the line shape for two-magnon scattering in $RbMnF_3$, the magnon-interaction seems to improve the agreement, which is now quantitative. Note that no ambiguous parameter is included in the magnon interactions are already mentioned previously in the case of two-magnon absorptions.

Finally we should mention that the intensity of two-magnon scattering depends on the spin ordering. For example, the largest term with $P_{j\ell}$, etc., is certainly predominant in anti-ferro- and ferrimagnets, while it is unimportant in ferromagnets unless magnetic anisotropy is extremely large. This situation is similar to the case of two-magnon absorptions.

2.5. Temperature dependence.

As already mentioned in 2.4.1, the temperature dependence of the light scattering by linear spin fluctuations is reduced to that of the wave-vector and frequency-dependent magnetic susceptibilities, and may be treated at any temperatures. Particularly, the critical scattering has been discussed for ferro- and antiferromagnets with some hopeful results for observation.[10)]

As for the two-magnon scattering or the scattering by quadratic spin fluctuations, no theory on its temperature dependence has been worked out so far. On the other hand, experimental studies on NiF_2 and $RbNiF_3$ have shown that substantial scattering intensity persists much above their Curie temperatures.[18, 19)] Strong scattering intensity which survives above the Curie temperature may be expected when ΔE_0 or ΔE in Eq. (7) is small. The line shape and its temperature dependence may be studied by statistical mechanical theory of short range order in the spin systems; quadratic spin correlation functions may be calculated, for example, by using some proper approximation methods in the Green function theory.

2.6. Cases where orbital angular momentum is not quenched.

Theoretical arguments explained above apply to transition metal compounds with well quenched orbital angular momentum. The symmetry arguments apply also to rare earth compounds if we replace $\underset{\sim}{S}_j$, etc. by $\underset{\sim}{J}_j = \underset{\sim}{S}_j + \underset{\sim}{L}_j$, etc. For rare earth ions, the polarizability associated with the total angular momentum of a single ion is generally large as compared with that for transition metal ions, since the spin-orbit coupling is larger than the crystalline field effect in the formers. On the other hand, the polarizability associated with a pair of rare earth ions is expected to be small compared with that for transition metal ions because of generally weak exchange interaction between rare earth ions.

When the orbital angular momentum is partially quenched, we have to extend the theory to include the degrees of freedom for orbital moments. This is actually the case for Co^{2+} and Fe^{2+} ions under a nearly octahedral crystalline electric field. We make use of an effective orbital angular momentum $\underset{\sim}{\ell}$ ($\ell = 1$) to represent the degree of freedom for three-fold degenerate t_{2g} states. The magnons and excitons arising from this $(2\ell+1) \times (2S+1)$ manifold may be treated by extending the spin Hamiltonian to include $\underset{\sim}{\ell}$. Similarly, light scattering by magnons and excitons may be treated by using the $\underset{\sim}{\ell}$, $\underset{\sim}{S}$-dependent electric polarizability. Such a calculation is under way.[20]

2.7 Recent experiments.

Light scattering from localized spin waves has been observed recently [21, 22] in MnF_2 doped with Ni^{2+} and Fe^{2+} . The observed scattering lines have been identified as, simultaneous excitation of a pair of magnon-modes on and around the impurity. Interaction between magnons is essential here to interpret the observed line positions.

One-magnon scattering has been observed in a ferromagnetic semiconductor $CdCr_2Se_4$.[23] This crystal has a spinel structure and has one accoustial $[\Gamma_1^+(A_{1g})]$ and one optical triply degenerate $\Gamma_{25}^+(F_{2g})]$ spin-wave mode at the Brillouin zone center. The observed line at 168cm^{-1} is assigned to the latter.

Scattering of infrared radiation by two-magnon processes has been observed in ferrimagnetic yttrium-ion garnet.[24]
The detection has been made by studying the population variation, due to the infrared radiation, of a coherent spinwave mode excited parametrically by parallel pumping. Possible study of light scattering by coherent spinwaves, excited parametrically, has been proposed earlier[25] with an order estimation for one-magnon processes

3. METALS

3.1. Spin-dependent polarizability and light scattering in metals.

The above discussed approach by using the spin-dependent electric polarizability applies to metals, with some modifications. Here it is essential to take account of dielectric properties of metals which give rise to the skin effect. We assume, for brevity, this dielectric constant $\varepsilon(\omega)$ to be isotropic, i.e., scalar. The metal surface is given by a plane: $z=0$ and the half space: $z > 0$ is a vacuum.

The plane of incidence is taken to be the xz plane.

The electric field of the incident radiation ($z > 0$):

$$E_0(r,t) = E_0 \exp[i(k_0 \cdot r - \omega_0 t)]$$

generates the refracted component ($z < 0$):

$$E_2(r,t) = \Gamma \cdot E_0 \exp[i(k_{0x}x - \omega_0 t) + \gamma_0 z] \tag{7}$$

with

$$\gamma_0 = [-(\omega_0/c)^2 \epsilon(\omega_0) - k_{0x}^2]^{1/2}, \quad Re\gamma_0 > 0, \tag{8}$$

and

$$\Gamma_{\mu\nu} = \delta_{\mu\nu}\Gamma_\mu,$$

$$\Gamma_x = 2\sqrt{\epsilon - \sin^2\theta_0}/(\sqrt{\epsilon - \sin^2\theta_0} + \epsilon\cos\theta_0),$$

$$\Gamma_y = 2\cos\theta_0/(\sqrt{\epsilon - \sin^2\theta_0} + \cos\theta_0), \tag{9}$$

$$\Gamma_z = 2\cos\theta_0/(\sqrt{\epsilon - \sin^2\theta_0} + \epsilon\cos\theta_0),$$

where θ_0 is the angle of incidence and $\epsilon(\omega_0)$ is abbreviated to ϵ.

The spin-dependent current induced in the metal may be given by

$$j(r,t) = -i(\omega_0/c)\theta(-z)\,\alpha(r,t) \cdot E_2(r,t), \tag{10}$$

where $\theta(x) = 1$ for $x > 0$ and 0 for $x < 0$, $\alpha(r, t)$ is the spin-dependent electric polarizability, and we consider only the frequency components close to ω_0 for which the frequency dependence of α is neglected.

The spin-dependent polarizability has a clear meaning in rare earth metals where localized atomic moments are well defined. For transition metals definition is less clear; we assume here that the polarizability is expanded phenomenologically in terms of spin density.

The vector potential for the scattered radiation may be obtained from the following Maxwell equation:

$$[\text{curl curl} + \{1+(\epsilon-1)\theta(-z)\}c^{-2}(\partial^2/\partial t^2)]\underset{\sim}{A}(\underset{\sim}{r},t) = (4\pi/c)\underset{\sim}{j}(\underset{\sim}{r},t). \quad (11)$$

This is solved by using the Green function defined by

$$[\text{curl curl} + \{1+(\epsilon-1)\theta(-z)\}c^{-2}(\partial^2/\partial t^2)].D(\underset{\sim}{r},\underset{\sim}{r}'; t-t') = -(4\pi/c)\underset{\sim}{1}\delta(t-t')\delta(\underset{\sim}{r}-\underset{\sim}{r}') \quad (12)$$

where $\underset{\sim}{1}$ is a unit matrix. The scattered radiation is given by

$$\underset{\sim}{A}^s(\underset{\sim}{r},t) = -\iint_{z'<0} d\underset{\sim}{r}'dt'\, D(\underset{\sim}{r},\underset{\sim}{r}'; t-t').\underset{\sim}{j}(\underset{\sim}{r}'t'). \quad (13)$$

The solution for the Eq. (12) for the Green function ($z>0$, $z'<0$) is as follows:[26)]

$$D(\underset{\sim}{r},\underset{\sim}{r}';t) = \int \frac{d\omega}{2\pi} e^{-i\omega t} \int \frac{d\underset{\sim}{k}_{\parallel}}{(2\pi)^2}(4\pi/\omega)D(\underset{\sim}{k}_{\parallel},\omega)\exp[i\underset{\sim}{k}_{\parallel}\cdot(\underset{\sim}{r}-\underset{\sim}{r}') + ik_z z + \gamma_s z'], \quad (14)$$

$$D(\underset{\sim}{k}_{\parallel},\omega) = (\omega/c\gamma_s)[\gamma_s^2 k_z(i\gamma_s k^2-\gamma^2 k_z)^{-1}\hat{\xi}\hat{\xi} + \gamma_s(ik_z-\gamma_s)^{-1}\hat{\eta}\hat{\eta} - ik_{\parallel}^2\gamma_s(i\gamma_s k^2-\gamma^2 k_z)^{-1}\hat{z}\hat{z} + i\gamma_s k_z k_{\parallel}(i\gamma_s k^2-\gamma^2 k_z)^{-1}\hat{\xi}\hat{z} - \gamma_s^2 k_{\parallel}(i\gamma_s k^2-\gamma^2 k_z)^{-1}\hat{z}\hat{\xi}], \quad (15)$$

with

$$\gamma_s^2 = -(\omega/c)^2\epsilon + k^2, \quad Re\gamma_s > 0,$$
$$\gamma^2 = -(\omega/c)^2\epsilon, \quad (16)$$

where $\underset{\sim}{k}_{\parallel}$ is the two-dimentional vector in the xy plane, $\hat{\xi}$, $\hat{\eta}$, and $\hat{z}$ are unit vectors parallel to $\underset{\sim}{k}_{\parallel}$, perpendicular to $\hat{\xi}$ and the z axis, and along the z axis, respectively. When $|\gamma| >> |k|$

$|kz|$, as is frequently the case in metals, Eq. (15) reduces to

$$D(\underset{\sim}{k}_{\parallel},\omega) \simeq (\omega/c\gamma_s)[-(1-\hat{z}\hat{z}) + (\gamma_s^2 k_{\parallel}/\gamma^2 k_z)\,\hat{z}\hat{\xi}]. \tag{17}$$

From this and Eq. (9) we see that the components of the spin-dependent polarizability parallel to the boundary surface (xx, yy, xy, yx-components) are most effectively studied by light scattering.

Finally from Eqs. (7), (10), (13) and (14) we get

$$\underset{\sim}{A}^s(\underset{\sim}{r}, t) = \int \frac{d\omega}{2\pi} \int \frac{dk}{(2\pi)^2} \underset{\sim}{A}^s(\underset{\sim}{k},\omega)\, e^{i(\underset{\sim}{k}\cdot\underset{\sim}{r}-\omega t)} \delta(k_z - \sqrt{(\omega/c)^2 - k_{\parallel}^2}), \tag{18}$$

$$\underset{\sim}{A}^s(\underset{\sim}{k}_{\parallel},\omega) = iS4\pi E_0\, D(\underset{\sim}{k}_{\parallel},\omega)\cdot \int_{-\infty}^{\infty} dt'\, \alpha(\underset{\sim}{k}_{\parallel} - \underset{\sim}{k}_{0\parallel}, \gamma_s + \gamma_0; t') \cdot \Gamma \cdot \underset{\sim}{e}_0 \exp[i(\omega - \omega_0)t'], \tag{19}$$

with

$$\alpha(\underset{\sim}{q}_{\parallel},\gamma;t) = S^{-1} \int_{z<0} d\underset{\sim}{r}\, \alpha(\underset{\sim}{r},t) \exp[-i\underset{\sim}{q}_{\parallel}\cdot\underset{\sim}{r} + \gamma z], \tag{20}$$

where S is the area of the metal surface, $E_0 = |\underset{\sim}{E}_0|$ and $\underset{\sim}{e}_0 = \underset{\sim}{E}_0/E_0$. For large $r = |\underset{\sim}{r}|$, an asymptotic expansion of Eq. (18) leads to

$$\underset{\sim}{A}^s(\underset{\sim}{r}, t) \sim \frac{\cos\theta_s}{ir} \int \frac{d\omega}{(2\pi)^2} \frac{\omega}{c} \underset{\sim}{A}^s(\underset{\sim}{k}_{r\parallel},\omega)\, e^{i\omega[(r/c)-t]}, \tag{21}$$

where

$$\underset{\sim}{k}_r = (\omega/c)(\underset{\sim}{r}/r),$$

and Θ_s is the angle between $\underset{\sim}{r}$, the direction of scattered radiation, and the z axis. In what follows we abbreviate $\underset{\sim}{k}_r$ to $\underset{\sim}{k}$.

The scattering efficiency (the number of photons with the frequency $\omega\sim\omega+d\omega$ scattered into a solid angle divided by incident number of photons) may be obtained as follows:

$$\varphi_0^{-1}d^2\varphi/d\Omega d\omega = (2/\pi)S(\omega_0\omega^3/c^4)[\cos^2\theta_s/\cos\theta_0]$$
$$\times\int_{-\infty}^{\infty} d\tau\ \exp[i(\omega-\omega_0)\tau]\langle \underset{\sim}{e}_0{}^*.\Gamma^*.\tilde{\alpha}^*(\underset{\sim}{k}_{\parallel}-\underset{\sim}{k}_{0\parallel},\gamma_s+\gamma_0;t).$$
$$.\tilde{D}^*(\underset{\sim}{k}_{\parallel},\omega).D(\underset{\sim}{k}_{\parallel},\omega).\alpha(\underset{\sim}{k}_{\parallel}-\underset{\sim}{k}_{0\parallel},\gamma_s+\gamma_0;t+\tau).\Gamma.\underset{\sim}{e}_0\rangle, \tag{22}$$

where $\langle\ \rangle$ means the statistical average, $\sim$ to transpose the matrix and $*$ to take complex conjugate. From Eq. (15) we get

$$\tilde{D}^*(\underset{\sim}{k}_{\parallel},\omega).D(\underset{\sim}{k}_{\parallel},\omega) = |\epsilon-\sin^2\theta_s|^{-1}\{|\gamma_s/(ik_z-\gamma_s)|^2\hat{\eta}\hat{\eta}$$
$$+|\gamma_s|^2k^2|i\gamma_sk^2-\gamma^2k_z|^{-2}[|\gamma_s|^2\hat{\xi}\hat{\xi}+k_{\parallel}{}^2\hat{z}\hat{z}+ik_{\parallel}(\gamma_s{}^*\hat{\xi}\hat{z}-\gamma_s\hat{z}\hat{\xi})]\}. \tag{23}$$

For small Θ_s Eq. (23) may be approximated by

$$\tilde{D}^*(\underset{\sim}{k}_{\parallel},\omega).D(\underset{\sim}{k}_{\parallel},\omega) \simeq |i-\sqrt{-\epsilon}|^{-2}(1-\hat{z}\hat{z}), \tag{24}$$

where $\sqrt{-\epsilon}$ is taken to have a positive real part.

3.2. Rare earth metals.

In the case of rare earth metals, where the localized atomic moment due to 4f electrons is well defined, the spin dependent electric polarizability may be defined in the same way as in insulators. Let us here consider the light scattering by linear spin fluctuations which is described by the polarizability:

$$\alpha(\underset{\sim}{r},t) = \sum_f\sum_\mu \alpha_\mu{}^{(f)}(\omega)\ S_\mu{}^{(f)}(\underset{\sim}{r},t), \tag{25}$$

where f specifies magnetic atoms in a unit cell, and μ stands for x, y, and z. From Eq. (22) and (20) the scattering efficiency may be calculated as

$$\varphi_0^{-1} d^2\varphi/d\Omega d\omega = \pi^{-2}(\omega_0\omega^3/c^4)[\cos^2\theta_s/\cos\theta_0] \sum_{ff'} \sum_{\mu\mu'}$$

$$\underset{\sim}{e}_0^* \cdot \Gamma^* \cdot \tilde{\alpha}_\mu^{(f)*}(\omega) \cdot \tilde{D}^*(\underset{\sim}{k}_\parallel, \omega) \cdot D(\underset{\sim}{k}_\parallel, \omega) \cdot \alpha_{\mu'}^{(f')}(\omega) \cdot \Gamma \cdot \underset{\sim}{e}_0$$

$$\times \int_{-\infty}^{\infty} d\tau \exp[i(\omega - \omega_0)\tau] \int dq_z \langle S_\mu^{(f)}(-\underset{\sim}{q}_\parallel, q_z; t) S_{\mu'}^{(f')}(\underset{\sim}{q}_\parallel, -q_z; t+\tau)\rangle$$

$$\times |\gamma_s + \gamma_0 - iq_z|^{-2}, \tag{26}$$

with

$$\underset{\sim}{S}^{(f)}(\underset{\sim}{q}_\parallel, q_z; t) = (SL)^{-1/2} \int d\underset{\sim}{r}\, \underset{\sim}{S}^{(f)}(\underset{\sim}{r}, t) \exp[-i(\underset{\sim}{q}_\parallel \cdot \underset{\sim}{r} + q_z z)],$$

$$\underset{\sim}{q}_\parallel = \underset{\sim}{k}_\parallel - \underset{\sim}{k}_{0\parallel}, \tag{27}$$

where S is the area of the metal surface and L its thickness and we assume that $\langle S_\mu^{(f)}(-\underset{\sim}{q}_\parallel, q_z; t) S_{\mu'}^{(f')}(\underset{\sim}{q}_\parallel, -q_z'; t+\tau)\rangle = 0$, unless $qz = qz'$.*

*Under a general boundary condition at the surface (z = 0), we should take account of $\langle S_\mu^{(f)}(\underset{\sim}{q}_\parallel, q_z; t)\, S_{\mu'}^{(f')}(-\underset{\sim}{q}_\parallel, qz; t+\tau)\rangle$, too. The following qualitative discussions are not influenced by this.

From Eq. (26) we see that the modes with $|q_z| \lesssim \gamma$ contribute to the scattering. When the q-dependence of the spin correlation function is weak, as is usually the case for magnons in anisotropic ferromagnets, the last integral in Eq. (26) may be approximated by $\langle S_\mu^{(f)}(O, O; t) S_{\mu'}^{(f')}(O, O; t+\tau)\rangle \cdot \pi / Re(\gamma_s + \gamma)$.

Thus the scattering efficiency is reduced in metals, as compared with insulators, by the order of the following factor:

$$|\Gamma|^2 \tilde{D}^*(k_{\parallel}, \omega) \cdot D(k_{\parallel}, \omega)(\pi/2)[Re(\gamma_s + \gamma_0)L]^{-1} \sim |\epsilon|^{-2}(\delta / L), \tag{28}$$

where δ is the skin depth. This reduction factor in usual metals ranges about 10^{-6} — 10^{-8}. For Nd, for example, we estimate this factor as 0.3×10^{-6} for a wave length 4500 Å. Since the scattering extinction coefficient in insulator compounds of rare earth metals is expected to be about 10^{-3} — 10^{-6}, and the value for $\alpha_\mu^{(f)}(\omega)$ is expected to be approximately the same in metals and insulators*, there is a fair possibility of observing one-magnon scattering in rare earth metals. The use of frequencies near the excitation energy: 4f Fermi level would be advantageous. Also, one can greatly increase the scattering intensity, through the factor in Eq. (28), by working with frequencies near the reflection minimum, i.e., near the phasma frequency.[26)]

---

* Although we have 6s-5d band in metals, the excitation energy for 4f → 5d (empty band) would not be very much different from that in insulators.[27)]

It should be noted here that the z component (perpendicular to the surface) of spin fluctuation is most effectively seen by light scattering, owing to the gyrotropic nature of the polarizability tensor and the skin effect in metals (see the paragraph following Eq. (17) Therefore, it is advantageous to magnetize the sample parallel to the surface. Possible light scattering study of magnons in helical states would also be very interesting.

3.3. Transition metals.

Elementary excitations with spin-flips in ferromagnetic transition metals are spin waves and the Stoner excitations: individual electron excitation with spin-flip. In the long wave limit the former is regarded as a motion of the spin system as a whole, while the latter is approximately a spin-flip of an individual electron and the total spin quantum number is changed by 1. Therefore, a radiation of long wavelength couples mainly to the spinwave mode, provided all the electrical couplings between individual electron spins and the radiation field are the same.

The order of magnitudes for the one-magnon scattering efficiency may be estimated[28] by using Faraday rotation data.[29] For iron, for example, where the specific rotation is about 10^4, a roughly estimated value of the scattering efficiency is about 10^{-13} - 10^{-14} ($\sim$5000Å). The observation may thus be possible if one choose more favorable conditions than the above.

Light scattering by the Stoner excitations may be observed, if the electron spin-radiation coupling has a significant varience depending on the orbital states. It may also be required that the line broadening due to electron interactions is not too large.

We may also expect some possibility of studying magnons in antiferromagnetic metals including static spin density wave states. The intensity for light scattering by individual electron excitations through the energy gap in chromium has recently been estimated to be of significant magnitude.[30)]

Finally, the writer would like to thank Drs. A. Kawabata and M. Inoue for their helpful discussions on the subject discussed in 3.

REFERENCES:

1) F. G. Bass and M. I. Kaganov, Zh. Eksp. i Teor. Fiz. 37, 1390(1959) translation: Soviet Physics - JETP 10, 986 (1960) .

2) R. J. Elliott and R. Loudon, Phys. Letters 3, 189 (1963).

3) J. T. Hougen and S. Singh, Phys. Rev. Letters 10, 406 (1963).

4) Y. R. Shen and N. Bloembergen, Phys. Rev. 143, 372 (1966).

5) P. A. Fleury, S. P. S. Porto, L. E. Cheesman, and H. J. Guggenheim, Phys. Rev. Letters 17, 84 (1966).

6) P. A. Fleury, S. P. S. Porto, and R. Loudon, Phys. Rev. Letters, 18, 658 (1967).

7) Y. Tanabe, T. Moriya, and S. Sugano, Phys. Rev. Letters 15, 1023 (1965).

8) Y. R. Shen, J. appl. Phys. 38, 1490 (1967).

9) T. Moriya, J. Phys. Soc. Japan 23, 490 (1967).

10) T. Moriya, J. appl. Phys. 39, 1042 (1968).

11) P. A. Fleury and R. Loudon, Phys. Rev. 166, 514 (1968).

12) R. J. Elliott, M. F. Thorpe, G. F. Imbusch, R. Loudon, and J. B. Parkinson, Phys. Rev. Letters 21, 147 (1968).

13) V. S. L'vov, Zh. Eksp. i Teor. Fiz. 53, 163 (1967) translation: Soviet Physics - JETP 26, 113 (1968) .

14) P. A. Fleury and S. P. S. Porto, J. appl. Phys. 39, 1035 (1968).

15) P. A. Fleury, Phys. Rev. Letters 21, 151 (1968).

16) See for example: R. Kubo, J. Phys. Soc. Japan 12, 570 (1957); L. D. L. Landan and E. M. Lishitz, "Statistical Physics" (Pergamon Press, Inc., New York, 1960).

17) This connection between the Faraday notation and the spin-Raman effect has been noted by P. S. Pershan, J. appl. Phys. 38, 1482 (1967).

17a) A. P. Cracknell, J. Phys. C Proc. Phys. Soc. (London), 2, 2, 500 (1969).

18) P. A. Fleury, Proc. International Conference on Light Scattering Spectra in Solids, New York, 1968, to be published.

19) S. R. Chinn and H. J. Zeiger, Phys. Rev. Letters 21, 1589 (1968).

20) A. Ishikawa and T. Moriya, private communication.

21) A. Oseroff and P. S. Pershan, Phys. Rev. Letters 21, 1593 (1968).

22) P. Moch, G. Parisot, R. E. Dietz, and H. J. Guggenheim, Phys. Rev. Letters 21, 1596 (1968).

23) G. Harbeke and E. F. Steigmeier, Solid State Communications 6, 747 (1968).

24) H. Le Gall, J. P. Jamet and V. Cagan, Solid State Communications 7, 27 (1969).

25) V. S. L'vov and S. S. Starobinets, Zh. Exsp. i Teor. Fiz. Pis'ma 5, 242 (1967) translation: JETP letters 5, 194 (1967).

26) The same Green function has been used in the case of phonon scattering by D. L. Mills, A. A. Maradudin, and E. Burstein.

Phys. Rev. Letters 21, 1178 (1968).

27) For optical properties of rare earth metals see: "Optical Properties and Electronic Structure of Metals and Alloys" (edited by F. Abeles, North Holland, 1966) pp. 221 - 256.

28) A detailed theoretical study of light scattering in ferro-magnetic transition metals is under way: A. Kawabata, private communication.

29) J. F. Dillon, J. Appl. Phys. 39, 922 (1968).

30) P. C. Kwok, J. W. F. Woo and S. S. Jha, Phys. Letters 28A, 691 (1969).